The R Series

Implementing Reproducible Research

Edited by

Victoria Stodden
Columbia University
New York, New York, USA

Friedrich Leisch
University of Natural Resources and Life Sciences
Institute of Applied Statistics and Computing
Vienna, Austria

Roger D. Peng
Johns Hopkins University
Baltimore, Maryland, USA

CRC Press
Taylor & Francis Group
Boca Raton   London   New York

CRC Press is an imprint of the
Taylor & Francis Group an **informa** business

A CHAPMAN & HALL BOOK

Chapman & Hall/CRC
The R Series

Series Editors

John M. Chambers
Department of Statistics
Stanford University
Stanford, California, USA

Torsten Hothorn
Division of Biostatistics
University of Zurich
Switzerland

Duncan Temple Lang
Department of Statistics
University of California, Davis
Davis, California, USA

Hadley Wickham
Department of Statistics
Rice University
Houston, Texas, USA

Aims and Scope

This book series reflects the recent rapid growth in the development and application of R, the programming language and software environment for statistical computing and graphics. R is now widely used in academic research, education, and industry. It is constantly growing, with new versions of the core software released regularly and more than 4,000 packages available. It is difficult for the documentation to keep pace with the expansion of the software, and this vital book series provides a forum for the publication of books covering many aspects of the development and application of R.

The scope of the series is wide, covering three main threads:
- Applications of R to specific disciplines such as biology, epidemiology, genetics, engineering, finance, and the social sciences.
- Using R for the study of topics of statistical methodology, such as linear and mixed modeling, time series, Bayesian methods, and missing data.
- The development of R, including programming, building packages, and graphics.

The books will appeal to programmers and developers of R software, as well as applied statisticians and data analysts in many fields. The books will feature detailed worked examples and R code fully integrated into the text, ensuring their usefulness to researchers, practitioners and students.

Published Titles

Analyzing Baseball Data with R, *Max Marchi and Jim Albert*

Customer and Business Analytics: Applied Data Mining for Business Decision Making Using R, *Daniel S. Putler and Robert E. Krider*

Dynamic Documents with R and knitr, *Yihui Xie*

Event History Analysis with R, *Göran Broström*

Implementing Reproducible Research, *Victoria Stodden, Friedrich Leisch, and Roger D. Peng*

Programming Graphical User Interfaces with R, *Michael F. Lawrence and John Verzani*

R Graphics, Second Edition, *Paul Murrell*

Reproducible Research with R and RStudio, *Christopher Gandrud*

Statistical Computing in C++ and R, *Randall L. Eubank and Ana Kupresanin*

CRC Press
Taylor & Francis Group
6000 Broken Sound Parkway NW, Suite 300
Boca Raton, FL 33487-2742

© 2014 by Taylor & Francis Group, LLC
CRC Press is an imprint of Taylor & Francis Group, an Informa business

No claim to original U.S. Government works

Printed on acid-free paper
Version Date: 20131025

International Standard Book Number-13: 978-1-4665-6159-5 (Hardback)

Library of Congress Cataloging-in-Publication Data

Implementing reproducible research / [edited by] Victoria Stodden, Friedrich Leisch, and Roger D. Peng.
    pages cm. -- (Chapman & Hall/CRC the R series)
    Includes bibliographical references and index.
    ISBN 978-1-4665-6159-5
    1. Reproducible research. 2. Research--Statistical methods. I. Stodden, Victoria, editor of compilation. II. Leisch, Friedrich, editor of compilation. III. Peng, Roger D., editor of compilation.

Q180.55.S7I47 2013
507.2--dc23
                                                                2013036694

Visit the Taylor & Francis Web site at
http://www.taylorandfrancis.com

and the CRC Press Web site at
http://www.crcpress.com

Contents

Part I Tools

Part II Practices and Guidelines

Part III   Platforms

Preface

Science moves forward when discoveries are replicated and reproduced. In general, the more frequently a given relationship is observed by independent scientists, the more trust we have that such a relationship truly exists in nature. Replication, the practice of independently implementing scientific experiments to validate specific findings, is the cornerstone of discovering scientific truth. Related to replication is reproducibility, which is the calculation of quantitative scientific results by independent scientists using the original datasets and methods. Reproducibility can be thought of as a different standard of validity from replication because it forgoes independent data collection and uses the methods and data collected by the original investigator (Peng et al., 2006). Reproducibility has become an important issue for more recent research due to advances in technology and the rapid spread of computational methods across the research landscape.

Much has been written about the rise of computational science and the complications computing brings to the traditional practice of science (Bailey et al. 2013; Birney et al. 2009; Donoho et al. 2009; Peng 2011; Stodden 2012; Stodden et al. 2013; Yale Roundtable 2010). Large datasets, fast computers, and sophisticated statistical modeling make a powerful combination for scientific discovery. However, they can also lead to a lack of reproducibility in computational science findings when inappropriately applied to the discovery process. Recent examples show that improper use of computational tools and software can lead to spectacularly incorrect results (e.g., Coombes et al. 2007). Making computational research reproducible does not guarantee correctness of all results, but it allows for quickly building on sound results and for rapidly rooting out unsound ones.

The sharing of analytic data and the computer codes used to map those data into computational results is central to any comprehensive definition of reproducibility. Except for the simplest of analyses, the computer code used to analyze a dataset is the only record that permits others to fully understand what a researcher has done. The traditional materials and methods sections in most journal publications are simply too short to allow for the inclusion of critical details that make up an analysis. Often, seemingly innocuous details can have profound impacts on the results, particularly when the relationships being examined are inherently weak. Some concerns have been raised over the sharing of code and data. For example, the sharing of data may allow other competing scientists to analyze the data and scoop the scientists who originally published the data, or the sharing of code may lead to the inability to monetize software through proprietary versions of the code. While these concerns are real and have not been fully resolved by the scientific community, we do not dwell on them in this book.

This book is focused on a simple question. Assuming one agrees that reproducibility of a scientific result is a good thing, _how do we do it_? In computational science, reproducibility requires that one make code and data available to others so that they may analyze the original data in a similar manner as in the original publication. This task requires that the analysis be done in such a way that preserves the code and data, and permits their distribution in a format that is generally readable, and a platform be available to the author on which the data and code can be distributed widely. Both data and code need to be licensed permissively enough so that others can reproduce the work without a substantial legal burden.

In this book, we cover many of the ingredients necessary for conducting and distributing reproducible research. The book is divided into three parts that cover the three principal areas: tools, practices, and platforms. Each part contains contributions from leaders in the area of reproducible research who have materially contributed to the area with software or other products.

Tools

Literate statistical programming is a concept introduced by Rossini, which builds on the idea of literate programming as described by Donald Knuth. With literate statistical programming, one combines the description of a statistical analysis and the code for doing the statistical analysis into a single document. Subsequently, one can take the combined document and produce either a human-readable document (i.e., PDF) or a machine-readable code file. An early implementation of this concept was the Sweave system of Leisch, which uses R as its programming language and LaTeX as its documentation language. Yihui Xie describes his knitr package, which builds substantially on Sweave and incorporates many new ideas developed since the initial development of Sweave. Along these lines, Tanu Malik and colleagues describe the Science Object Linking and Embedding framework for creating interactive publications that allow authors to embed various aspects of computational research in a document, creating a complete research compendium.

There have been a number of systems developed recently that are designed to track the provenance of data analysis outputs and to manage a researcher's workflow. Juliana Freire and colleagues describe the VisTrails system for open source provenance management for scientific workflow creation. VisTrails interfaces with existing scientific software and captures the inputs, outputs, and code that produced a particular result, even presenting this workflow in flowchart form. Andrew Davison and colleagues describe the Sumatra toolkit for reproducible research. Their goal is to introduce a tool for reproducible research that minimizes the disruption to scientists' existing workflows, therefore maximizing the uptake by current scientists. Their tool serves as a kind of "backend" to keep track of the code, data, and dependencies as a researcher works. This allows for easily reproducing specific analyses and for sharing with colleagues.

Philip Guo takes the "backend tracking" idea one step further and describes his Code, Data, Environment (CDE) package, which is a minimal "virtual machine" for reproducing the environment as well as the analysis. This package keeps track of all files used by a given program (i.e., a statistical analysis program) and bundles everything, including dependencies, into a single package. This approach guarantees that all requirements are included and that a given analysis can be reproduced on another computer.

Peter Murray-Rust and Dave Murray-Rust introduce The Declaraton, a tool for the precise mapping of mathematical expressions to computational implementations. They present an example from materials science, defining what reproducibility means in this field, in particular for unstable dynamical systems.

Practices and Guidelines

Conducting reproducible research requires more than the existence of good tools. Ensuring reproducibility requires the integration of useful tools into a larger workflow that is rigorous in keeping track of research activities. One metaphor is that of the lab notebook, now extended to computational experiments. Jarrod Millman and Fernando Pérez raise important points about how computational scientists should be trained, noting that many are not formally trained in computing, but rather pick up skills "on the go." They detail skills and tools that may be useful to computational scientists and describe a web-based notebook system developed in IPython that can be used to combine text, mathematics, computation, and results into a reproducible analysis. Titus Brown discusses tools that can be useful in the area of bioinformatics as well as good programming practices that can apply to a broad range of areas.

Holger Hoeing and Anthony Rossini present a case study in how to produce reproducible research in a commercial environment for large-scale data analyses involving teams of investigators, analysts, and stakeholders/clients. All scientific practice, whether in academia or industry, can be informed by the authors' experiences and the discussion of tools they used to organize their work.

Closely coupled with the idea of reproducibility is the notion of "open science," whereby results are made available to the widest audience possible through journal publications or other means. Luis Ibanez and colleagues give some thoughts on open science and reproducibility and trends that are either encouraging or discouraging it. Bill Howe discusses the role of cloud computing in reproducible research. He describes how virtual machines can be used to replicate a researcher's entire software environment and allow researchers to easily transfer that environment to a large number of people. Other researchers can then copy this environment and conduct their own research without having to go through the difficult task of reconstructing the environment from scratch.

Members of the Open Science Collaboration outline the need for reproducibility in all science and detail why most scientific findings are rarely reproduced. Reasons include a lack of incentives on the part of journals and investigators to publish reproductions or null findings. He describes the Reproducibility Project, whose goal is to estimate the reproducibility of scientific findings in psychology. This massive undertaking represents a collaboration of over 100 scientists to reproduce a sample of findings in the psychology literature. By spreading the effort across many people, the project overcomes some of the disincentives to reproducing previous work.

Platforms

Related to the need for good research practices to promote reproducibility is the need for software and methodological platforms on which reproducible research can be conducted and distributed. Mikio Braun and Cheng Soon Ong discuss the area of machine learning and place it in the context of open source software and open science. Aspects of the culture of machine learning have led to many open source software packages and hence reproducible methods.

Christophe Hurlin and colleagues, in Chapter 14, describe the RunMyCode platform for sharing reproducible research. This chapter addresses a critical need in the area of reproducible research, which is the lack of central infrastructure for distributing results. A key innovation of this platform is the use of cloud computing to allow research findings to be reproduced through the RunMyCode web interface, or on the user's local system via code and data download.

Perhaps the oldest "platform" for distributing research is the journal. Iain Hrynaszkiewicz and colleagues describe some of the infrastructure available for publishing reproducible research. In particular, they review how journal policies and practices in the growing field of open access journals encourage reproducible research.

Victoria Stodden provides a primer on the current legal and policy framework for publishing reproducible scientific work. While the publication of traditional articles is rather clearly covered by copyright law, the publication of data and code treads into murkier legal territory. Stodden describes the options available to researchers interested in publishing data and code and summarizes the recommendations of the Reproducible Research Standard.

Summary

We have divided this book into three parts: Tools, Practices and Guidelines, and Platforms. These mirror the composition of research happening in reproducibility today. Even just over the last two years, tool development for computational science has taken off. An early conference at Applied Mathematics Perspectives called "Reproducible Research: Tools and Strategies for Scientific Computing" in July of 2011 sought to encourage the nascent community of research tool builders (Stodden 2012). Recently, in December of 2012, a workshop entitled "Reproducibility in Computational and Experimental Mathematics" was held as part of the ICERM workshop series (ICERM Workshop 2012). A summary of the tools presented is available on the workshop wiki (Stodden 2012), and the growth of the field is evident. Additional material, including code and data, is available from the editor's website: www.ImplementingRR.org.

Journals are continuing to raise standards to ensure reproducibility in computational science (*Nature* Editorial 2013; Marcia 2014) and funding agencies have recently been instructed by the White House to develop plans for the open dissemination of data arising from federally funded research (Stebbins 2013). We feel that a book documenting available tools, practices, and dissemination platforms could not come at a better time.

References

Bailey, D.H., Borwein, J.M., LeVeque, R.J., Rider, B., Stein, W., and Stodden, V. (2012). Reproducibility in computational and experimental mathematics, in *ICERM Workshop*, December 10–14, 2012. http://icerm.brown.edu/tw12-5-.rcem.

Bailey, D.H., Borwein, J.M., Stodden, V., Set the default to 'Open,' notices of the American Mathematical Society, June/July 2013. http://www.ams.org/notices/201306/rnoti-p679.pdf.

Birney, E., Hudson, T. J., Green, E. D., Gunter, C., Eddy, S., Rogers, J., Harris, J. R. et al. (2009), Prepublication data sharing, *Nature*, 461, 168–170.

Coombes, K., Wang, J., and Baggerly, K. (2007), Microarrays: Retracing steps, *Nat. Med.*, 13, 1276–1277.

Donoho, D.L., Maleki, A., Rahman I.U., Shahram, M., and Stodden V., Reproducible research in computational harmonic analysis, computing in science and engineering, *IEEE Comput. Sci. Eng.*, 11(1), 8–18, Jan/Feb 2009, doi:10.1109/MCSE.2009.15.

McNutt, M. (2014, January 17), Reproducibility, *Science*, 343(6168), 229. http://www.sciencemag.org/content/343/6168/229.summary.

Nature Editorial (2013, April 24), Announcement: Reducing our irreproducibility, *Nature*, 496. http://www.nature.com/news/announcement-reducing-our-irreproducibility-1.12852.

Peng, R. D. (2011), Reproducible research in computational science, *Science*, 334, 1226–1227.

Peng, R. D., Dominici, F., and Zeger, S. L. (2006), Reproducible epidemiologic research, *Am. J. Epidemiol.*, 163, 783–789.

Stebbins, M. (2013), Expanding public access to the results of federally funded research, February 22, 2013. http://www.whitehouse.gov/blog/2013/02/22/expanding-public-access-results-federally-funded-research.

Stodden, V. (2012), Reproducible research: Tools and strategies for scientific computing, *Comput. Sci. Eng.*, 14(4), 11–12 July/August 2012. http://www.computer.org/csdl/mags/cs/2012/04/mcs2012040011-abs.html.

Stodden, V., Borwein, J., and Bailey, D.H. (2013), Setting the default to reproducible in computational science research, *SIAM News*, June 3, 2013. http://www.siam.org/news/news.php?id=2078.

Yale Roundtable (2010), Reproducible research: Addressing the need for data and code sharing in computational science, *IEEE Comput. Sci. Eng.*, 12, 8–13.

MATLAB® is a registered trademark of The MathWorks, Inc. For product information, please contact:

The MathWorks, Inc.
3 Apple Hill Drive
Natick, MA 01760-2098 USA
Tel: 508 647 7000
Fax: 508-647-7001
Email: info@mathworks.com
Web: www.mathworks.com

Acknowledgment

The editors acknowledge the generous support of Awards R01ES019560 and R21ES020152 from the National Institute of Environmental Health Sciences (Peng), NSF Award 1153384 "EAGER: Policy Design for Reproducibility and Data Sharing in Computational Science" (Stodden), and the Sloan Foundation Award "Facilitating Transparency in Scientific Publishing" (Stodden). The content is solely the responsibility of the authors and does not necessarily represent the official views of the National Institute of Environmental Health Sciences, the National Institutes of Health, the National Science Foundation, or the Alfred P. Sloan Foundation.

Editors

Victoria Stodden is an assistant professor of statistics at Columbia University and affiliated with the Columbia University Institute for Data Sciences and Engineering, New York City, New York. Her research centers on the multifaceted problem of enabling reproducibility in computational science. This includes studying adequacy and robustness in replicated results, designing and implementing validation systems, developing standards of openness for data and code sharing, and resolving legal and policy barriers to disseminating reproducible research. She is the developer of the award-winning "Reproducible Research Standard," a suite of open licensing recommendations for the dissemination of computational results.

Friedrich Leisch is head of the Institute of Applied Statistics and Computing at the University of Natural Resources and Life Sciences in Vienna. He is a member of the R Core Team, the original creator of the Sweave system in R, and has published extensively about tools for reproducible research. He is also a leading researcher in the area of high-dimensional data analysis.

Roger D. Peng is an associate professor in the Department of Biostatistics at the Johns Hopkins Bloomberg School of Public Health, Baltimore, Maryland. He is a prominent researcher in the areas of air pollution and health risk assessment and statistical methods for environmental health data. Dr. Peng is the associate editor for reproducibility for the journal *Biostatistics* and is the author of numerous R packages.

Contributors

Mikio L. Braun
Department of Computer Science
Technical University of Berlin
Berlin, Germany

C. Titus Brown
Department of Computer Science
and Engineering
and
Department of Microbiology and
Molecular Genetics
Michigan State University
East Lansing, Michigan

Fernando Chirigati
Department of Computer Science
and Engineering
Polytechnic Institute of New York
University
Brooklyn, New York

Andrew P. Davison
Unité de Neurosciences, Information
& Complexité
Centre National de la Recherche
Scientifique
Gif sur Yvette, France

Scott Edmunds
Beijing Genomics Institute
Beijing, People's Republic of China

Juliana Freire
Department of Computer Science
and Engineering
Polytechnic Institute of New York
University
Brooklyn, New York

Philip J. Guo
University of Rochester
Rochester, New York

Marcus D. Hanwell
Kitware, Inc.
Clifton Park, New York

Holger Hoefling
Novartis, Pharma
Basel, Switzerland

Bill Howe
Scalable Data Analytics
University of Calabria
Rende, Italy

and

eScience Institute
and
Department of Computer Science
and Engineering
University of Washington
Seattle, Washington

Iain Hrynaszkiewicz
Outreach Director
Faculty of 1000
London, United Kingdom

Christophe Hurlin
Department of Economics
University of Orléans
Orléans, France

Luis Ibanez
Kitware, Inc.
Clifton Park, New York

David Koop
Department of Computer Science
   and Engineering
Polytechnic Institute of New York
   University
Brooklyn, New York

Peter Li
Giga Science
Beijing Genomics Institute
Beijing, People's Republic of China

Michele Mattioni
European Molecular Biology
   Laboratory
European Bioinformatics Institute
Hinxton, United Kingdom

K. Jarrod Millman
Division of Biostatistics
School of Public Health
University of California, Berkeley
Berkeley, California

Dave Murray-Rust
Department of Informatics
University of Edinburgh
Edinburgh, Scotland

Peter Murray-Rust
Department of Chemistry
University of Cambridge
Cambridge, United Kingdom

Cheng Soon Ong
Bioinformatics Group
National ICT Australia
University of Melbourne
Melbourne, Victoria, Australia

Open Science Collaboration
Charlottesville, Virginia

Fernando Pérez
Henry H. Wheeler Jr. Brain Imaging
   Center
Helen Wills Neuroscience Institute
University of California, Berkeley
Berkeley, California

Christophe Pérignon
Finance Department
Hautes études commerciales de
   Paris
Paris, France

Likit Preeyanon
Department of Microbiology and
   Molecular Genetics
Michigan State University
East Lansing, Michigan

Alexis Black Pyrkosz
Avian Disease and Oncology
   Laboratory
East Lansing, Michigan

Anthony Rossini
Novartis, Pharma
Basel, Switzerland

Dmitry Samarkanov
Ecole Centrale de Lille
Lille University of Science and
   Technology
Villeneuve-d'Ascq, France

William J. Schroeder
Kitware, Inc.
Clifton Park, New York

Cláudio T. Silva
Polytechnic Institute of New York
   University
Brooklyn, New York

Victoria Stodden
Department of Statistics
Columbia University
New York City, New York

Bartosz Teleńczuk
Unité de Neurosciences, Information
    & Complexité
Centre National de la Recherche
    Scientifique
Gif sur Yvette, France

and

Institute for Theoretical Biology
Humboldt University
Berlin, Germany

Yihui Xie
Department of Statistics
Iowa State University
Ames, Iowa

Part I

Tools

1

knitr: A Comprehensive Tool for Reproducible Research in R

Yihui Xie

CONTENTS

Reproducibility is the ultimate standard by which scientific findings are judged. From the computer science perspective, reproducible research is often related to literate programming [13], a paradigm conceived by Donald Knuth, and the basic idea is to combine computer code and software documentation in the same document; the code and documentation can be identified by different special markers. We can either compile the code and mix the results with documentation or extract the source code from the document. To some extent, this implies reproducibility because everything is generated automatically from computer code, and the code can reflect all the details about computing.

Early implementations like WEB [12] and Noweb [20] were not directly suitable for data analysis and report generation, which was partly overcome by later tools like Sweave [14]. There are still a number of challenges that were not solved by existing tools; for example, Sweave is closely tied to LaTeX and hard to extend. The **knitr** package [28,29] was built upon the ideas of previous tools with a framework redesigned, enabling easy and fine control of many aspects of a report. Sweave can be regarded as a subset of **knitr** in terms of the features.

In this chapter, we begin with a simple but striking example that shows how reproducible research can become natural practice to authors given a simple and appealing tool. We introduce the design of the package in Section 1.2 and how it works with a variety of document formats, including LaTeX, HTML, and Markdown. Section 1.3 lists the features that can be useful to data analysis such as the cache system and graphics support. Section 1.4 covers advanced features that extend **knitr** to a comprehensive environment for data analysis; for example, other languages such as Python, awk, and shell scripts can also be integrated into the **knitr** framework. We will conclude with a few significant examples, including student homework, data reports, blog posts, and websites built with **knitr**.

The main design philosophy of **knitr** is to make reproducible research easier and more enjoyable than the common practice of cut-and-paste results. This package was written in the R language [11,19]. It is freely available on CRAN (Comprehensive R Archive Network) and documented in its website http://yihui.name/knitr/; the development repository is on Github: https://github.com/yihui/knitr, where users can file bug reports and feature requests and participate in the development.

There are obvious advantages of writing a literate programming document over copying and pasting results across software packages and documents. An overview of literate programming applied to statistical analysis can be found in [22]; [8] introduced general concepts of literate programming documents for statistical analysis, with a discussion of the software architecture; [7] is a practical example based on [8], using an R package **GolubRR** to distribute reproducible analysis; and [2] revealed several problems that may arise with the standard practice of publishing data analysis results, which can lead to false discoveries due to lack of enough details for reproducibility (even with datasets supplied). Instead of separating results from computing, we can actually put everything in one document (called a *compendium* in [8]), including the computer code and narratives. When we compile this document, the computer code will be executed, giving us the results directly.

This is the central idea of this chapter—we go from the source code to the report in one step, and everything is automated by the source code.

1.1 Web Application

R Markdown (referred to as *Rmd* hereafter) is one of the document formats that **knitr** supports, and it is also the simplest one. Markdown [10] is a both easy-to-read and easy-to-write language that was designed primarily for writing web content easily and can be translated to HTML (e.g., `**text**` translates to `<strong>text</strong>`). What follows is a trivial example of how Rmd looks like:

```
# First section

Description of the methods.

```{r brownian-motion, fig.height=4, fig.cap='Brownian Motion'}
x <- cumsum(rnorm(100))
plot(x)
```

The mean of x is `r mean(x)`.
```

We can compile this document with **knitr**, and the output will be an HTML web page containing all the results from R, including numeric and graphical results. This is not only easier for authors to write a report but also guarantees a report is reproducible since no cut-and-paste operations are involved. To compile the report, we only need to load the **knitr** package in R and call the `knit()` function:

```
library(knitr)
knit("myfile.Rmd")   # suppose we saved the above file as
   myfile.Rmd
```

Based on this simple idea, **knitr** users have contributed hundreds of reports to the hosting website RPubs (http://rpubs.com) within a few months since it was launched, ranging from student homework, data analysis reports, HTML5 slides, and class quizzes. Traditionally, literate programming tools often choose LaTeX as the authoring environment, which has a steep learning curve for beginners. The success of R Markdown and RPubs

shows that one does not have to be a typesetting expert in order to make use
of literate programming and write reproducible reports.

1.2 Design

The package design consists of three components: parser, evaluator, and
renderer. The parser identifies and extracts computer code from the source
document; the evaluator executes the code; and the renderer generates the
final output by appropriately marking up the results according to the output
format.

1.2.1 Parser

To include computer code into a document, we have to use special patterns
to separate it from normal texts. For instance, the Rmd example in Section
1.1 has an R code chunk that starts with ``` ```{r} ``` and ends with ``` ``` ```.

Internally, **knitr** uses the object `knit_patterns` to set or get the pattern rules, which are essentially regular expressions. Different document
formats use different sets of regular expressions by default, and all built-in
patterns are stored in the object `all_patterns` as a named list. For example, `all_patterns$rnw` is a set of patterns for the Rnw format, which has
an R code embedded in a LaTeX document using the Noweb syntax. Similarly, **knitr** has default syntax patterns for other formats like Markdown
(`md`), HTML (`html`), and reStructuredText (`rst`). We take the Rnw syntax
for example.

```
library(knitr)
names(all_patterns)  # all built-in document formats
```

```
## [1] "rnw"  "brew" "tex"  "html" "md"   "rst"
```

```
all_patterns$rnw[c("chunk.begin", "chunk.end", "inline.code")]
```

```
## $chunk.begin
## [1] "^\\s*<<(.*)>>="
##
## $chunk.end
## [1] "^\\s*@\\s*(%+.*|)$"
##
## $inline.code
## [1] "\\\\Sexpr\\{([^}]+)\\}"
```

In the pattern list for the Rnw format, there are three major elements as shown earlier: `chunk.begin`, `chunk.end`, and `inline.code`, which are regular expressions indicating the patterns for the beginning and ending of a code chunk, and inline code, respectively. For example, the regular expression `^\s*<<(.*)>>=` means the pattern for the beginning of a code chunk is: in the beginning (`^`) of this line, there are at most some white spaces (`\s*`), then the chunk header starts with `<<`; inside the chunk header, there can be some texts denoting chunk options (`(.*)`), which can be regarded as metadata for a chunk (e.g., `fig.height=4` means the figure height will be 4 in. for this chunk); the chunk header is closed by `>>=`. The code chunk is usually closed by `@` (white spaces are allowed before it and TEX comments are allowed after it), and we can also write inline code inside the pseudo TEX command `\Sexpr{}`. What follows is an example of a fragment of an Rnw document:

```
\section{First section}

Description of the methods.

<<brownian-motion, fig.height=4, fig.cap='Brownian  Motion'>>=
x <- cumsum(rnorm(100))
plot(x)
@

The mean of x is \Sexpr{mean(x)}.
```

Based on the Rnw syntax, **knitr** will find out the code chunk, as well as the inline code `mean(x)`. Anything else in the document will remain untouched and will be mixed with the results from the computer code eventually. To show the parser can be easily generalized, we take a look at the Rmd syntax as well:

```
str(all_patterns$md[c("chunk.begin", "chunk.end",
  "inline.code")])

## List of 3
##  $ chunk.begin: chr "^\\s*`{3,}\\s*\\{r(.*)\\}\\s*$"
##  $ chunk.end  : chr "^\\s*`{3,}\\s*$"
##  $ inline.code: chr "`r +([^`\n]+)\\s*`"
```

Roughly speaking, the three major patterns are changed to ```` ```{r *} ```` (beginning), ```` ``` ```` (ending), and `` `r *` `` (inline), respectively. If we want to specify our own syntax, we can use the `knit_patterns$set()` function, which will override the default syntax, for example:

```
knit_patterns$set(chunk.begin = "^<<r(.*)", chunk.end =
   "^r>>$", inline.code = "\\{\\{([^}]+)\\}\\}")
```

Then, we will be able to parse a document like this with the custom syntax:

```
<<r brownian-motion, fig.height=4, fig.cap='Brownian  Motion'
x <- cumsum(rnorm(100))
plot(x)
r>>

The mean of x is {{mean(x)}}.
```

In practice, however, this kind of customization is often unnecessary. It is better to follow the default syntax, otherwise additional instructions will be required in order to compile a literate programming document. Table 1.1 shows all the document formats that are currently supported by **knitr**.

Among all chunk options, there is a special option called the chunk label. It is the only chunk option that does not have to be of the form `option = value`. The chunk label is supposed to be a unique identifier of a code chunk, which will be used as the filename for figure files, cache files, and also ids for chunk references. We will mention these later in Section 1.3.

1.2.2 Evaluator

Once we have the code chunks and inline code expressions extracted from the document, we need to evaluate them. The **evaluate** package [26] is used to execute code chunks, and the `eval()` function in base R is used to execute the inline R code. The latter is easy to understand and is made possible by

TABLE 1.1

Code Syntax for Different Document Formats

Format	Start	End	Inline	Output
Rnw	<<*>>=	@	\Sexpr x	TEX
Rmd	```{r *}	```	'r x'	Markdown
Rhtml	<!-begin.rcode *	end.rcode->	<!-rinline x->	HTML
Rrst	.. {r *}	.. ..	:r:'x'	reST
Rtex	% begin.rcode *	% end.rcode	\rinlinex	TEX
brew			<% x %>	text

* Denotes local chunk options, for example, `<<label, eval=FALSE>>=`; x denotes inline R code, for example, `<% 1+2 %>`.

the power of "computing on the language" [18] of R. Suppose we have a code fragment 1+1 as a character string, we can parse and evaluate it as an R code:

```
eval(parse(text = "1+1"))
```

```
## [1] 2
```

For code chunks, it is more complicated. The **evaluate** package takes a piece of R source code, evaluates it, and returns a list containing the results of six possible classes: `character` (normal text output), `source` (source code), `warning`, `message`, `error`, and `recordedplot` (plots).

```
library(evaluate)
res <- evaluate(c("'hello world!'", "1:2+1:3"))
str(res, nchar.max = 37)
```

```
## List of 5
##  $ :List of 1
##  ..$ src: chr "'hello world!'\n"
##  ..- attr(*, "class")= chr "source"
##  $ : chr "[1] \"hello world!\"\n"
##  $ :List of 1
##  ..$ src: chr "1:2+1:3"
##  ..- attr(*, "class")= chr "source"
##  $ :List of 2
##  ..$ message: chr "longer object length is not a
##     multip"| __truncated__
##  ..$ call    : language 1:2 + 1:3
##  ..- attr(*, "class")= chr [1:3] "simpleWarning"
##     "warning" "condition"
##  $ : chr "[1] 2 4 4\n"
```

An internal S3 generic function *wrap()* in **knitr** is used to deal with different types of output using output hooks defined in the object `knit_hooks`, which constitutes the renderer. Before the final output is rendered, we may have to postprocess the output from **evaluate** according to the chunk options. For example, if the chunk option is `echo=FALSE`, we need to remove the source code. This is one advantage of using the **evaluate** package because we can easily filter out the result elements that we do not want according to the classes of the elements. Continuing the aforementioned example, we can remove the source code by

```
## filter out elements which are not source
res <- Filter(Negate(is.source), res)
str(res, nchar.max = 37)
```

```
## List of 3
##  $ : chr "[1] \"hello world!\"\n"
##  $ :List of 2
##   ..$ message: chr "longer object length is not a
     multip"| __truncated__
##   ..$ call   : language 1:2 + 1:3
##   ..- attr(*, "class")= chr [1:3] "simpleWarning" "warning"
     "condition"
##  $ : chr "[1] 2 4 4\n"
```

Similarly, we can process other elements according to the chunk options; for instance, `warning=FALSE` means to remove warning messages, and `results='hide'` means to remove elements of the class `character`; **knitr** has a large number of chunk options to tweak the output, which are documented at http://yihui.name/knitr/options.

One notable feature of the **evaluate** package that may be surprising to most R users is that it does not stop on errors by default. This is to mimic the behavior of R when we copy and paste R code in the console (or terminal): If an error occurs in a previous R expression, the rest of the code will still be pasted and executed. To completely stop on errors, we need to set a chunk option in **knitr**:

```
opts_chunk$set(error = FALSE)
```

1.2.3 Renderer

Unlike other implementations such as Sweave, **knitr** makes almost everything accessible to the users, including every piece of results returned from **evaluate**. The users are free to write these results in any formats they like via output hook functions. Consider the following simple example:

```
1 + 1
```

```
## [1] 2
```

There are two parts in the returned results: the source code 1 + 1 and the output [1] 2. Users may define a hook function for the source code like this to use the `lstlisting` environment in LaTeX:

```
knit_hooks$set(source = function(x, options) {
    paste("\\begin{lstlisting}\n", x, "\\end{lstlisting}\n",
        sep = "")
})
```

Or put it inside the `<pre>` tag with a CSS class `source` in HTML:

```
knit_hooks$set(source = function(x, options) {
    paste("<pre class='source'>", x, "</pre>", sep = "")
})
```

Here, the name of the hook function corresponds to the class of the element returned from **evaluate**; see Table 1.2 for the mapping between the two sets of names. The argument `x` of the hook denotes the corresponding output (a character string), and `options` is a list of chunk options for the current code chunk, for example, `options$fig.width` is a numeric value that determines the width of figures in the current chunk. Note that there are two additional output hooks called `chunk` and `document`. The chunk hook takes the output of the whole chunk as input, which has been processed by the previous six output hooks; the `document` hook takes the output of the whole document as input and allows further postprocessing of the output text.

Like the parser, **knitr** also has a series of default output hooks for different document formats, so users do not have to rewrite the renderer in most cases.

TABLE 1.2

Output Hook Functions and the Object Classes of Results from the **evaluate** Package

Class	Output Hook	Arguments
source	source	x, options
character	output	x, options
recordedplot	plot	x, options
message	message	x, options
warning	warning	x, options
error	error	x, options
	chunk	x, options
	document	x

1.3 Features

The **knitr** package borrowed features such as TikZ graphics [25] and cache from **pgfSweave** [3] and **cacheSweave** [16], respectively, but the implementations are completely different. New features like code reference from an external R script, as well as output customization, are also introduced. The feature of hook functions in Sweave was reimplemented and hooks have extended power now. Special emphasis was put on graphics: there can be any number of plots per chunk, there are more than 20 graphical devices to choose from (PDF, PNG, and Cairo devices), and it is also easy to specify the size and alignment of plots via chunk options.

There are several other small features that were motivated from the experience of using Sweave. For example, a progress bar is provided when knitting a file so we more or less know how long we still need to wait; output from inline R code (e.g., \Sexpr{x[1]}) is automatically formatted in scientific notation (like 1.2346×10^8) if the result is numeric (this applies to all document formats), and we will not get too many digits by default (the default number in R is 7, which is too long).

As we emphasize the ease of use, the concept of an "R Notebook" was also introduced in this package, which enables one to write a pure R script to create a report, and **knitr** will take care of the details of formatting and compilation.

1.3.1 Code Decoration

Syntax highlighting comes by default in **knitr** (chunk option highlight= TRUE) since we believe it enhances the readability of the source code. The **formatR** [27] is used to reformat R code (option tidy=TRUE), for example, add spaces and indentation, break long lines into shorter ones, and automatically replace the assignment operator = to <-; see the manual of **formatR** for details.

For LaTeX output, the **framed** package is used to decorate code chunks with a light gray background (as we can see in this document). If this LaTeX package is not found in the system, a version will be copied directly from **knitr**. The output for HTML documents is styled with CSS, which looks similar to LaTeX (with gray shadings and syntax highlighting).

The prompt characters are removed by default because they mangle the R source code in the output and make it difficult to copy the R code. The R output is masked in comments by default based on the same rationale. In fact, this was largely motivated from my experience of grading homework; with the default prompts, it is difficult to verify the results in the homework because it is so inconvenient to copy the source code. Anyway, it is easy to revert to the output with prompts (set option prompt=TRUE), and we will

quickly realize the inconvenience to the readers if they want to run the code
in the output document:

```
> x <- rnorm(5)
> x
[1] -0.56048 -0.23018  1.55871  0.07051  0.12929
> var(x)
[1] 0.6578
```

The example below shows the effect of tidy=TRUE/FALSE:

```
## option tidy=FALSE
for(k in 1:10){j=cos(sin(k)*k^2)+3;print(j-5)}
```

```
## option tidy=TRUE
for (k in 1:10) {
    j <- cos(sin(k) * k^2) + 3
    print(j - 5)
}
```

While this may seem to be irrelevant to reproducible research, we would
argue that it is of great importance to design styles that look appealing and
helpful at the first glance, which can encourage users to write reports in
this way.

1.3.2 Graphics

Graphics is an important part of reports, and several enhancements have
been made in **knitr**. For example, **grid** graphics [15] may not need to be
explicitly printed as long as the same code can produce plots in the R console
(in some cases, however, they have to be printed, e.g., in a loop, because we
have to do so in an R console); what follows is a chunk of code that will
produce a plot in both the R console and the **knitr**:

```
library(ggplot2)
p <- qplot(carat, price, data = diamonds) + geom_hex()
p  # no need to print(p)
```

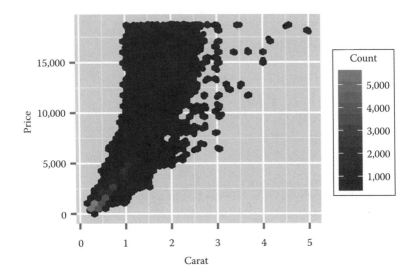

1.3.2.1 Graphical Devices

Over a long time, a frequently requested feature for Sweave was the support for other graphics devices, which has been implemented since R 2.13.0. Instead of using several logical options like png or jpeg, **knitr** uses a single option dev (like grdevice in Sweave), which has support for more than 20 devices. For instance, dev='png' will use the png() device in the **grDevices** package in base R, and dev='CairoJPEG' uses the CairoJPEG() device in the add-on package **Cairo** (it has to be installed first, of course). Here are the possible values for dev:

```
##  [1] "bmp"            "postscript"    "pdf"           "png"
##  [5] "svg"            "jpeg"          "pictex"        "tiff"
##  [9] "win.metafile"   "cairo_pdf"     "cairo_ps"      "quartz_pdf"
## [13] "quartz_png"     "quartz_jpeg"   "quartz_tiff"   "quartz_gif"
## [17] "quartz_psd"     "quartz_bmp"    "CairoJPEG"     "CairoPNG"
## [21] "CairoPS"        "CairoPDF"      "CairoSVG"      "CairoTIFF"
## [25] "Cairo_pdf"      "Cairo_png"     "Cairo_ps"      "Cairo_svg"
## [29] "tikz"
```

If none of these devices is satisfactory, we can provide the name of a customized device function, which must have been defined in this form before it is used:

```
custom_dev <- function(file, width, height, ...) {
    # open the device here, e.g. pdf(file, width, height, ...)
}
```

Then, we can set the chunk option dev='custom_dev'.

1.3.2.2 *Plot Recording*

All the plots in a code chunk are first recorded as R objects and then "replayed" inside a graphical device to generate plot files. The **evaluate** package will record plots per *expression* basis; in other words, the source code is split into individual complete expressions and **evaluate** will examine the possible plot changes in snapshots after each single expression has been evaluated. For example, the following code consists of three expressions, out of which two are related to drawing plots, therefore **evaluate** will produce two plots by default:

```
par(mar = c(3, 3, 0.1, 0.1))
plot(1:10, ann = FALSE, las = 1)
text(5, 9, "mass $\\rightarrow$ energy\n$E=mc^2$")
```

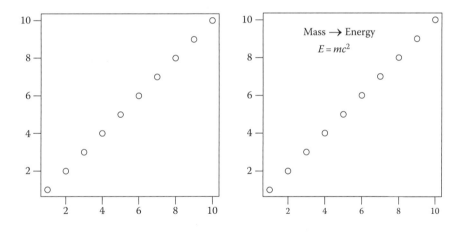

This brings a significant difference with traditional tools in R for dynamic report generation since low-level plotting changes can also be recorded. The option `fig.keep` controls which plots to keep in the output; `fig.keep='all'` will keep low-level changes in separate plots; by default (`fig.keep='high'`), **knitr** will merge low-level plot changes into the previous high-level plot, like most graphics devices do. This feature may be useful for teaching R graphics step by step. Note, however, that low-level plotting commands in a single expression (a typical case is a loop) will not be recorded cumulatively, but high-level plotting commands, regardless of where they are, will always be recorded. For example, this chunk will only produce 2 plots instead of 21 plots because there are 2 complete expressions:

```
plot(0, 0, type = "n", ann = FALSE)
for (i in seq(0, 2 * pi, length = 20)) points(cos(i), sin(i))
```

But this will produce 20 plots as expected:

```
for (i in seq(0, 2 * pi, length = 20)) {
    plot(cos(i), sin(i), xlim = c(-1, 1), ylim = c(-1, 1))
}
```

We can discard all previous plots and keep the last one only by
`fig.keep='last'`, or keep only the first plot by `fig.keep='first'`, or
discard all plots by `fig.keep='none'`.

1.3.2.3 Plot Rearrangement

The chunk option `fig.show` can decide whether to hold all plots while
evaluating the code and "flush" all of them to the end of a chunk
(`fig.show='hold'`; see the previous plot example), or just insert them to
the places where they were created (by default `fig.show='asis'`). Here is
an example of `fig.show='asis'` for two plots in one chunk:

```
contour(volcano)   # contour lines
```

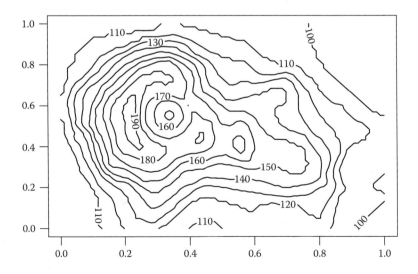

```
filled.contour(volcano)   # fill contour plot with colors
```

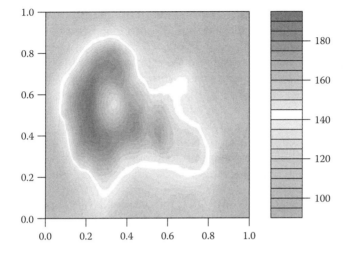

Besides 'hold' and 'asis', the option fig.show can take a third value, 'animate', which makes it possible to insert animations into the output document. In LaTeX, the package **animate** is used to put together image frames as an animation. For animations to work, there must be more than one plot produced in a chunk. The option interval controls the time interval between animation frames; by default it is 1 s. Note that we have to add \usepackage{animate} in the LaTeX preamble because **knitr** will not add it automatically. Animations in the PDF output can only be viewed in Adobe Reader. There are animation examples in both the main manual and graphics manual of **knitr**, which can be found on the package website.

We can specify the figure alignment via the chunk option fig.align ('left', 'center', and 'right'). The plot example in the previous section used fig.align='center' so the two plots were centered.

1.3.2.4 Plot Size

The fig.width and fig.height options specify the size of plots in the graphics device (units in inches), and the real size in the output document can be different (specified by out.width and out.height). When there are multiple plots per code chunk, it is possible to arrange multiple plots side by side. For example, in LaTeX, we only need to set out.width to be less than half of the current line width, for example, out.width='.49\\linewidth'.

1.3.2.5 Tikz Device

Besides PDF, PNG, and other traditional R graphical devices, **knitr** has special support to TikZ graphics via the **tikzDevice** package [24], which is similar to the feature of **pgfSweave**. If we set the chunk option dev='tikz', the tikz() device in **tikzDevice** will be used to generate plots. The options sanitize (for escaping special TEX characters) and external are related to the tikz device: see the documentation of tikz() for details. Note that external=TRUE in **knitr** has a different meaning with **pgfSweave**— it means standAlone=TRUE in tikz(), and the TikZ graphics output will be compiled to PDF *immediately* after it is created, so the "externalization" does not depend on the official but complicated externalization commands in the **tikz** package in LATEX. To maintain consistency in (font) styles, **knitr** will read the preamble of the input document and pass it to the tikz device so that the font style in the plots will be the same as the style of the whole LATEX document.

Besides consistency of font styles, the tikz device also enables us to write arbitrary LATEX expressions into R plots. A typical use is to write math expressions. The traditional approach in R is to use an expression() object to write math symbols in the plot, and for the tikz device, we only need to write normal LATEX code. What follows is an example of a math expression $p(\theta|\mathbf{x}) \propto \pi(\theta)f(\mathbf{x}|\theta)$ using the two approaches, respectively:

```
plot(0, type = "n", ann = FALSE)
text(0, expression(p(theta ~ "|" ~ bold(x)) %prop% pi(theta)
    * f(bold(x) ~ "|" ~ theta)), cex = 2)
```

$$p(\theta \mid \mathbf{x}) \propto \pi(\theta)f(\mathbf{x} \mid \theta)$$

With the tikz device, it is both straightforward (if we are familiar with LATEX) and more beautiful:

```
plot(0, type = "n", ann = FALSE)
text(0, "$p(\\theta|\\mathbf{x})\\propto\\pi(\\theta)
    f(\\mathbf{x}|\\theta)$", cex = 2)
```

$$p(\theta|\mathbf{x}) \propto \pi(\theta)\,f(\mathbf{x}|\theta)$$

One disadvantage of the tikz device is that LATEX may not be able to handle too large tikz files (it can run out of memory). For example, an R plot with tens of thousands of graphical elements may fail to compile in LATEX if we use the tikz device. In such cases, we can switch to the PDF or PNG device, or reconsider our decision on the type of plots, for example, a scatter plot with millions of points is usually difficult to read, and a contour plot or a hexagon plot showing the 2D density can be a better alternative (they are smaller in size).

We emphasized the uniqueness of chunk labels in Section 1.2.1, and here is one reason why it has to be unique: the chunk label is used in the filenames of plots; if there are two chunks that share the same label, the latter chunk will override the plots generated in the previous chunk. The same is true for cache files in the next section.

1.3.3 Cache

The basic idea of cache is that we directly load results from a previous run instead of recompute everything from scratch if nothing has been changed since the last run. This is not a new idea—both **cacheSweave** [16] and **weaver** [6] have implemented it based on Sweave, with the former using **filehash** [17] and the latter using .RData images; **cacheSweave** also supports lazy-loading of objects based on **filehash**. The **knitr** package directly uses internal base R functions to save (`tools:::makeLazyLoadDB()`) and lazy-load objects (`lazyLoad()`). The **cacheSweave** vignette has clearly explained lazy-loading; roughly speaking, lazy-loading means an object will not be really loaded into memory unless it is really used somewhere. This is very useful for cache; sometimes, we read a large object and cache it, then take a subset for analysis and this subset is also cached; in the future, the initial large object will not be loaded into R if our computation is only based on the subset object.

The paths of cache files are determined by the chunk option `cache.path`; by default all cache files are created under a directory cache/ relative to the current working directory, and if the option value contains a directory (e.g., `cache.path='cache/abc-'`), cache files will be stored under the directory cache/ (automatically created if it does not exist) with a prefix abc-. The cache is invalidated and purged on any changes to the code chunk, including both the R code and chunk options; this means previous cache files of this chunk are removed (filenames are identified by the chunk label) and a new set of cache files will be written. The change is detected by verifying if the MD5 hash of the code and options has changed, which is calculated from the **digest** package [5].

Two new features that make **knitr** different from other packages are as follows: cache files will never accumulate since old cache files will always be removed, and **knitr** will also try to preserve side effects such as printing and loading add-on packages. However, there are still other types of side effects like setting `par()` or `options()`, which are not cached. Users should be aware of these special cases and make sure to clearly divide the code that is not meant to be cached into other chunks that are not cached, for example, set all global options in the first chunk of a document and do not cache that chunk.

Sometimes, a cached chunk may need to use objects from other cached chunks, which can bring a serious problem—if objects in previous chunks have changed, this chunk will not be aware of the changes and will still use old cached results, unless there is a way to detect such changes from other chunks. There is an option called `dependson` in **cacheSweave**, which does this job. In **knitr**, we can also explicitly specify which other chunks this chunk depends on by setting an option like `dependson=c('chunkA',` `'chunkB')` (a character vector of chunk labels). Each time the cache of a chunk is rebuilt, all other chunks that depend on this chunk will lose cache, hence their cache will be rebuilt as well.

There are two alternative approaches to specify chunk dependencies: `dep_auto()` and `dep_prev()`. For the former, we need to turn on the chunk option `autodep` (i.e., set `autodep=TRUE`), then put `dep_auto()` in the first chunk in a document. This is an experimental feature borrowed from **weaver** that frees us from setting chunk dependencies manually. The basic idea is, if a latter chunk uses any objects created from a previous chunk, the latter chunk is said to depend on the previous one. The function `findGlobals()` in the **codetools** package is used to find out all global objects in a chunk, and according to its documentation, the result is an approximation. Global objects roughly mean the ones that are not created locally, for example, in the expression `function() {y <- x}`, x should be a global object, whereas y is local. Meanwhile, we also need to save the list of objects created in each cached chunk so that we can compare them to the global objects in latter chunks. For example, if chunk A created an object x and chunk B uses this object, chunk B must depend on A, that is, whenever A changes, B must also be updated. When `autodep=TRUE`, **knitr** will write out the names of objects created in a cached chunk as well as those global objects in two files named __objects and __globals, respectively; later we can use the function `dep_auto()` to analyze the object names to figure out the dependencies automatically. For `dep_prev()`, it is a very conservative approach that sets the dependencies so that a cached chunk will depend on *all* of its previous chunks, that is, whenever a previous chunk is updated, all later chunks will be updated accordingly; similarly, this function needs to be called in the first code chunk in a document.

1.3.4 Code Externalization

It can be more convenient to write R code in a separate file rather than mixing it into a literate programming document; for example, we can run R code successively in a pure R script from one chunk to the other without jumping through other text chunks. This may not sound important for some editors that support interaction with R, such as RStudio (http://www.rstudio.com/ide) or Emacs with ESS [21], since we can send

R code chunks directly from the editor to R, but for other editors like LYX (http://www.lyx.org), we can only compile the whole report as a batch job, which can be inconvenient when we only want to know the results of a single chunk.

The second reason for the feature of code externalization is to be able to reuse code across different documents. Currently the setting is like this: the external R script also has chunk labels for the code in it (marked in the form ## @knitr chunk-label by default); if the code chunk in the input document is empty, **knitr** will match its label with the label in the R script to input external R code. For example, suppose this is a code chunk labeled as Q1 in an R script named mycode.R, which is under the same directory as the source document:

```
## @knitr Q1
#' find the greatest common divisor of m and n
gcd <- function(m, n) {
    while ((r <- m%%n) != 0) {
        m <- n
        n <- r
    }
    n
}
```

In the source document, we can first read the script using the function read_chunk(), which is available in **knitr**:

```
read_chunk("mycode.R")
```

This is usually done in an early chunk, and we can use the chunk Q1 later in the source document (e.g., an Rnw document):

```
<<Q1, echo=TRUE, tidy=TRUE>>=
@
```

Different documents can read the same R script, so the R code can be reusable across different input documents. In a large project, however, this may not be an ideal approach to organizing code since there are too many code fragments. We may consider an R package to organize functions, which can be easier to call and test.

1.3.5 Chunk Reference

Code externalization is one way to reuse code chunks across documents, and for a single document, all its code chunks are also reusable in this document.

and `spin()`. The idea of "stitch" is we fit an R script into a predefined template in **knitr** (choices of templates include LATEX, HTML, and Markdown) and compile the mixed document to a report; all the code in the script will be put into one single chunk. The idea of "spin" is to write a specially formatted script, with normal texts masked in roxygen comments (i.e., after #′) and chunk options after #+. Here is an example for `spin()`:

```
#' This is a report.
#'
#+ chunkA, eval=TRUE
# generate data
x <- rnorm(100)
#'
#' The report is done.
```

This script will be parsed and translated to one of the document formats that **knitr** supports (Table 1.1), and then compiled to a report. This can be done through a single click in RStudio or we can also call the functions manually in R:

```
library(knitr)
stitch("mycode.R")   # stitch it, or spin it
spin("mycode.R")
```

1.4 Extensibility

The **knitr** package is highly extensible. We have seen in Section 1.2 that both the syntax patterns and output hooks can be customized. In this section, we introduce two new concepts: chunk hooks and language engines.

1.4.1 Hooks

A chunk hook (not to be confused with the output hooks) is a function to be called when a corresponding chunk option is not `NULL`, and the returned value of the function is written into the output if it is `character`. All chunk hooks are also stored in the object `knit_hooks`.

One common and tedious task when using R base graphics is we often have to call `par()` to set graphical parameters. This can be abstracted into a chunk hook, so that before a code chunk is evaluated, a set of graphical parameters can be automatically set. A chunk hook can be arbitrarily named as long as it does not conflict with existing hooks in `knit_hooks`. For example, we create a hook named `pars`:

```
knit_hooks$set(pars = function(before, options, envir) {
    if (before)
        par(options$pars)
})
```

Now we can pass a list of parameters to the `pars` option in a chunk, for example, `<<pars = list(col = 'gray', mar = c(4, 4, .1, .1), pch = 19)>>=`. Because this list is obviously not `NULL`, **knitr** will run the chunk hook `pars`. In this hook, we specified that `par()` is called *before* a chunk is evaluated (i.e., what `if (before)` means), and `options` argument in the hook function is a list of current chunk options, so the value of `options$pars` is just the list we passed to the chunk option `pars`. As we can see, the name of the hook function and the name of the chunk option should be the same, and that is how **knitr** knows which hook function to call based on a chunk option. What follows is a code chunk testing the `pars` hook:

```
plot(rnorm(100), ann = FALSE)
```

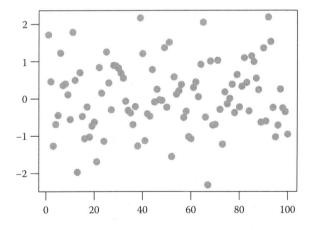

We see a scatter plot with solid gray points, which means `par()` was indeed called (the default of R is black open circles), although it did not show up in the source code. Because the hook function does not return character results, nothing else is written in the output. Now we show another example on how to save **rgl** plots [1] using a built-in chunk hook `hook_rgl()` in **knitr**. Note this function returns a character string depending on the output format, for example if it is LATEX, it returns a character string like `\includegraphics{filename}` where `filename` is the filename of the **rgl** plot captured by **knitr**.

```
knit_hooks$set(rgl = hook_rgl)
head(hook_rgl, 7)  # the hook function is defined as this

##
## 1 function (before, options, envir)
## 2 {
## 3     library(rgl)
## 4     if (before || rgl.cur() == 0)
## 5         return()
## 6     name = fig_path("", options)
## 7     par3d(windowRect = 100 + options$dpi * c(0, 0,
##         options$fig.width,
```

Then, we only have to set the chunk option `rgl` to a non-NULL value, for example, <<rgl=TRUE, dev='png'>>= (when dev='png', we record the plot using `rgl.snapshot()` in **rgl** to capture the snapshot as a PNG image):

```
library(rgl)
demo("bivar", package = "rgl", echo = FALSE)
par3d(zoom = 0.7)
```

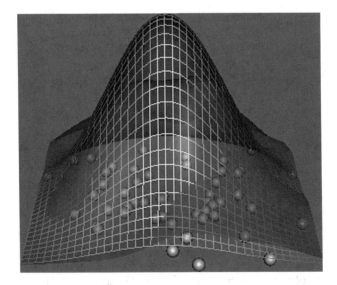

In all, chunk hooks help us do additional tasks before or after the evaluation of code chunks, and we can also use them to write additional content to the output document.

1.4.2 Language Engines

Although **knitr** was created in R, it also supports other languages like Python, Perl, awk, and shell scripts. For the time being, the interface is still very preliminary: it is a call to external programs via the `system()` function in R, and the results are collected as character strings.

The chunk option `engine` is used to specify the language engine, which is `'R'` by default. It can be `'python'`, `'perl'`, `'awk'`, `'haskell'`, and `'bash'`. Although the interface is naive, the design is very general. For example, these engines can be used for all the document formats, and appropriate renderers have been set up for them. For example, we can call Python in this LATEX document:

```
x = 'hello python from knitr'
print x.split(' ')
['hello', 'python', 'from', 'knitr']
```

As all other components of **knitr**, language engines can be customized as well. The object that controls the engines is `knit_engines`, for example, we can call `knit_engines$get('python')` to check how the Python engine was defined, or `knit_engines$set(python = ...)` to override the default engine. See the documentation in the package for more details.

A data analysis project often involves multiple tools other than R—we may use a shell script to decompress the data, awk to preprocess the data, and R to read the data. By integrating all tools into one framework, a project can be more tight in the sense that all the relevant code lives in the same document. It will be easy to redo the whole analysis without worrying if a certain part of the project is not up-to-date.

1.5 Discussion

A few future directions about tools for reproducible research were outlined in [8], including multilanguage compendiums, conditional chunks, and interactivity. All of these have been made possible in the **knitr** framework. For example, modern web technologies have enabled us to interact with web pages easily. RPubs mentioned in Section 1.1 is a good example: we can publish reports to the web from RStudio with a single mouse click; besides, we can also write interactive content into the web page based on **knitr** and other tools like JavaScript: http://rpubs.com/jverzani/1143 is an interactive quiz for R; the questions and answers were generated dynamically from **knitr**. Another application is the **googleVis** package [9]:

http://rpubs.com/gallery/googleVis (we are able to interact with tables and Google maps there).

We observed a lot of homework submissions on RPubs (e.g., http://rpubs.com/kaz_yos/1519), and we believe this is a good indication from the educational point of view. When students are trained to write homework in a reproducible manner, it should have more positive impact on scientific research in the future.

It is debatable which authoring environment is ideal for reproducible research (e.g., [8] suggested XML), and we would argue that a wide list of choices should be made available. LaTeX is a perfect typesetting tool for experts, but it is very likely that beginners can get stuck. Markdown is much less frustrating, and the most important thing is, users can step into the paradigm of reproducible research really quickly rather than spending most of their time figuring out typesetting problems. As one example, an RPubs user published a data analysis about the hurricane "Sandy" almost immediately after it hit the east coast of the United States: http://rpubs.com/JoFrhwld/sandy.

Everything is moving to the "cloud" nowadays, and lots of applications are developed and deployed on the server side. OpenCPU is a platform that provides the service of R through a set of APIs that can be programmed in JavaScript; **knitr** has a simple application there that allows one to write a report in the web browser: http://public.opencpu.org/apps/knitr. The computing is done on OpenCPU, and nothing is required on the client side except a web browser. This could be one of the future directions of statistical computing and report generation. A similar platform sponsored by RStudio is the Shiny [23] server, and a **knitr** example can be found at http://glimmer.rstudio.com/yihui/knitr/.

Web applications may also have an impact on publications related to data analysis because it is convenient to collaborate with other people, fast to publish reports, and get feedback. Vistat (http://vis.supstat.com) is an attempt to build a collaborative and reproducible website featuring statistical graphics like a journal. It is based on Github and R Markdown; authors can submit new articles through the version control tool GIT and reviewers can make comments online. All the graphics will be verified independently, hence it requires the author(s) to submit a detailed source document for other people to reproduce the results.

There are a number of important issues when implementing the software package for reproducible research. For example, cache may be handy because it can save us a lot of time, but we have to be cautious about when to invalidate the cache. Even if the code and chunk options are not changed, do we need to purge the cache and recompute everything after we have upgraded R from version 2.15.1 to 2.15.2? To incorporate with this kind of questions, **knitr** provides additional approaches to invalidate the cache, for example we can add a chunk option `cache.extra=R.version.string` so that whenever the R version has changed, the cache will be rebuilt. Besides

R itself, there can also be problems with add-on packages. In **knitr**, there is a convenience function `write_bib()` that can automatically write the citation information about R packages in the current R session into a BibTEX database; this guarantees that the version information of packages is always up-to-date. We illustrate one more issue as a potential problem: When we distribute our analysis, how are we supposed to include external materials such as the figure files? For LATEX, this is not a problem since images are embedded in PDF; for Markdown/HTML, **knitr** uses the R package **markdown** to encode images as base64 strings and embed the character strings into HTML so that a web page is self-contained (i.e., no extra files are required to publish it). However, it can be difficult, if not impossible, to embed everything in a single document, for example, how should we disseminate datasets and unit tests? A potential media is an R package as proposed by [8], which has a nice structure of a project (source code, documentation, vignettes, tests, and datasets). In this case, **knitr** will be one part of a reproducible project. In fact, this has been made possible since R 3.0.0—we can build package vignettes with **knitr** (traditionally only Sweave was allowed) and the document formats can be LATEX, HTML, and Markdown.

The **knitr** package has gained support in many editors that make it easy to write the source documents; at the moment, RStudio has the most comprehensive support. We can also use LYX, Emacs/ESS, WinEdt, Eclipse, and Tinn-R. All of them support the compilation of the source document with one mouse click or keyboard shortcut.

We emphasized graphics but not tables in this chapter because tables are essentially text output and can be supported by other packages such as **xtable** [4]; in **knitr**, we just need to use the chunk option `results='asis'` when we want a table in the chunk output. Put it another way, tables are orthogonal to **knitr**'s design.

In all, we have mainly introduced one comprehensive tool for reproducible research, namely **knitr**, in this chapter. It has a flexible design to allow customization and extension in several aspects from the input to the output. The major functionality of this package has stabilized, and the future work will be primarily bug fixes and improving existing features such as the language engines. A much more detailed introduction of this package can be found in the book [28].

Acknowledgments

First, I thank Friedrich Leisch for the seminal work on Sweave, which deserves credits of the design and many features in **knitr**. As I mentioned in Section 1.3, the ideas of cache and TikZ graphics were from **cacheSweave**

(Roger Peng), **pgfSweave** (Cameron Bracken and Charlie Sharpsteen), and **weaver** (Seth Falcon); syntax highlighting was inspired by Romain Francois from his **highlight** package. I thank all these package authors as well as Hadley Wickham for his unpublished **decumar** package, which greatly influenced the initial design of **knitr**. There have been a large number of users giving me valuable feedbacks in the mailing list https://groups.google.com/group/knitr and on Github, and I really appreciate the communications. I thank the authors and contributors of open-source editors such as LyX and RStudio for the quick support. I thank my advisors Di Cook and Heike Hofmann for their guidance. Last but not least, I thank the R Core Team for providing such a wonderful environment for both data analysis and programming. There are a few nice functions in R that introduced very useful features into **knitr**, such as `recordPlot()` and `lazyLoad()`.

References

1. D. Adler and D. Murdoch. rgl: 3D visualization device system (OpenGL), 2013. R package version 0.93.929/r929.
2. K.A. Baggerly, J.S. Morris, and K.R. Coombes. Reproducibility of SELDI-TOF protein patterns in serum: Comparing datasets from different experiments. *Bioinformatics*, 20(5):777–785, 2004.
3. C. Bracken and C. Sharpsteen. pgfSweave: Quality speedy graphics compilation and caching with Sweave, 2012. R package version 1.3.0.
4. D.B. Dahl. xtable: Export tables to LaTeX or HTML, 2013. R package version 1.7-1.
5. D. Eddelbuettel. digest: Create cryptographic hash digests of R objects, 2013. R package version 0.6.3.
6. S. Falcon. weaver: Tools and extensions for processing Sweave documents, 2013. R package version 1.24.0.
7. R. Gentleman. Reproducible research: A bioinformatics case study. *Statistical Applications in Genetics and Molecular Biology*, 4(1):1034, 2005.
8. R. Gentleman and D. Temple Lang. Statistical analyses and reproducible research. Bioconductor Project Working Papers, 2004.
9. M. Gesmann and D. de Castillo. googleVis: Interface between R and the Google chart tools, 2013. R package version 0.4.2.
10. J. Gruber. The Markdown project, 2004. http://daringfireball.net/projects/markdown/
11. R. Ihaka and R. Gentleman. R: A language for data analysis and graphics. *Journal of Computational and Graphical Statistics*, 5(3):299–314, 1996.
12. D.E. Knuth. The WEB system of structured documentation. Technical report, Department of Computer Science, Stanford University, Stanford, CA, 1983.

13. D.E. Knuth. Literate programming. *The Computer Journal*, 27(2):97–111, 1984.
14. F. Leisch. Sweave: Dynamic generation of statistical reports using literate data analysis. In *COMPSTAT 2002 Proceedings in Computational Statistics*, Vol. 69, pp. 575–580. Physica Verlag, Heidelberg, Germany, 2002.
15. P. Murrell. *R Graphics*, 2nd ed. Chapman & Hall/CRC, Boca Raton, FL, 2011.
16. R.D. Peng. cacheSweave: Tools for caching Sweave computations, 2012. R package version 0.6-1.
17. R.D. Peng. filehash: Simple key-value database, 2012. R package version 2.2-1.
18. R Core Team. *R Language Definition. R Foundation for Statistical Computing*. Vienna, Austria, 2012.
19. R Core Team. *R: A Language and Environment for Statistical Computing*. R Foundation for Statistical Computing. Vienna, Austria, 2013. ISBN 3-900051-07-0.
20. N. Ramsey. Literate programming simplified. *IEEE Software*, 11(5): 97–105, 1994.
21. A.J. Rossini, R.M. Heiberger, R.A. Sparapani, M. Maechler, and K. Hornik. Emacs speaks statistics: A multiplatform, multipackage development environment for statistical analysis. *Journal of Computational and Graphical Statistics*, 13(1):247–261, 2004.
22. A. Rossini. Literate statistical analysis. In *Proceedings of the 2nd International Workshop on Distributed Statistical Computing*, Hornik, K. and Leisch, F. (Eds.), Technische Universität Wien, Vienna, Austria, 2001. http://www.ci.tuwien.ac.at/Conferences/DSC-2001/Proceedings/
23. RStudio, Inc. shiny: Web Application Framework for R, 2013. R package version 0.4.0.99.
24. C. Sharpsteen and C. Bracken. tikzDevice: R graphics output in LaTeX format, 2012. R package version 0.6.3/r49.
25. T. Tantau. The TikZ and PGF packages, 2008. http://sourceforge.net/projects/pgf/
26. H. Wickham. evaluate: Parsing and evaluation tools that provide more details than the default, 2013. R package version 0.4.3.
27. Y. Xie. formatR: Format R code automatically, 2012. R package version 0.7.2.
28. Y. Xie. *Dynamic Documents with R and knitr*. Chapman & Hall/CRC, Boca Raton, FL, 2013. ISBN 978-1482203530.
29. Y. Xie. knitr: A general-purpose package for dynamic report generation in R, 2013. R package version 1.1.8.

2

Reproducibility Using VisTrails

Juliana Freire, David Koop, Fernando Chirigati, and Cláudio T. Silva

CONTENTS

2.1 Introduction

Science has long placed an emphasis on revisiting and reusing past results: reproducibility is a core component of the scientific process. Testing and extending published results are *standard* activities that lead to practical progress: science moves forward using past work and allowing scientists to "stand on the shoulders of giants." In natural science, long tradition requires experiments to be described in enough detail so that they can be reproduced by other researchers. This standard, however, has not been widely applied for computational experiments. Researchers often have to rely on tables, plots, and figure captions included in papers. Consequently, it is difficult to verify and reproduce many published results [43], and this has led to a credibility crisis in computational science [17].

Scientific communities in different domains have started to act in an attempt to address this problem. Prestigious conferences such as *SIGMOD* [58] and *VLDB* [71], and journals such as *PNAS* [52], *Biostatistics* [7], the *IEEE Transactions on Signal Processing* [65], *Nature*, and *Science*, to name a few, have been encouraging—and sometimes requiring—that published results be accompanied by the necessary data and code needed to reproduce them. However, it can be difficult and time-consuming for authors to make their experiment reproducible and for reviewers to verify the results. Authors need to encapsulate the whole experiment (data, parameter settings, source code, and environment) to guarantee that the same results are generated. Even when an experiment compendium is available, reviewers may have difficulties reproducing the experiments due to missing libraries or dependences on a specific operating system version to run (or compile) the experiment. We posit that by planning for reproducibility and through the use of systems that systematically capture provenance of the scientific exploration process, researchers will not only create results that are reproducible but they can also streamline many of the tasks they have to carry out. With this in mind, we have built a framework that supports the life cycle of computational experiments [25,38]. This framework has been implemented and is currently released as part of VisTrails [20,70], an open-source, workflow-based data exploration and visualization system. VisTrails relies on a provenance management component to automatically and transparently capture the necessary metadata to allow experiments to be reproduced, including executable specification of computational processes (i.e., the workflow structure), parameter settings, input and output data, library versions, and code. It implements mechanisms that leverage the provenance information to support the exploratory process [37,39,60], which is common in data-intensive science [30]. These mechanisms also make it easier for reviewers to run and verify the results.

In this chapter, we describe the VisTrails reproducibility infrastructure. We start in Section 2.2 with a definition for computational reproducibility and reproducibility levels. We also give an overview of workflow systems and discuss their benefits and limitations for the creation of reproducible experiments. The VisTrails system is described in Section 2.3 and in Section 2.4, we present specific components we have added to the system to support both authors and reviewers of reproducible experiments. Some limitations of our reproducibility framework and directions for future work are discussed in Section 2.5. We review related work in Section 2.6, and conclude in Section 2.7.

2.2 Reproducibility, Workflows, and Provenance

2.2.1 Anatomy of a Reproducible Experiment

A computational experiment that has been developed at time t on hardware/operating system s on data d is reproducible if it can be executed at time t' on system s' on data d' that is similar to (or potentially the same as) d with consistent results [22]. For this to be possible, the description of the experiment must be sufficiently precise and include

- A description of the input data, either in extension (the actual data) or in intention (e.g., a script that derives the data)
- Detailed information about the system where the experiments were run, including hardware and software configuration
- An executable specification for the experiment that describes the steps followed to derive the result

The components of a reproducible paper are illustrated in Figure 2.1. This paper investigates Galois conjugates of quantum double models [19]. Each figure in the paper is accompanied by its provenance, consisting of the workflow used to derive the plot, the underlying libraries invoked by the workflow, and links to the input data, that is, simulation results stored in an archival site. This provenance information allows all results in the paper to be reproduced. In the PDF version of the paper,* the figures are active, and when clicked on, the corresponding workflow is loaded into the VisTrails and executed on the reader's machine. The reader may then modify the workflow, change parameter values, and input data.

2.2.1.1 Levels of Reproducibility

While full reproducibility is desirable, it can be hard or impossible to attain. Therefore, it is important to consider different levels of reproducibility. Freire et al. [22] have introduced three criteria to characterize the level of reproducibility of experiments:

1. The *depth* evinces how much of an experiment is made available. The default today is to include a set of figures in a manuscript. Higher depths can be obtained by including the script (or spreadsheet file) used to generate the figures in the paper together with the appropriate datasets; the raw data and intermediate results derived during the experiments; the set of experiments (system configuration and

---

* This paper can be downloaded from http://arxiv.org/abs/1106.3267.

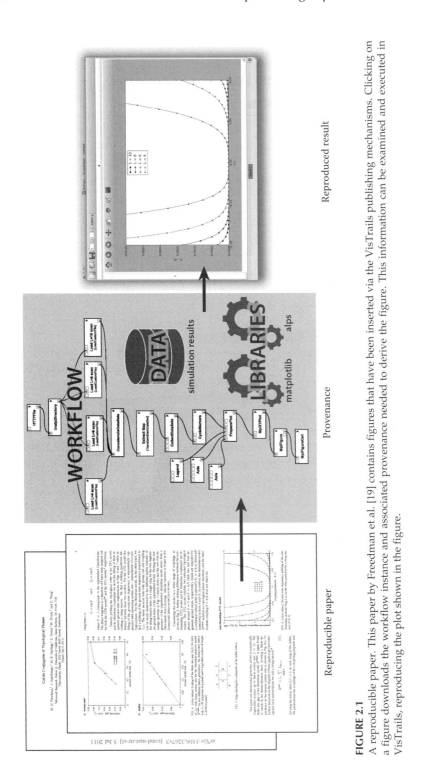

FIGURE 2.1

A reproducible paper. This paper by Freedman et al. [19] contains figures that have been inserted via the VisTrails publishing mechanisms. Clicking on a figure downloads the workflow instance and associated provenance needed to derive the figure. This information can be examined and executed in VisTrails, reproducing the plot shown in the figure.

initialization, scripts, workload, measurement protocol) used to produce the raw data; and the software system as a white box (source, configuration files, build environment) or black box (executable) on which the experiments are performed.

2. The level of *portability* indicates whether the experiments can be reproduced (a) on the original environment (basically the author of the experiment can replay it on his or her machine), (b) on a similar environment (e.g., same OS but different machines), or (c) on a different environment (i.e., on a different OS or machine).

3. *Coverage* specifies how much of an experiment can be reproduced. Coverage can be partial or full. For example, considering an experiment that requires special hardware to derive data, partial reproducibility can be obtained by providing the data produced by the hardware and the analysis processes used to derive the plots included in the paper.

2.2.2 Describing Computations as Workflows

Workflows are widely used to represent and execute computational experiments, as evidenced by the emergence of several workflow-based systems, such as Apple's Mac OS X Automator, Yahoo! Pipes, Galaxy, NiPype, VisTrails, Kepler, and Taverna, to name a few. Workflow systems have features that make them suitable as tools to create reproducible experiments. Notably, (1) workflow specifications provide an explicit representation of the structure of the experiments, (2) workflows automate repetitive tasks and computations, and (3) workflow systems can transparently capture provenance information.

In a workflow, computational steps are represented by modules, and there is a connection between two modules if there is a dependency relation between them. The dependency relation can be either control or data driven. When the dependencies are data driven, workflows are referred to as *dataflows*. Dataflows can be naturally represented as directed-acyclic graphs (DAGs) where the connections (or edges) correspond to data flowing between modules. Dataflows are the underlying model for the major scientific workflow systems [35,62,70] and also for many workflow-based systems used for data processing and visualization.

Dataflows have several advantages over scripts or programs written in high-level languages [25]. They provide a simple programming model where a sequence of tasks is composed by connecting the outputs of one task to the inputs of another. This enables the use of visual interfaces that are suitable for users without programming expertise. The explicit DAG structure supports useful manipulations, including the ability to query workflows and update them in a programmatic fashion [57]. Workflows can also be represented at different levels of *abstraction* [23]. As illustrated in Figure 2.2,

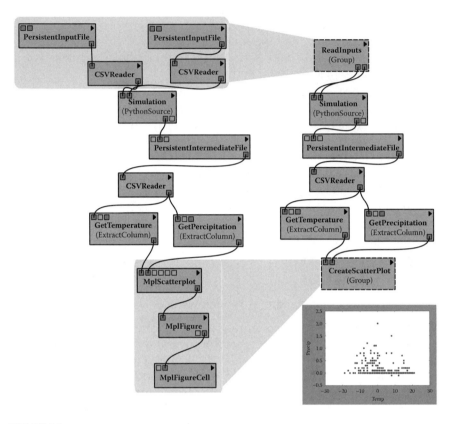

FIGURE 2.2
A scientific workflow for visualizing temperature and precipitation data. Besides providing an executable specification for the derivation of the scatterplot (bottom right), through the use of abstraction, the workflow can also provide a specification that hides unnecessary details and is easier to understand.

a series of modules can be grouped to hide unnecessary details. Abstraction can be used to make the specification easier to understand and more amenable for publication.

A given workflow instance embodies not only the structure of the experiment but also its configuration, that is, the input data and parameters used to produce a result. Having a workflow instance associated with a published result simplifies reproducibility. In addition, because the workflow specification is executable and has an explicit structure, users can easily perform parameter sweeps and run experiments varying the input data, while ensuring the same configuration is used across the different runs. For example, the VisTrails system provides a mechanism for parameter exploration and allows users to compare the results side by side [24]. Figure 2.3 shows a series of scatter plots showing temperature and precipitation derived for multiple years: the same workflow is run varying the input files for

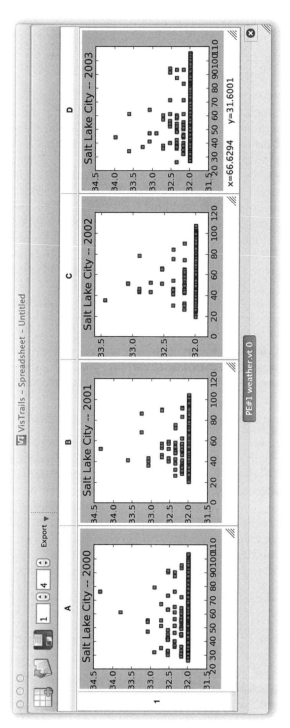

FIGURE 2.3
The VisTrails parameter exploration mechanism. A workflow is used to derive scatterplots for temperature and precipitation varying the input data to analyze the behavior in different years.

different years. Such a mechanism is useful, for instance, to verify results and perform sensitivity analyses, tasks that are essential for reviewers.

2.2.3 Provenance in Workflow Systems

In a script or a program, unless specified by the programmer, it is difficult to record the steps taken, the inputs consumed, and the tools called throughout the execution. Because workflow systems control the execution of computational processes, they can systematically *capture their provenance*. As provenance is a key ingredient for reproducibility [23,25,60], workflows systems are well suited as a platform to create reproducible experiments.

There are three main types of provenance captured by workflow systems: prospective, retrospective, and workflow evolution [23]. *Prospective provenance* embodies the description of an experiment—the specification of the workflow structure, including modules, connections, and inputs. *Retrospective provenance* captures information about the execution of the workflow, what actually happened when the workflow was run. *Workflow evolution* captures the history of a workflow, that is all the different versions of the workflow. Evolution provenance is especially useful for data-intensive tasks, where workflows are iteratively refined, for example, to experiment with different algorithms or simulation codes, and different input data. Interfaces can be created to allow users to navigate over workflow versions in an intuitive way, undo changes without losing results, visually compare multiple workflows, and show their results side by side (see Figure 2.4).

2.2.4 Workflows and Reproducibility

Workflow systems capture both prospective and retrospective provenance, as such, they can provide a high depth of reproducibility: a detailed account of how a result was derived. Because the prospective provenance, that is, the workflow specification, is executable, the experiments can be reproduced. One caveat is the fact that workflows may not be portable: it may not be possible to run a workflow in an environment different from the one where it was created. This can be due to a number of factors, including hard-coded file names, missing libraries, and OS incompatibility. In Section 2.4, we discuss how we have extended the VisTrails system to deal with these limitations.

2.3 VisTrails System

VisTrails (http://www.vistrails.org) is an open-source provenance management and scientific workflow system that was designed to support the scientific discovery process [20,24,25]. VisTrails provides unique support

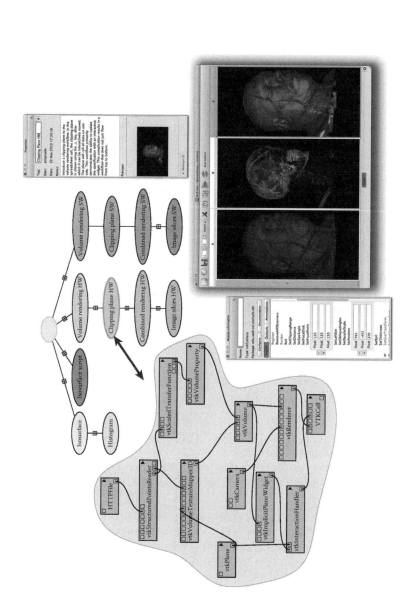

FIGURE 2.4

The VisTrails system consists of layers to manage computations, provenance, and data. The interface includes a version tree that allows users to navigate over past versions, a workflow view where users can create or modify workflows, and a visual spreadsheet where results can be compared side by side. Users may also add tags and annotations to the different workflow versions.

for data analysis and visualization, a comprehensive provenance infrastructure, and a user-centered design. The system combines and substantially extends useful features of visualization and scientific workflow systems. Similar to visualization systems [33,36,41,66], VisTrails makes advanced visualization techniques available to users, allowing them to explore and compare different visual representations of their data; and similar to scientific workflow systems [35,50,62,69], VisTrails enables the composition of workflows that combine specialized libraries, distributed computing infrastructure, and web services. As a result, users can create complex workflows that encompass important steps of scientific discovery, from data gathering and manipulation to complex analyses and visualizations, all integrated in one system.

Whereas workflows have been traditionally used to automate repetitive tasks, for applications that are exploratory in nature, such as simulations, data analysis, and visualization, very little is repeated—change is the norm. As a scientist generates and evaluates hypotheses about data under study, a series of different, albeit related, workflows are created as they are adjusted in an iterative process. VisTrails was designed to manage these rapidly evolving workflows. Another distinguishing feature of VisTrails is a *comprehensive provenance infrastructure* that maintains detailed history information about the steps followed and data derived in the course of an exploratory task [24]: VisTrails maintains provenance of data products (e.g., visualizations, plots), the workflows that derive these products, and their executions. The system also provides extensive annotation capabilities that allow users to enrich the automatically captured provenance. This information is persisted as XML files or in a relational database. Besides enabling reproducible results, VisTrails *leverages provenance information through a series of operations and intuitive user interfaces that aid users to collaboratively analyze data.* Notably, the system supports reflective reasoning by storing temporary results, by providing users the ability to reason about these results and to follow chains of reasoning backward and forward [47]. Users can navigate workflow versions in an intuitive way, undo changes but not lose any results, visually compare multiple workflows and show their results side by side in a visual spreadsheet, and examine the actions that led to a result [6,24,60]. In addition, the system has native support for parameter sweeps, whose results can also be displayed on the spreadsheet [24].

VisTrails addresses important usability issues that have hampered a wider adoption of workflow and visualization system. It provides a series of operations and user interfaces that simplify workflow design and use, including the ability to create and refine workflows by analogy, to query workflows by example, and a recommendation system that automatically suggests workflow completions as users interactively construct their workflows [40,57]. The system also supports the creation of mashups—customized and simplified applications that can be more easily deployed to scientists [55,56].

A beta version of the VisTrails system was first released in January 2007. Since then, the core system has been downloaded over 40,000 times. The VisTrails wiki has had over 1.7 million page views, and Google Analytics reports that visitors to the site come from 70 different countries. VisTrails has been adopted in several scientific projects, both nationally and internationally, and in different areas, including environmental science [4,12,31,32], psychiatry [3], astronomy [63], cosmology [2], high-energy physics [16], molecular modeling [29], quantum physics [5,19], earth observation [14,68], and habitat modeling [46]. Besides being a stand-alone system, VisTrails has been used as a key component of domain-specific tools. One notable example is UV-CDAT, a new toolset for large-scale climate data analysis [54,67].

A number of groups have contributed to the project, some directly—by checking in code into our git repository—and others by sharing packages that add functionality to VisTrails, for example, ALPS (ETH Zurich) [1], control flow [11] (Federal University of Rio de Janeiro, Brazil), ITK [34] (University of Utah), GridFields [28] (University of Washington), vtDV3D [72] (NASA), and SAHM [46,53] (USGS). Researchers at the Council for Scientific and Industrial Research (CSIR) in South Africa have added spatial–temporal data access and data preprocessing capabilities to VisTrails [18]. The system has been used by many projects, including DataONE [15] and STC-CMOP [12]. VisTrails has also been successfully used as a tool for teaching, having been adopted at universities in the United States and abroad [59].

2.4 Reproducing and Publishing Results with VisTrails

Because it was designed as a provenance-aware system, VisTrails natively stores much of the information necessary for users to revisit or extend existing work. To provide better support for result publication, we have added new functionality to VisTrails including support for portability (i.e., the ability to run a workflow in an environment different from the one where it was created) and to connect published results to their provenance [25,38]. The new functionality both integrates with and complements the core provenance features. At the same time, they are not required for users of VisTrails nor are the ideas strictly dependent on VisTrails—they could be implemented on other systems.

2.4.1 Reproducibility Support

VisTrails captures both the provenance of workflow executions and the provenance of the workflow specifications. The system captures what happens when a workflow is run—the dependencies and properties of intermediate data, and how workflows are modified from run to run—which parameters are changed, which modules are connected together. Workflow evolution is a key element in maintaining reproducible results, capturing the changes that are made from initial explorations, tests, and extensions to the final published results. Because no workflow versions are deleted or replaced, all results can be retrieved, reproduced, and compared against. At the same time, knowing how a computation was built can help others understand the process, extend the results, or tweak them in meaningful ways.

2.4.1.1 Provenance-Rich Results

Not only does VisTrails capture provenance, but it also makes this information available to users to help them organize and understand past work. This can be especially useful in collaborative settings, where multiple users are contributing ideas and making changes. In addition to a visual workflow builder, VisTrails provides a version tree that displays all of the workflow versions users have created and their relationships. A version tree is shown in Figure 2.4. Each node in the tree corresponds to a workflow version; an edge corresponds to an action or sequence of actions applied to transform the parent node into the child. For example, the node tagged `Clipping Plane HW` was created by modifying the node `Volume Rendering HW` to include a clipping plane. Unlike standard undo stacks, VisTrails captures and maintains all of the steps a user has taken, regardless of whether they may have been unproductive. A user can switch to any version by selecting the appropriate node in the version tree. Having access to this workflow evolution information allows users to investigate any idea they wish without worrying about explicitly saving versions along the way. In addition, as illustrated in Figure 2.4 (top right), each version is associated to metadata that includes an optional label (tag) that describes the version, information about the user who created the version, the time/date of creation, and free-text notes that users can add to provide further details about their findings and to better document the exploratory process. Because both the workflow specification and metadata can be searched (and queried), users also do not have to worry about meticulously labeling each version with the exact parameters and inputs used. This reduces the burden on users to maintain all of the information needed for reproducibility.

In addition to workflow evolution provenance, VisTrails also allows users to view the provenance of workflow executions (see Figure 2.5). This provenance information, containing timing information, any errors encountered, and which system was used, can be valuable in diagnosing possible issues or determining more efficient methods of execution. It can also be used to determine which outputs used a particular input, making it possible to highlight results that may be invalidated by, for example, a malfunctioning sensor. Such information is also important for reproducibility because it is possible that the system being used or any previous errors can inform

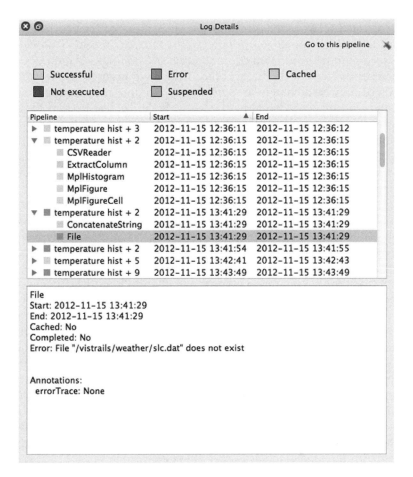

FIGURE 2.5
Retrospective provenance in VisTrails. Detailed information is kept about the execution of workflows and their modules, including when and for how long they ran and whether their execution completed or failed.

those who attempt to emulate the original work later. For this reason, a VisTrails contains a set of related workflows, the changes differentiating one from another, and the provenance of any executions.

2.4.1.2 Workflow Upgrades

As time progresses, workflows can become stale for a number of reasons, notably due to the fact that the software the workflow relies upon may change. This can happen both during the scientific exploration or after results are published. When possible, we wish to retain the original steps so that given a compatible system (e.g., a virtual machine), they can be repeated exactly. However, for later use, it is more convenient to be able to work with the original computations in a current system. This requires both a recognition of outdated computations and an upgrade path.

There are a variety of changes that may have occurred to warrant an upgrade. It may be a change in algorithm where the interface (parameters, input and output ports) remains the same. It could also be that the ports were changed to provide new parameters or outputs, possibly with relabeling. Finally, a module (or network of modules) may have been replaced or removed due to a reorganization or reimplementation. All of these types of upgrades can be handled, but some can be defaulted to automatic steps while others require developer or user involvement.

VisTrails detects when upgrades are necessary by comparing the version of a module in a given workflow with the currently available version. If the module is out-of-date, it tries to upgrade the old module. Upgrade paths are either determined automatically or specified by the module developer [39]. When the interface has not changed, VisTrails replaces the old module with the new version. When it has, VisTrails will attempt to replace the old module, and when ports have only been added or a changed port has not been used, it will succeed. When it does not, it alerts the user so he or she can determine a next step. However, for nontrivial upgrades, developers are encouraged to provide explicit upgrade paths for the modules in their packages. VisTrails passes along any modules that the developer has designated for special handling, and the developer can write a set of changes (the normal VisTrails actions like "add module" and "change parameter") to be applied. These changes can be based on the current parameters being sent to the module as well as any neighboring modules. When automated upgrades do not work and developers have not specified an upgrade path, VisTrails alerts the user who can then make the necessary changes.

Most importantly, any upgrades are recorded with the same change-based provenance as a normal action. This means the original version of the workflow is always retained, and anyone looking at the steps in the future can see exactly how the original workflow was modified. One could, then, re-execute the original version in a virtual machine and compare it to an upgraded version to check if the behavior is unchanged or if a specific bug has been fixed.

2.4.1.3 Managing Data and Their Versions

Even if authors have maintained the specification of the computation needed to reproduce their results, reproducibility also requires the data used in the work. The provenance of a computation may indicate the file name used, but if this file is moved, deleted, or modified, reproducing that work is not possible. One value we have added to VisTrails provenance is a hash of a file's contents, allowing users to later check if the provided data does indeed match the one used in the original computation. In addition, we have developed a *persistence package* that allows users to store input, output, and intermediate data in a versioned repository [37]. This repository not only ensures that data can be accessed later, but it also automatically tracks changes in data, ensuring that different versions of the input data can be recovered in the future. If a user generated a figure based on a dataset that was later updated, the user can go back to the original version and reproduce the original run. This mechanism also creates strong links between output data and the computations used to generate them: users can match output data and computations by comparing hashes with the repository and finding provenance traces that match the given identifiers.

Scientific workflows may also access external data sources such as relational databases as part of an experiment. For example, a module may have to remotely query a large dataset and use the results for the remaining computation, or it may also have to update information in the dataset. As with file-based references, accessing relational databases in workflows may also lead to problems for reproducibility. For example, when workflow consumes data from a database, its results may depend on the data. When the database is updated, the results of the workflow may no longer be reproduced, as database updates may have changed the data that is used by the workflow. Reproducibility, in this case, is more challenging because there is an inherent mismatch between workflow and database models: while workflows are stateless and deterministic, databases are stateful—new states reflect the changes applied to the database.

To address this issue, we have implemented in VisTrails a model that integrates workflow and database provenance and enables reproducibility for workflows that interact with relational databases [9]. We rely on a transaction temporal model that is currently supported by commercial RDMS—in our implementation, we used Oracle [48] and its Total Recall functionality [49]. The states of the database are systematically captured and added to the workflow provenance, and this information is used by VisTrails to communicate with the database and go back to a previous state so the workflow can be correctly reproduced.

2.4.1.4 Using Provenance for Future Work

Provenance allows users to go back to previous work and possibly extend it, but it can also be used to help create or inform future work. For example, if a user performs some modification to a workflow like adding a data filtering step, he or she may wish to add a similar step to other workflows as well. This process could be very tedious, but because the VisTrails provenance contains the steps needed to transform the workflow, it can usually be automated. VisTrails workflow analogies [57] allow users to modify a set of workflows based on the changes they have made to a single workflow; the technique uses a flexible matching algorithm that allows the changes to be applied to workflows that have different structures. While analogies are useful for users that have well-defined changes, provenance can also be mined to derive common workflow patterns. Similar to completion techniques often seen in text entry boxes, VisComplete [40] uses a collection of workflows to build ranked sets of possible completions for a partially constructed workflow. This allows users to save time spent rebuilding common substructures or searching examples. These techniques highlight the use of provenance not only as a record of past work but also as the starting point for future exploration.

2.4.2 Publishing Results

One common problem with published work is that the caption of a figure, table, or other result does not provide the whole story about the origin of that figure. For example, after plotting raw data, an author may restrict the dataset, transform the data to emphasize a particular aspect, or just refine the graphical presentation. Sometimes, these steps are not recorded, and later, readers (or even the original authors) are hard pressed to determine all the steps used to generate the result. Our approach is to encode the actual computation as part of the paper so that upon generating the final PDF, the result is recalculated if necessary and a link to the exact computation included.

Our implementation of this hard link between the results displayed in a paper and the computation uses VisTrails, LaTeX, and a LaTeX package that defines a new command to reference the underlying computation and passes the information to VisTrails for computation, placing the result from Vis-Trails directly into the paper [38]. The `vistrails` LaTeX package defines the base command for inserting a result as well as options for adding links to a web-hosted definition of the workflow or an interactive version of the result. The user can specify the workflow used to generate a result using a tag or a unique identifier. With tags, the computation can be edited, and the paper, upon recompilation, will include the updated result automatically. We have also implemented mechanisms to include results derived by VisTrails into web pages, wikis, Word documents, and PowerPoint presentations. Having these mechanisms reduces the burden on authors to manually update results and mitigates the problem of losing previous results.

2.4.3 Publishing Interactive Results on the Web

The web has opened the possibility of publishing much more than a printable PDF, but the same issues that arise in traditional publishing must be addressed. Specifically, authors must make sure that the published result accurately reflects the underlying computation, and the idea of reproducibility as inspection and further exploration is not met simply by having an interactive visualization. As with papers, being able to download, execute, and modify a result is preferred over a static result locked into a particular

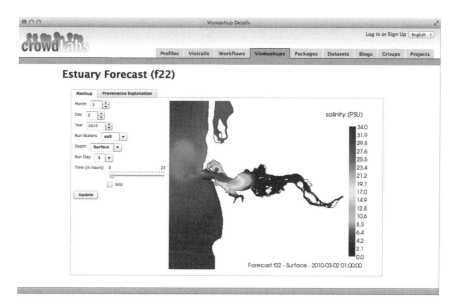

FIGURE 2.6
Authors may publish interactive results on the web. Here, we show how VisTrails results can be displayed and interacted with on the crowdLabs site.

site. That said, there are benefits of server-hosted results, and many of the goals of reproducibility can be met in such an environment. Because users do not need to install additional software or download large datasets, the burden of exploring results is lessened. Furthermore, the interactive possibilities on the web allow authors to publish results that permit recomputation with user-defined inputs and parameters.

With VisTrails, we have explored two alternatives for web-based publishing: wikis and interactive social websites. Our approach for wikis is very similar to publishing for traditional PDFs: there is a specific `vistrails` tag that allows a user to specify the workflow that should be computed to generate the output. As with LaTeX, a user can indicate the workflow by a tag or a specific, unique identifier. After editing a wiki page, there is a MediaWiki extension that processes the `vistrail` tag to re-execute, if necessary, the workflow and insert the result directly into the wiki output.

We have also developed the crowdLabs website as a place where users can host and share their workflows and results [45]. In addition, crowdLabs supports VisTrails mashups, which allow workflow creators to easily specify higher-level interfaces to workflows [56]; an example is shown in Figure 2.6. Such interfaces make it easy for users to quickly explore different parameter settings for a given computation. Users can upload an entire version tree (with all its tagged workflows), a specific workflow, datasets, or a mashup directly from the VisTrails application. The computations can then be executed on the crowdLabs server and the results made available via a web browser. For mashups, a user can modify inputs to generate new results on the fly without having VisTrails installed locally. This server-side execution allows some amount of reproducibility without requiring a reader to download the required packages and data. Furthermore, crowdLabs allows users to comment on others' results and approaches, enabling conversations about the computations that can help further future extensions.

2.5 Challenges and Opportunities

In the previous section, we have described our efforts in the area of computational reproducibility. The infrastructure we have presented, along with all its features, has already been successfully used by different research groups [38]. However, the infrastructure is by no means comprehensive. Through our collaborations with scientists, we have gathered a set of requirements that guided our design. In the long term, our goal is to build a general system where different components and methods can be mixed and used to achieve reproducibility for many different domains and scenarios.

Our infrastructure is built on top of the VisTrails system, but this tight integration is not always desirable. Although there are significant advantages for using workflows (Section 2.2.2), it may be time consuming to wrap the experiment into a workflow system if scientists are already using another approach for the execution. Besides, the experiment may involve interactive tools, which cannot be wrapped into a workflow. Nonetheless, the key contribution here are the core ideas and functionalities on which the infrastructure is based—the infrastructure itself is a proof-of-concept implementation of our efforts to simplify the creation, review, and reuse of reproducible experiments. A direction we are currently pursuing is to break this infrastructure into components that can be more easily integrated with other systems. In addition, we have developed a plug-in mechanism that leverages the VisTrails provenance management subsystem to provenance support to other tools. Examples of such tools are VisIt, Autodesk's Maya, and ParaView [8].

2.6 Related Work

There are many scientific workflow systems, besides VisTrails, that represent computations as dataflows, including Swift [61], Taverna [62], Kepler [35], Triana [64], and Pegasus/Wings [50]. While these systems support reproducibility, they do not have support for portability as discussed in Section 2.2.4.

There are also a number of other tools that provide support for reproducibility. Madagascar [44] is a software package for geophysics that allows scientists to generate computational results and include these on reproducible documents by making use of SCons, a software construction tool, and LaTeX. Sweave [42] is a tool that embeds source code from R, the statistical computing software, in LaTeX documents. In this way, every time data and analysis change, the document is automatically updated, creating *dynamic reports*, which supports reproducible research. ReproZip is a tool that automatically captures the provenance of existing experiments and packs all the components necessary to reproduce the results in different environments [10]. Reviewers can then unpack and run the experiments in their environment without having to install any additional software. ReproZip also generates a workflow specification for the experiment, which reviewers can use to explore the experiment and try different configurations.

The idea of linking data to publications is also explored by SOLE, Collage, SHARE, and VCR. SOLE [51] is a system that defines command-line tools to create scientific objects, such as source code, annotations in PDFs and virtual machine images hosted on a cloud, and link those objects to a paper in HTML format. Collage [13] is a framework, integrated by the publisher Elsevier, that was developed to create executable papers, where authors include their code and data. SHARE [27] is a web portal that allows authors to create and share remote virtual machines. These machines can be cited in research papers, and readers can access them and fully reproduce the experiment. Lastly, verifiable computational results (VCRs) [26] are computational results that have an identifier, known as verifiable result identifier (VRI), which is a URL that points to a repository where the results and the computational process behind it are located. VCRs can then be published in papers, and reviewers and readers may follow the VRIs to possibly reproduce the results.

2.7 Conclusion

End-to-end and long-term reproducibility of a scientific result is hard to achieve due to the factors that include the use of specialized hardware, proprietary data, and inevitable changes in hardware and software environments. Nonetheless, with the infrastructure we have built, it is possible to accurately document the processes through provenance capture, as well as to attain reproducibility for important subcomponents of a result, for example, the analysis and visualization of data derived from simulations run on special hardware.

As reproducibility becomes more widely adopted, the availability of repositories that contain fully documented experiments will open up new opportunities for scientific sharing and progress. These repositories have the potential to streamline scientific discovery by allowing researchers to search through and more easily reuse existing work [21].

Acknowledgments

We thank the VisTrails team for making this work possible, especially Emanuele Santos, Tommy Ellqvist, and Huy T. Vo. We also thank our many collaborators and users that have provided us feedback on our reproducibility infrastructure, particularly Philippe Bonnet, Dennis Shasha, Joel Tohline, and Matthias Troyer. The research and development of the VisTrails system has been funded by the National Science Foundation under grants CNS-1229185, IIS-1139832, IIS-1142013, IIS-0905385, IIS-1050422, IIS-0844572, ATM-0835821, IIS-0844546, IIS-0746500, CNS-0751152, IIS-0713637, OCE-0424602, IIS-0534628, CNS-0514485, IIS-0513692, CNS-0524096, CCF-0401498, OISE-0405402, CCF-0528201, CNS-0551724, the Department of Energy SciDAC (VACET and SDM centers), and IBM Faculty Awards.

References

1. The ALPS project. http://alps.comp-phys.org (Accessed March, 2013).
2. E.W. Anderson, J.P. Ahrens, K. Heitmann, S. Habib, and C.T. Silva. Provenance in comparative analysis: A study in cosmology. *Computing in Science and Engineering*, 10(3):30–37, 2008.
3. E.W. Anderson, G.A. Preston, and C.T. Silva. Towards development of a circuit based treatment for impaired memory: A multidisciplinary approach. In *IEEE EMBS Neural Engineering*, Kohala Coast, HI, pp. 302–305, 2007.
4. A. Baptista, B. Howe, J. Freire, D. Maier, and C.T. Silva. Scientific exploration in the era of ocean observatories. *Computing in Science and Engineering*, 10(3):53–58, 2008.
5. B. Bauer et al. The ALPS project release 2.0: Open source software for strongly correlated systems. *Journal of Statistical Mechanics: Theory and Experiment*, 2011(05):P05001, 2011.
6. L. Bavoil, S. Callahan, P. Crossno, J. Freire, C. Scheidegger, C.T. Silva, and H.T. Vo. VisTrails: Enabling interactive multiple-view visualizations. In *Proceedings of IEEE Visualization*, Minneapolis, MN, pp. 135–142, 2005.
7. *Biostatistics Journal*. http://biostatistics.oxfordjournals.org (Accessed March, 2013).
8. S.P. Callahan, J. Freire, C.E. Scheidegger, C.T. Silva, and H.T. Vo. Towards provenance-enabling ParaView. In J. Freire, D. Koop, and

L. Moreau, editors, *Provenance and Annotation of Data and Processes*, Lecture Notes in Computer Science, pp. 120–127. Springer-Verlag, Salt Lake City, UT, 2008.

9. F. Chirigati and J. Freire. Towards integrating workflow and database provenance. In P. Groth and J. Frew, editors, *Provenance and Annotation of Data and Processes*, Lecture Notes in Computer Science, pp. 11–23. Springer, Berlin, Germany, 2012.

10. F. Chirigati, D. Shasha, and J. Freire. Reprozip: Packing experiments for sharing and publication. In *Proceedings of the 2013 ACM SIGMOD International Conference on Management of Data, SIGMOD'13*, 2013 ACM, New York, 2013.

11. F.S. Chirigati, R. Dahis, S. Manuel Serra da Cruz, J. Freire, C.T. Silva, and M. Mattoso. Desenvolvimento de estruturas de controle explícito para o SGWfC VisTrails. In *Brazilian Symposium on Databases (SBBD)*, Fortaleza, Brazil, 2009 (Best poster award).

12. NSF Center for Coastal Margin Observation and Prediction (CMOP). http://www.stccmop.org (Accessed March, 2013).

13. Collage: Authoring Environment for Executable Publications. https://collage.elsevier.com/ (Accessed March, 2013).

14. Council for Scientific and Industrial Research (CSIR) in South Africa. http://www.csir.co.za (Accessed March, 2013).

15. The Data Observation Network for Earth (DataONE). https://dataone.org/ (Accessed March, 2013).

16. A. Dolgert, L. Gibbons, C.D. Jones, V. Kuznetsov, M. Riedewald, D. Riley, G.J. Sharp, and P. Wittich. Provenance in high-energy physics workflows. *Computing in Science and Engineering*, 10(3):22–29, 2008.

17. D.L. Donoho, A. Maleki, I.U. Rahman, M. Shahram, and V. Stodden. Reproducible research in computational harmonic analysis. *Computing in Science and Engineering*, 11(1):8–18, Jan–Feb 2009.

18. EO4VisTrails—Earth Observation Capabilities for VisTrails. http://code.google.com/p/eo4vistrails (Accessed March, 2013).

19. M.H. Freedman, J. Gukelberger, M.B. Hastings, S. Trebst, M. Troyer, and Z. Wang. Galois conjugates of topological phases. *Physical Review B*, 85:045414, Jan 2012.

20. J. Freire, D. Koop, E. Santos, C. Scheidegger, C.T. Silva, and H.T. Vo. *The Architecture of Open Source Applications*, Brown, A. and Wilson, G. (Eds.) Lulu Publishing Inc., 2011. http://www.aosabook.org/en/vistrails.html

21. J. Freire, P. Bonnet, and D. Shasha. Exploring the coming repositories of reproducible experiments: Challenges and opportunities. *PVLDB*, 4(12):1494–1497, 2011.

22. J. Freire, P. Bonnet, and D. Shasha. Computational reproducibility: State-of-the-art, challenges, and database research opportunities. In *Proceedings of the 2012 ACM SIGMOD International Conference on Management of Data, SIGMOD'12*, pp. 593–596, ACM, New York, 2012.

23. J. Freire, D. Koop, E. Santos, and C.T. Silva. Provenance for computational tasks: A survey. *Computing in Science and Engineering*, 10(3):11–21, May 2008.

24. J. Freire, C.T. Silva, S. Callahan, E. Santos, C. Scheidegger, and H.T. Vo. Managing rapidly-evolving scientific workflows. In *International Provenance and Annotation Workshop (IPAW)*, LNCS 4145, pp. 10–18. Springer Verlag, Chicago, IL, 2006.

25. J. Freire and C.T. Silva. Making computations and publications reproducible with VisTrails. *Computing in Science and Engineering*, 14(4):18–25, 2012.

26. M. Gavish and D. Donoho. A universal identifier for computational results. *Procedia Computer Science*. In *Proceedings of the International Conference on Computational Science (ICCS)*, Singapore, 4:637–647, 2011.

27. P. Van Gorp and S. Mazanek. SHARE: A web portal for creating and sharing executable research papers. *Procedia Computer Science*, 4:589–597, 2011. In *Proceedings of the International Conference on Computational Science (ICCS)*. Singapore.

28. GridFields. http://code.google.com/p/gridfields

29. R. Heiland, M. Swat, B. Zaitlen, J. Glazier, and A. Lumsdale. Workflows for parameter studies of multi-cell modeling (HPC). In *Proceedings of the ACM High Performance Computing Symposium*, Orlando, FL, pp. 94:1–94:6, 2010.

30. T. Hey, S. Tansley, and K. Tolle, editors. *The Fourth Paradigm: Data-Intensive Scientific Discovery*. Microsoft Research, Redmond, WA, 2009.

31. B. Howe, P. Lawson, R. Bellinger, E. Anderson, E. Santos, J. Freire, C. Scheidegger, A. Baptista, and C.T. Silva. End-to-end escience: Integrating workflow, query, visualization, and provenance at an ocean observatory. In *IEEE International Conference on eScience*, Indianapolis, IN, pp. 127–134, 2008.

32. B. Howe, C. Silva, and J. Freire. A science cloud on your desktop: VisTrails + GridFields, 2009. http://clue.cs.washington.edu (Accessed March, 2013).

33. IBM. OpenDX. http://www.research.ibm.com/dx (Accessed March, 2013).

34. The Insight Toolkit. http://www.itk.org (Accessed March, 2013).

35. The Kepler Project. http://kepler-project.org (Accessed March, 2013).

36. Kitware. ParaView. http://www.paraview.org (Accessed March, 2013).

37. D. Koop, E. Santos, B. Bauer, M. Troyer, J. Freire, and C.T. Silva. Bridging workflow and data provenance using strong links. In *SSDBM*, Heidelberg, Germany, pp. 397–415, 2010.

38. D. Koop, E. Santos, P. Mates, H.T. Vo, P. Bonnet, B. Bauer, B. Surer, M. Troyer, D.N. Williams, J.E. Tohline, J. Freire, and C.T. Silva. A provenance-based infrastructure to support the life cycle of executable papers. *Procedia Computer Science*, 4:648–657, 2011. In *Proceedings of the International Conference on Computational Science (ICCS)*. Singapore.

39. D. Koop, C. Scheidegger, J. Freire, and C.T. Silva. The provenance of workflow upgrades. In *IPAW*, Troy, NY, pp. 2–16, 2010.
40. D. Koop, C.E. Scheidegger, S.P. Callahan, J. Freire, and C.T. Silva. VisComplete: Automating suggestions for visualization pipelines. *IEEE Transactions on Visualization and Computer Graphics*, 14(6):1691–1698, 2008.
41. Lawrence Livermore National Laboratory. VisIt: Visualize It in Parallel Visualization Application. https://wci.llnl.gov/codes/visit, March 29, 2008. (Accessed March, 2013).
42. F. Leisch. Sweave: Dynamic generation of statistical reports using literate data analysis. In *Compstat*, pp. 575–580, 2002.
43. R.J. LeVeque. Python tools for reproducible research on hyperbolic problems. *Computing in Science and Engineering*, 11(1):19–27, Jan–Feb 2009.
44. Madagascar. http://www.ahay.org/wiki/Main_Page (Accessed March, 2013).
45. P. Mates, E. Santos, J. Freire, and C.T. Silva. CrowdLabs: Social analysis and visualization for the sciences. In *SSDBM*, Portland, OR, pp. 555–564, 2011.
46. J. Morisette, C. Jarnevich, T. Holcombe, C. Talbert, D. Ignizio, M. Talbert, C.T. Silva, D. Koop, A. Swanson, and N. Young. VisTrails SAHM: Visualization and workflow management for ecological niche modeling. *Ecography*, 36:129–135, 2013.
47. D.A. Norman. *Things That Make Us Smart: Defending Human Attributes in the Age of the Machine*. Addison Wesley, Reading, MA, 1994.
48. Oracle Database 11g Release 2. http://www.oracle.com/technetwork/database/enterprise-edition/overview (Accessed March, 2013).
49. Oracle Total Recall with Oracle Database 11g Release 2. http://www.oracle.com/technetwork/database/application-development/total-recall-1667156.html (Accessed March, 2013).
50. The Pegasus Project. http://pegasus.isi.edu/ (Accessed March, 2013).
51. Q. Pham, T. Malik, I. Foster, R. Di Lauro, and R. Montella. SOLE: Linking research papers with science objects. In P. Groth and J. Frew, editors, *Provenance and Annotation of Data and Processes*, vol. 7525 of Lecture Notes in Computer Science, pp. 203–208. Springer, Berlin, Germany, 2012.
52. PNAS Submission Guidelines. http://www.pnas.org/site/misc/iforc.shtml\#submission (Accessed March, 2013).
53. Software for Assisted Habitat Modeling Package for VisTrails (SAHM: VisTrails). http://www.fort.usgs.gov/products/software/sahm (Accessed March, 2013).
54. E. Santos, D. Koop, T. Maxwell, C. Doutriaux, T. Ellqvist, G. Potter, J. Freire, D. Williams, and C.T. Silva. Designing a provenance-based climate data analysis application. In P. Groth and J. Frew, editors, *Provenance and Annotation of Data and Processes*, vol. 7525 of Lecture Notes in Computer Science, pp. 214–219. Springer, Berlin, Germany, 2012.

55. E. Santos, D. Koop, H.T. Vo, E.W. Anderson, J. Freire, and C.T. Silva. Using workflow medleys to streamline exploratory tasks. In *SSDBM*, New Orleans, LA, pp. 292–301, 2009.

56. E. Santos, L. Lins, J. Ahrens, J. Freire, and C.T. Silva. VisMashup: Streamlining the creation of custom visualization applications. *IEEE Transactions on Visualization and Computer Graphics*, 15(6):1539–1546, 2009.

57. C.E. Scheidegger, H.T. Vo, D. Koop, J. Freire, and C.T. Silva. Querying and creating visualizations by analogy. *IEEE Transactions on Visualization and Computer Graphics*, 13(6):1560–1567, 2007.

58. SIGMOD Experimental Repeatability. http://www.sigmod2011.org/calls_papers_sigmod_research_repeatability.shtml (Accessed March, 2013).

59. C.T. Silva, E. Anderson, E. Santos, and J. Freire. Using VisTrails and provenance for teaching scientific visualization. *Computer Graphics Forum*, 30(1):75–84, 2011.

60. C.T. Silva, J. Freire, and S.P. Callahan. Provenance for visualizations: Reproducibility and beyond. *Computing in Science and Engineering*, 9(5):82–89, September 2007.

61. The Swift System. http://www.ci.uchicago.edu/swift (Accessed March, 2013).

62. The Taverna Project. http://www.taverna.org.uk/ (Accessed March, 2013).

63. J.E. Tohline, J. Ge, W. Even, and E. Anderson. A customized python module for CFD flow analysis within VisTrails. *Computing in Science and Engineering*, 11(3):68–73, 2009.

64. The Triana Project. http://www.trianacode.org (Accessed March, 2013).

65. *IEEE Transactions on Signal Processing*—Reproducible Research. http://www.signalprocessingsociety.org/publications/periodicals/tsp/ (Accessed March, 2013).

66. C. Upson et al. The application visualization system: A computational environment for scientific visualization. *IEEE Computer Graphics and Applications*, 9(4):30–42, 1989.

67. Ultrascale Visualization—Climate Data Analysis Tools (UV-CDAT). http://uv-cdat.llnl.gov (Accessed March, 2013).

68. T.L. Van Zyl, G. McFerren, and A. Vahed. Earth observation scientific workflows in a distributed computing environment. Technical Report 7727, CSIR, 2011. http://hdl.handle.net/10204/5435 (Accessed March, 2013).

69. VDS—The GriPhyN Virtual Data System. http://www.ci.uchicago.edu/wiki/bin/view/VDS/VDSWeb/WebMain (Accessed March, 2013).

70. VisTrails. http://www.vistrails.org (Accessed March, 2013).

71. VLDB Experimental Reproducibility. http://www.vldb.org/2013/experimental_reproducibility.html (Accessed March, 2013).

72. vtDV3D VisTrails Package. http://portal.nccs.nasa.gov/DV3D/vtDV3D/_build/html/index.html (Accessed March, 2013).

3

Sumatra: A Toolkit for Reproducible Research

Andrew P. Davison, Michele Mattioni, Dmitry Samarkanov, and Bartosz Teleńczuk

CONTENTS

3.1 Introduction

Lack of replicability in computational studies is, at base, a problem of shortcomings in record keeping. In laboratory-based experimental science, the tradition is to write down all experimental details in a paper notebook. This approach is no longer viable for many computational studies as the number of details that could have an impact on the final result is so large. Automated or semiautomated tools for keeping track of all the experimental details—the scientist's own code, input and output data, supporting software, the computer hardware used, etc.—are therefore needed.

For the busy scientist, the time investment needed to learn to use these tools, or to adapt their workflow so as to make use of them, may be one they are reluctant to make, especially since the problems of lack of reproducibility often take some time to manifest themselves. To achieve wide uptake among computational scientists, therefore, tools to support reproducible research should aim to minimize the effort required to learn, adopt, and use them (see [1] for a more detailed version of this argument).

Sumatra is a software tool to support reproducible computational research, which aims to make reproducible computational science as easy to achieve (or easier) than nonreproducible research, largely by automating the process of capturing all the experimental details. In practice, this means that using Sumatra should require minimal changes to existing workflows and, given the wide diversity in workflows for computational science, Sumatra should be easy to adapt to different computational environments.

This chapter is intended for two groups of people:

- Scientists who are interested in using Sumatra to track the details of their own research
- Developers who are interested in using Sumatra as a library in their own software for reproducible research

The first section is an extended case study, illustrating how Sumatra may be of use in day-to-day research. This is followed by an in-depth explanation of Sumatra's architecture, including examples of how to use Sumatra as a Python library and how to extend and customize Sumatra.

3.2 Using Sumatra

We will illustrate one way to use Sumatra, and why you might want to use Sumatra, with a story about Alice and Bob. Bob is a graduate student in Alice's lab. When Alice was a graduate student herself, she kept track of the evolution of her code by giving each significant version a different file name, and she included the file name as a label in every figure she generated. Alice used to be quite confident she could, if it were ever necessary, go back and recreate the results from her earlier papers since she has the original data carefully archived on CD-ROMs. However, after her recent experience with Charlie, she is not so sure. Charlie was a postdoc in Alice's lab, who got some great results, which they wrote up and submitted to a high-profile journal. The reviews were quite positive, but the reviewers asked for some new figures and a change to one of the existing figures. The problem was that when they tried to generate the modified figure, they could not get the results to match: the new graph looked significantly different, and no longer showed the effect they had found. Although Charlie had used the Subversion version control system for his code, he had not been so careful about keeping track of which version of the code had been used for each figure in the manuscript: several of the figures had originally been generated for a poster, and in the rush to get the poster finished in time to send to the printers, Charlie had not had time to keep such careful notes as usual, and had not always remembered to check-in changes in his code to the Subversion repository. Now Charlie has left science for a job with a major bank, and the manuscript is languishing in a drawer.

As a consequence of these experiences, Alice asked Bob, her new graduate student, to try out Sumatra. Sumatra automates the necessary but tedious and error-prone process of keeping track of which code version was used to produce which output. Bob has his code in a Mercurial version control repository. (For the purposes of this chapter, we will use a simplified version of Bob's code. If you would like to follow along, the repository is available at http://bitbucket.org/apdavison/ircr2013.) Bob downloaded and installed Sumatra according to the instructions at http://neuralensemble. org/sumatra.

Bob normally runs his analysis (of scanning electron microscope images of glass samples) as follows:

```
$ python glass_sem_analysis.py MV_HFV_012.jpg
1699.875 65.0
```

This analyses the image specified on the command line, generates some further images, and prints out some statistics (see the SciPy tutorial at http://scipy-lectures.github.com/ for more details). The output images are saved to a specific subdirectory labeled according to the day on which the code is run, and the individual files are labeled with a timestamp, for example, "Data/20121025/MV_HFV_012_163953_phases.png."

He creates a new Sumatra project in the same directory using the smt command-line tool:

```
$ smt init ProjectGlass
$ smt configure -e python -m glass_sem_analysis.py -i . -d Data
```

This creates a new project, and sets "python" as the default executable to be used, "glass_sem_analysis.py" as the default script file, the current directory (".") as the place to look for input data, and a subdirectory "Data" as the place to start looking for output files. (If Bob could not remember the various options to the "smt configure" command, "smt help configure" would tell him.)

"smt info" shows the current configuration of Bob's project. Note that it is using the already-existing Mercurial repository in his working directory:

```
$ smt info
Project name       : ProjectGlass
Default executable : Python (version: 2.6.7) at
                     /usr/bin/python
Default repository : MercurialRepository at
                     /home/bob/Projects/Glass
Default main file  : glass_sem_analysis.py
```

```
Default launch mode : serial
Data store (output) : ./Data
   .          (input) : .
Record store         : Django record store at
                       /home/bob/Projects/Glass/.smt/records
Code change policy   : error
Append label to      : None
```

Now to run the analysis using Sumatra:

```
$ smt run MV_HFV_012.jpg
1699.875 65.0
```

Since Bob has already specified the executable and script file, all he has to provide is the name of the input data file. The program runs as before and gives the same results, but in addition, Sumatra has captured a great deal of information about the *context* of the computation—exactly which version of the code was used, what the input and output data files were, what operating system and processor architecture were used, etc. Some of this information can be viewed in the console:

```
$ smt list -l
Label            : 20121025-170718
Timestamp        : 2012-10-25 17:07:18
Reason           :
Outcome          :
Duration         : 3.73256802559
Repository       : MercurialRepository at /home/bob/Projects/
                   Glass
Main_File        : glass_sem_analysis.py
Version          : 9d24b099b5f3
Script_Arguments : MV_HFV_012.jpg
Executable       : Python (version: 2.6.5) at /usr/bin/python
Parameters       :
Input_Data       : MV_HFV_012.jpg
                   (5d789282b10a0da7a91560f33f8baf7272f7543d)
Launch_Mode      : serial
Output_Data      : 20121025/MV_HFV_012_170722_phases.png
                   (c9955f84ca3c1912...
                 : 20121025/MV_HFV_012_170722_sand.png
                   (20bd5420d37ee589f3...
                 : 20121025/MV_HFV_012_170722_histogram.png
                   (e7884dc5f3e9c...
Tags             :
```

but in general it is better to use the built-in web browser-based interface, launched with the smtweb command—see Figure 3.1.

Two things in particular should be noted from this figure. The first is that the versions of not only the Python interpreter and Bob's own code but also the libraries on which Bob's code depends (NumPy, etc.), are captured. The

FIGURE 3.1
Record of a computation captured with Sumatra, displayed in the web browser interface.

second is that the path of each input and output data file is accompanied by a long hexadecimal string. This is the SHA1 digest, or hash, of the file contents (as used in crypographic applications, and also in version control systems such as Git and Mercurial). If the file contents are changed even slightly, the hash will change, which allows us to check for files being corrupted or accidentally overwritten.

Now Bob would like to investigate how his image analysis method is affected by changing its parameters. He thinks this will be easier to keep track of if the parameters are separated out into a separate file, so he modifies his script and adds a new file `default_parameters`. The script now expects two arguments, first the parameter file, second the input data, and would normally be run using

```
$ python glass_sem_analysis.py default_parameters
  MV_HFV_012.jpg
```

but Bob wants to run it with Sumatra:

```
$ smt run default_parameters MV_HFV_012.jpg
Code has changed, please commit your changes.
```

Bob has forgotten to commit his changes to the version control repository. Sumatra detects this and will then either refuse to run (the default, seen here) or will store the differences since the last commit. Bob commits and tries again.

```
$ hg commit -m 'Separated out parameters into separate file'
$ smt run -r 'test separate parameter file' default_parameters
  MV_HFV_012.jpg
1699.875 65.0
```

Note that he has also used the "-r" flag to note the reason for running this analysis, in case he forgets in future. Have Bob's modifications had any effect on his results? The output statistics are the same, and an inspection of the output data hashes in the web interface shows they have not changed either, so no, the results are unchanged.

We have seen already that Bob has less typing to do when running his analyses with Sumatra, as he has already specified the executable and script file as defaults. This is an example of how Sumatra tries to make it easier to use a tool for reproducible research than not to use one. Another example is the ability to specify parameters on the command line rather than having to edit the parameter file each time:

```
$ smt run -r 'No filtering' default_parameters MV_HFV_012.jpg
  filter_size=1
$ smt run -r 'Trying a different colourmap' default_parameters
                    MV_HFV_012.jpg phases_colourmap=hot
$ smt comment 'The default colourmap is nicer'
```

So far, Bob has been using Charlie's old computer, running Ubuntu Linux 10.04. The next day, he is excited to find that the new computer Alice ordered for him has arrived. He installs Ubuntu 12.04, together with all the latest versions of the Python scientific libraries. He also copies over his glass analysis data and migrates the Sumatra project. He tries to run the analysis script, but gets an error: in the latest version of NumPy, the return format of the `histogram()` function has changed. This is straightforward to fix (see https://bitbucket.org/apdavison/ircr2013/changeset/924a39a), so now Bob can commit and try again:

```
$ smt run -r 'Fixed to work with new histogram() function'
                    default_parameters MV_HFV_012.jpg
```

Has the upgrade affected Bob's results?

```
$ smt diff 20121025-172833 20121026-174545
Record 1                 : 20121025-172833
Record 2                 : 20121026-174545
Executable differs       : no
Code differs             : yes
   Repository differs    : no
   Main file differs     : no
   Version differs       : yes
   Non checked-in code   : no
   Dependencies differ   : yes
Launch mode differs      : no
Input data differ        : no
Script arguments differ  : no
Parameters differ        : no
Data differ              : yes
```

OK, Bob knew he had changed the code because of the new histogram() function, and he knew the dependencies had changed, because of the operating system upgrade, but it was a bit disappointing to see the output data are different. Using the web browser, we can look at the results from the two simulations (one from Ubuntu 10.04, one from Ubuntu 12.04) side by side (Figure 3.2)—visually there is no difference, just a tiny change in the margins, probably due to the upgraded matplotlib package.

Alice puts her head round the door to ask how Bob is getting on with Sumatra. So far, Bob is happy. His productive workflow has hardly changed—in fact, he has a little bit less to type, since Sumatra stores the names of the default executable and default script for him, and he can modify parameters quickly on the command line rather than having to open up the parameter file in his editor. The web browser interface lets him quickly browse and search through his results (Figure 3.3), and compare different runs side by side. And he feels much more confident that he will be able to replicate his results in the future.

Alice tries Sumatra out for herself the following week. Alice wants to use one of Bob's figures in a grant application, but Bob is on vacation, and she wants to make a few small changes to the figure. She copies Bob's Sumatra record store (which by default was created as the file .smt/records in a subdirectory of Bob's working directory) to the lab network file server, so that she can access Bob's records and Bob in turn will be able to see her results when he returns, and sets up a new project on her MacBook:

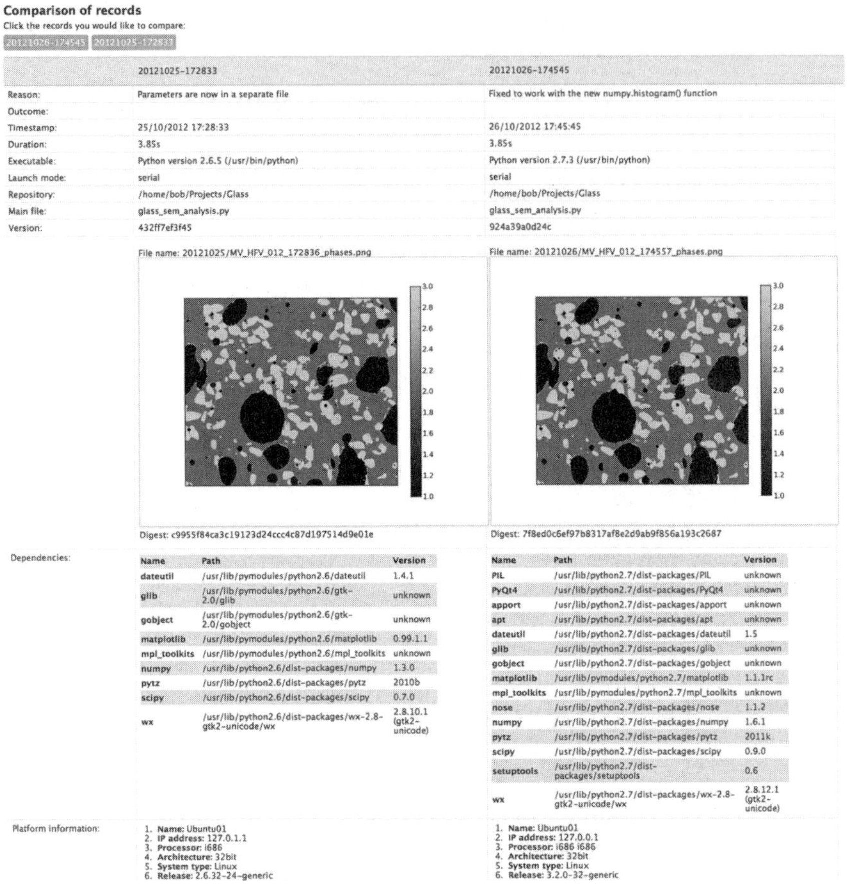

Comparison of records

Click the records you would like to compare:

`20121026-174545` `20121025-172833`

	20121025-172833	20121026-174545
Reason:	Parameters are now in a separate file	Fixed to work with the new numpy.histogram() function
Outcome:		
Timestamp:	25/10/2012 17:28:33	26/10/2012 17:45:45
Duration:	3.85s	3.85s
Executable:	Python version 2.6.5 (/usr/bin/python)	Python version 2.7.3 (/usr/bin/python)
Launch mode:	serial	serial
Repository:	/home/bob/Projects/Glass	/home/bob/Projects/Glass
Main file:	glass_sem_analysis.py	glass_sem_analysis.py
Version:	432ff7ef3f45	924a39a0d24c

File name: 20121025/MV_HFV_012_172836_phases.png

File name: 20121026/MV_HFV_012_174557_phases.png

Digest: c9955f84ca3c19123d24ccc4c87d197514d9e01e

Digest: 7f8ed0c6ef97b8317af8e2d9ab9f856a193c2687

Dependencies:

Name	Path	Version
dateutil	/usr/lib/pymodules/python2.6/dateutil	1.4.1
glib	/usr/lib/pymodules/python2.6/gtk-2.0/glib	unknown
gobject	/usr/lib/pymodules/python2.6/gtk-2.0/gobject	unknown
matplotlib	/usr/lib/pymodules/python2.6/matplotlib	0.99.1.1
mpl_toolkits	/usr/lib/pymodules/python2.6/mpl_toolkits	unknown
numpy	/usr/lib/python2.6/dist-packages/numpy	1.3.0
pytz	/usr/lib/python2.6/dist-packages/pytz	2010b
scipy	/usr/lib/python2.6/dist-packages/scipy	0.7.0
wx	/usr/lib/python2.6/dist-packages/wx-2.8-gtk2-unicode/wx	2.8.10.1 (gtk2-unicode)

Name	Path	Version
PIL	/usr/lib/python2.7/dist-packages/PIL	unknown
PyQt4	/usr/lib/python2.7/dist-packages/PyQt4	unknown
apport	/usr/lib/python2.7/dist-packages/apport	unknown
apt	/usr/lib/python2.7/dist-packages/apt	unknown
dateutil	/usr/lib/python2.7/dist-packages/dateutil	1.5
glib	/usr/lib/python2.7/dist-packages/glib	unknown
gobject	/usr/lib/python2.7/dist-packages/gobject	unknown
matplotlib	/usr/lib/pymodules/python2.7/matplotlib	1.1.1rc
mpl_toolkits	/usr/lib/pymodules/python2.7/mpl_toolkits	unknown
nose	/usr/lib/python2.7/dist-packages/nose	1.1.2
numpy	/usr/lib/python2.7/dist-packages/numpy	1.6.1
pytz	/usr/lib/python2.7/dist-packages/pytz	2011k
scipy	/usr/lib/python2.7/dist-packages/scipy	0.9.0
setuptools	/usr/lib/python2.7/dist-packages/setuptools	0.6
wx	/usr/lib/python2.7/dist-packages/wx-2.8-gtk2-unicode/wx	2.8.12.1 (gtk2-unicode)

Platform information:

1. Name: Ubuntu01
2. IP address: 127.0.1.1
3. Processor: i686
4. Architecture: 32bit
5. System type: Linux
6. Release: 2.6.32-24-generic

1. Name: Ubuntu01
2. IP address: 127.0.0.1
3. Processor: i686 i686
4. Architecture: 32bit
5. System type: Linux
6. Release: 3.2.0-32-generic

FIGURE 3.2

Excerpts from a side-by-side comparison of two computation records, one run on Ubuntu 10.04, the other on Ubuntu 12.04.

```
$ smt init -s /Volumes/shared/glass/smt_records ProjectGlass
$ smt configure -e python -m glass_sem_analysis.py -i .
  -d Data
```

Before starting her own modifications, she re-runs Bob's last analysis:

```
$ smt repeat 20121026-174545
The new record does not match the original. It differs as
  follows.
Record 1              : 20121026-174545
Record 2              : 20121026-174545_repeat
Executable differs    : no
Code differs          : yes
  Repository differs  : no
```

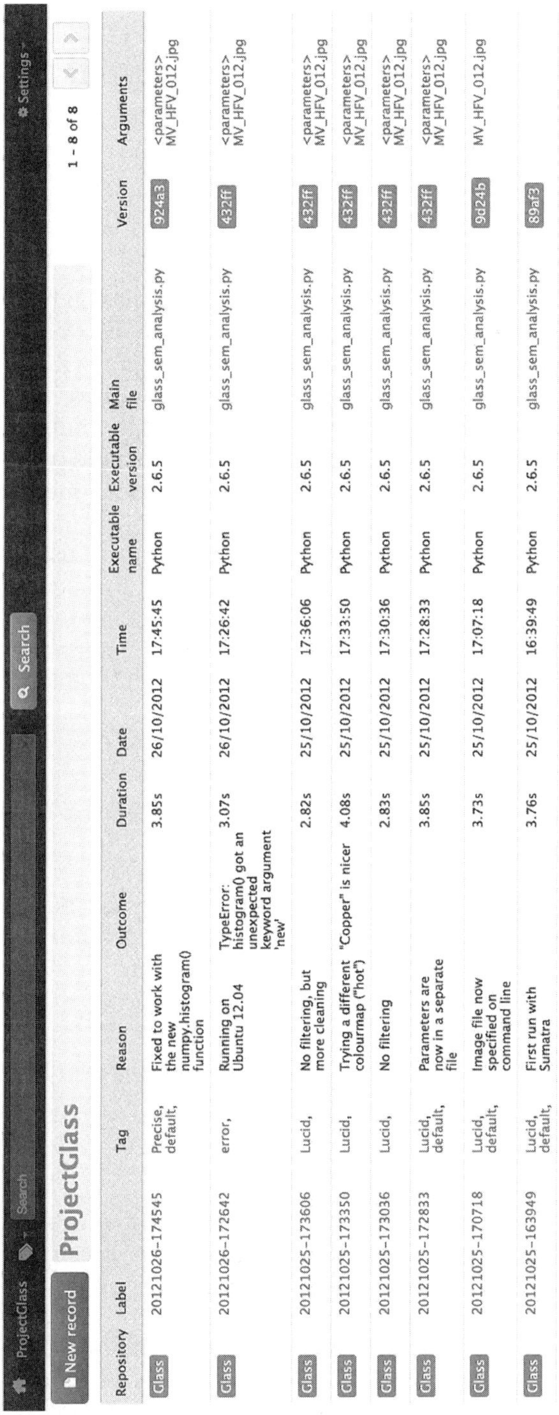

ProjectGlass   Search   Q Search   Settings

ProjectGlass

New record   1 – 8 of 8

Repository	Label	Tag	Reason	Outcome	Duration	Date	Time	Executable name	Executable version	Main file	Version	Arguments
Glass	20121026-174545	Precise, default,	Fixed to work with the new numpy.histogram() function		3.85s	26/10/2012	17:45:45	Python	2.6.5	glass_sem_analysis.py	924a3	<parameters> MV_HFV_012.jpg
Glass	20121026-172642	error,	Running on Ubuntu 12.04	TypeError: histogram0 got an unexpected keyword argument 'new'	3.07s	26/10/2012	17:26:42	Python	2.6.5	glass_sem_analysis.py	432ff	<parameters> MV_HFV_012.jpg
Glass	20121025-173606	Lucid,	No filtering, but more cleaning		2.82s	25/10/2012	17:36:06	Python	2.6.5	glass_sem_analysis.py	432ff	<parameters> MV_HFV_012.jpg
Glass	20121025-173350	Lucid,	Trying a different colourmap ("hot")	"Copper" is nicer	4.08s	25/10/2012	17:33:50	Python	2.6.5	glass_sem_analysis.py	432ff	<parameters> MV_HFV_012.jpg
Glass	20121025-173036	Lucid,	No filtering		2.83s	25/10/2012	17:30:36	Python	2.6.5	glass_sem_analysis.py	432ff	<parameters> MV_HFV_012.jpg
Glass	20121025-172833	Lucid, default,	Parameters are now in a separate file		3.85s	25/10/2012	17:28:33	Python	2.6.5	glass_sem_analysis.py	432ff	<parameters> MV_HFV_012.jpg
Glass	20121025-170718	Lucid, default,	Image file now specified on command line		3.73s	25/10/2012	17:07:18	Python	2.6.5	glass_sem_analysis.py	9d24b	MV_HFV_012.jpg
Glass	20121025-163949	Lucid, default,	First run with Sumatra		3.76s	25/10/2012	16:39:49	Python	2.6.5	glass_sem_analysis.py	89af3	

FIGURE 3.3

List of computation records in the Sumatra web browser interface.

```
Main file differs        : no
Version differs          : no
Non checked-in code      : no
Dependencies differ      : yes
Launch mode differs      : no
Input data differ        : no
Script arguments differ  : no
Parameters differ        : no
Data differ              : no
```

She has slightly different versions of the dependencies on her MacBook, but the results are unchanged. Alice can now proceed to reformat the figures, confident that her computing environment is consistent with that of her graduate student. Since the grant application is being written in LATEX, Alice can also use the sumatra LATEX package to automatically pull images from the Sumatra record store into her document, with automatic cross-checking of SHA1 hashes to ensure the image is indeed the correct one and has not been accidentally overwritten.

In conclusion, we hope to have demonstrated that by using Sumatra, Alice and Bob have improved the reproducibility of their computational experiments, enhanced communication within their lab, and increased the manageability of their projects, with minimal effort and minimal change to their existing workflow.

3.3 Design Criteria

In introducing the architecture of Sumatra so that others can build upon and extend it, we begin by describing the constraints we wish Sumatra to satisfy, before describing, in the following section, its current architecture.

The design of Sumatra is driven by two principles:

1. There is a huge diversity in computational science workflows.
2. Software to assist reproducibility must be very easy to use, or only the very conscientious will use it.

To elaborate on the first issue, of workflow diversity, different scientists may launch computations from the command line in interactive notebooks, in graphical interfaces, and in web-based tools. Computations may be launched serially as batch jobs or as distributed computations for immediate execution or queued for deferred execution on local machines, small clusters, supercomputers, grids, or in the cloud. Projects may be solo or collaborative efforts. Different workflows may be used for different components of a project or during different phases of a project (e.g., exploration vs. preparation of final published figures).

Given this diversity, it is unlikely that there is a single software tool to support reproducible research that will be optimal for all possible workflows. At the same time, there is a considerable amount of functionality that is required whatever the workflow, for example, unambiguous identification of exactly which code has been run. Sumatra is therefore designed as a core library of loosely coupled components for common functionality, easily extensible and customizable, so that people can adapt Sumatra to their own use cases and other people can build other tools on top of Sumatra.

Such a library is potentially useful to tool developers, but will not on its own promote reproducibility: it must be integrated into scientists' existing workflows, so that the barrier to adoption is as low as possible. Sumatra also, therefore, provides tools, built on top of the core library, that wrap around or work alongside widely used types of workflow. Three such tools are available at the time of writing: `smt`, which supports workflows built around running individual computations on the command line; `smtweb`, which provides a browser-based tool for browsing and querying the results of previous computations; and a LATEX package, which allows the automated inclusion of figures generated by a Sumatra-tracked computation in documents, with hyperlinks to the provenance information. The use of these tools was demonstrated in the previous section. In the future, further tools may be developed to support more interactive workflows.

Given the aforementioned constraints, Sumatra must enable a scientist to easily respond to the following questions:

- What code was run?
  - Which executable?
    - Name, location, version, compilation options
  - Which script?
    - Name, location, version
    - Options, parameters
    - Dependencies (name, location, version)
- What were the input data?
  - Name, location, content
- What were the outputs?
  - Data, logs, stdout/stderr
- Who launched the computation?
- When was it launched/when did it run? (queueing systems)
- Where did it run?
  - Machine name(s), other identifiers (e.g., IP addresses)
  - Processor architecture
  - Available memory
  - Operating system

- Why was it run?
- What was the outcome? (interpreted in terms of the ongoing project)
- Which project was it part of?

3.4 Architecture

This section gives an overview of Sumatra's architecture, intended for readers who may be interested in extending or building upon Sumatra, or applying some of its methods in their own approaches to replicability. More fine-grained detail is available in the online documentation at http://neuralensemble.org/sumatra. Sumatra has a modular design, with the coupling between modules made as loose as possible. Within modules, a common motif to provide flexibility and configurability is to use abstract base classes to define a common interface, which are then subclassed to provide different implementations of a given type of functionality (e.g., version control, data storage). The principal classes in the core Sumatra library, and their composition, are shown in Figure 3.4. More detail about the individual modules, classes, and their interactions is given in the following sections.

3.4.1 Code Versioning and Dependency Tracking

To ensure replication, we need to capture identifying information about all of the code that was run. Where code is modular, this means capturing the local file system path of each library/module/package that is included/imported by the "main" file (its "dependencies"), together with, if possible, the version of the module, so that (1) the environment could be recreated in future, (2) if failing to replicate with more up-to-date versions of libraries in future, we can investigate what has changed. This must be done recursively, of course, if a dependency itself has dependencies.

Finding the dependencies requires, in general, being able to parse the programming language used (although in future it may be possible to use a tool such as CDE [5,6] to determine which dependencies are loaded at run time). Sumatra therefore requires a "dependency finder" module to be provided for each programming language used. At the time of writing, such modules are all distributed within Sumatra, that is, as modules `dependency_finder.python`, `dependency_finder.matlab`, etc., but a plug-in architecture is planned so that users can easily extend Sumatra where the language they are using is not supported.

Version information may be provided in many ways, some of which are dependent on the programming language used, others independent. As an example of the former, Python modules often define a variable called `__version__`, `VERSION` or `version`, or a function called `get_version()`. Two examples of the latter are obtaining the version from a VCS and

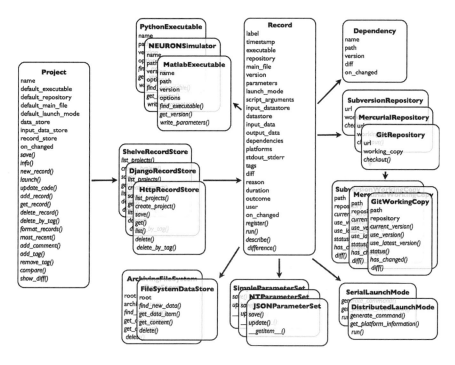

FIGURE 3.4

The principal classes in Sumatra with their attributes and methods. The arrows denote the relationship "contains an instance or instances of." Not shown, for reasons of space, are the classes DataKey, DataItem, RecordDifference, PlatformInformation, and Formatter. Not all subclasses are shown.

obtaining the version from a package management system (such as apt, on Debian). Sumatra's strategy, therefore, is that each `dependency_finder` module provides a list of functions, each implementing one heuristic for finding versions, for example, `find_versions_by_attribute()`, or `find_versions_by_version_control()`. Each of these is tried in turn, and the first version found is the one used (the order is important: generally a version obtained from a VCS is more reliable/precise than a version obtained from a variable defined within the code).

It may happen that some of the code under version control has been modified since the last commit. In this scenario, it is usually best to abort the computation and to commit the changes before proceeding. However, there may be good reasons for not wanting to commit, and so Sumatra also provides the option of storing the "diff" between the VCS working copy and the last commit.

Given the variety of VCSs in use, Sumatra's strategy is to wrap each VCS so as to provide a uniform interface. For each VCS supported by Sumatra, the `versioncontrol` module contains a submodule containing two classes—a subclass of `versioncontrol.base.Repository` and a subclass of `versioncontrol.base.WorkingCopy`. Sumatra does not require all the functionality of VCSs and is not intended to replace the normal methods of interacting with a VCS for code development. The `Repository` subclass has two roles: storing the repository URL and obtaining a fresh checkout/clone of the code base from a remote server (even the latter is not strictly necessary). The functionality required of the `WorkingCopy` subclass is more extensive: determine the current version; determine whether any of the code has been modified; determine the diff between the working code and the last commit; determine whether a given file is under version control; and change the working copy to an older or newer version (for replicating previous computations and then returning to the most recent state of the code base).

In general, the difference between distributed and centralized version control systems is not important for Sumatra. The only difference is that, for distributed VCSs, the repository used is always a local one, and it is therefore often useful, for the purposes of future replication and open science, to store the URL of the "upstream" repository, often a public repository on a remote server.

3.4.2 Data Handling

Replicability of a computational result requires knowing what the input data (if any) were, and it requires storing the output data so that future replication attempts can be checked against the original results. Inputs to a program can be subdivided into data and configuration/parameters. These can be generally distinguished in that data could be processed by a different program, while parameters are tightly tied to the code. Sumatra attempts to distinguish parameter/configuration files from input data files by the structure of the data; as a fall back, parameters will be treated as input data. Parameter file handling is described later.

Data may be stored in many ways: in individual files on a local or remote file system, in a relational database, and in a remote resource accessed over the Internet by some API. However it is stored, the most important thing to know about data is its content. However, it would be redundant for Sumatra to store a separate copy of each input and output data item, especially given the potentially enormous size of data items in many scientific disciplines. Sumatra therefore stores an identifier for each data item, which enables retrieval of the item from whichever data store—the file system, a relational database, etc.—is used. In the case of the file system, for example, the identifier consists of the file system path relative to a user-defined root directory together with the SHA1 hash of the file contents. The latter is needed to catch overwriting or corruption of files.

To handle different ways of storing data, Sumatra defines an abstract `DataStore` class, which is then subclassed: for example, the

`FileSystemDataStore` that is used to work with data stored on a local file system. The minimal functionality required of a `DataStore` subclass is: find new content, given a time stamp (used to link output data to a given computation); return a data item object, given the item's identifier ("key"); return the contents of a data item; and delete a data item. `DataItem` objects support obtaining the data contents and may also contain additional metadata, such as the mimetype.

It is straightforward to add extra functionality to a `DataStore` subclass. For example, the `ArchivingFileSystemDataStore` works the same as the plain `FileSystemDataStore`, but in addition copies all the output data files to an archive format. The `MirroredFileSystemDataStore` allows specifying a URL from which the data file can be retrieved (in addition to the local version). This supports, for example, using Dropbox (https://www.dropbox.com) with a public folder, or FTP, or FigShare (http://figshare.com) to make your data available online.

3.4.3 Storing Provenance Information

Once Sumatra has captured the context of your computational experiment, it needs to store all this information somewhere. For individual projects, a local database is probably the best way to do this. For collaborative projects, or if you often work while traveling, it may be necessary for this information to be stored in a remote database accessible over the Internet. To provide this flexibility, Sumatra defines an abstract `RecordStore` class, which is then subclassed.

Sumatra currently provides three `RecordStore` subclasses: `Shelve RecordStore`, which provides only basic functionality, but has the advantage of requiring no external libraries to be installed; `DjangoRecordStore`, which uses the Django web framework to store the provenance information in a relational database (SQLite by default, but MySQL, PostgreSQL, and others are also supported) and adds the ability to browse the record store using a web browser; and `HttpRecordStore`, which is a client for storing provenance information in a remote database accessed over HTTP using JSON as the transport format. The server for the `HttpRecordStore` is not distributed with Sumatra, but such a server is straightforward to implement. Two implementations currently exist—a Django-based implementation at https://bitbucket.org/apdavison/sumatra_server and a MongoDB-based version at https://github.com/btel/Sumatra-MongoDB.

The functionality required of a `RecordStore` subclass is: support multiple Sumatra projects; list all projects contained in the store; save a Sumatra `Record` object under a given project; list all the records in a project; retrieve a `Record` given its identifier (project+label); delete a `Record` given its identifier; delete all `Records` that have a given tag; return the most recent record; export a record in JSON format; import a record in the same format; and synchronize with another record store so that they both contain the same records for a given project.

3.4.4 Parameter Handling

It is a common practice in scientific computing to run a simulation or analysis with different parameters and to compare the results. Given this important use case, Sumatra allows parameters to be handled differently from other input data. If Sumatra is able to recognize a particular parameter file format, then (1) the parameters are available for future searching/querying/comparison and (2) Sumatra can add extra parameters. An important use case of the latter is that Sumatra can add the label/identifier for the current record, for use by the user's code in constructing file names, etc. Sumatra currently supports four parameter file formats, including simple "key=value" files, JSON, and config/ini-style formats. Implementing support for a new parameter file format is straightforward: define a `MyParameterSet` class whose constructor accepts either a filename or a text string containing the file contents. The class should also implement method `as_dict()`, which returns parameter names and values in a (possibly nested) Python dict; `update()`, which functions like `dict.update()`; and `save()`, which writes the parameter set to file in the given format.

3.4.5 Launching Computations

If your code is written in Python, then you can use Sumatra directly within your scripts and run your computation with Python as usual. If you are using other tools (or if using Python and you do not want to modify your code), then Sumatra needs to launch your computation in order to be able to capture the context. The challenge here is that there are so many different workflows ways of launching a computation: from the command line on the local machine and from the command line on a remote machine (e.g., using ssh); on a cluster, computing grid, or supercomputer using a job manager; as a parallel computation using MPI; or by clicking a button in a graphical interface.

To handle this variety, Sumatra follows the usual pattern of defining an abstract base class, `LaunchMode`, which is then subclassed to support different methods of launching computations. A `LaunchMode` subclass needs to define a method `generate_command()`, which should return a string that will be executed on the command line. The `LaunchMode` is also responsible for capturing information about the platform—the operating system, the processor architecture, etc. For computations run on the local machine, the base class takes care of this. For computations run on a remote machine or machines, the `LaunchModel` subclass must override the `get_platform_information()` method. Sumatra currently provides `SerialLaunchMode` and `DistributedLaunchMode` subclasses.

To generate the launch command, Sumatra may need extra information about the particular executable being used—particular arguments or flags that are needed in different circumstances. Similarly, there may be a build step or other preliminary that is needed before launching the computation. If this is the case, a user may define an `Executable` subclass that may define any of the attributes `pre_run`, `mpi_options`, `requires_script`, and may optionally redefine the method `_get_version()`. The user then calls the `programs.register_executable()` method to register the new subclass with Sumatra.

3.4.6 Putting It All Together

Tying all of the foregoing together are the `Record` class and the `Project` class. The `Record` class has two main roles: gathering provenance information when running a computation and acting as a container for provenance information. When launching a new computation, as diagrammed in Figure 3.5, a new `Record` object stores the identifiers of any input data, interacts with a `WorkingCopy` object to check that the code is at the requested version, uses the `dependency_finder` module to find the list of dependencies (and their versions), and then obtains platform information from the appropriate `LaunchMode`. It then runs any precursor tasks, such as building the executable, writes a modified parameter file, if necessary, and then passes control to the `LaunchMode`, which spawns a new process in which it runs the requested computation while capturing the standard output and standard error streams. Once this completes, the `Record` object calculates the time taken, stores stdout and stderr, asks the `DataStore` object to find any new data generated by the computation, and stores the identifiers of this output data.

The `Project` class has one main role: to simplify use of the Sumatra API by storing default values and providing shortcut functions for frequently performed tasks. Thus, for example, while creating a new `Record` object requires passing up to 16 arguments, the `Project.new_record()` method will often be called with just two—the parameter set and the list of input data items—since most of the others take default values stored by the `Project`. The `smt` command accesses Sumatra's functionality almost entirely through an instance of the `Project` class.

The precise division of responsibilities between the `Record` and `Project` class is not critical and could evolve in future versions of Sumatra to enhance usability of the API.

3.4.7 Search/Query/Reuse

So far, we have talked about the API from the perspective of capturing provenance information. We now consider the use cases of accessing, querying, and using the stored provenance information.

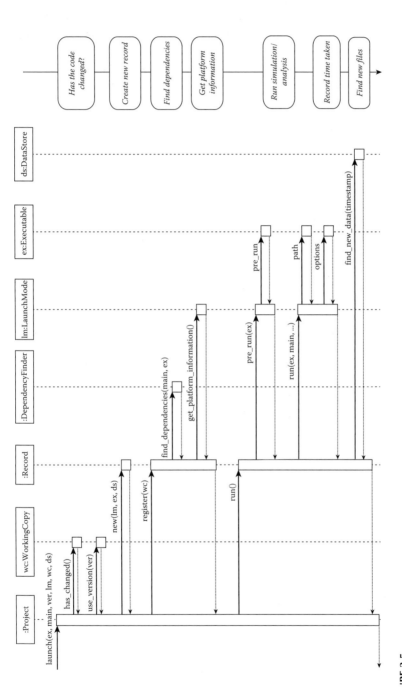

FIGURE 3.5

The flow of control between different Sumatra objects during a computation. Time flows from top to bottom. Each dashed vertical line represents the lifetime of an object, labeled at the top with the class and, in some cases, an instance name. Solid horizontal arrows represent method calls or attribute access.

As described earlier, this information is stored in a "record store," represented by a subclass of `RecordStore`, and whose backend may be a flat file, relational database, or web service. The common record store interface allows querying based on record identifiers (project + label) and on tags. Individual record store implementations may allow more sophisticated queries: for example, the `DjangoRecordStore` allows queries based on Django's object-relational-mapper or even using plain SQL.

The main use cases for accessing records of previous computations are (1) comparing the results of similar runs (e.g., examining the effects of parameter changes); (2) repeating a previous computation to check that the results are reproducible; and (3) further processing of results, for example, further analyses, visualization, and inclusion in manuscripts.

The first two of these use cases are supported by the `Project.compare()` method, which calls `Record.difference()`, that returns an instance of the `RecordDifference` class. This class has assorted methods that allow a precise dissection of the differences between two computations.

3.5 Discussion

In this chapter, we have presented Sumatra for two (although overlapping) audiences: the working computational scientist and the software developer or scientist-developer who may wish to extend or build upon Sumatra. In this book as a whole, a number of different tools to support reproducible research have been presented. For a scientist interested in ensuring their research is easily reproducible, when should you use Sumatra and when another tool?

Software for reproducible research can be divided into three general categories: tools for literate programming, workflow management systems, and tools for environment capture.

Literate programming* and the closely related "interactive notebook" approach[†] inextricably bind together code and the results generated by that code, which is clearly hugely beneficial for reproducible research. With some such systems, information about software versions, input data, and the computing environment can also be included in the final document. If your literate programming environment or interactive notebook supports Python, you could also use Sumatra via its API to provide this functionality.

---

* See, for example, Ref. [2], which explains the use of Sweave (http://www.statistik.lmu.de/ leisch/Sweave/) and Org-mode (http://orgmode.org) for reproducible research, and Ref. [11] in the current volume.

[†] For example, Mathematica (http://www.wolfram.com/mathematica/), Sage (http://www.sagemath.org), and IPython (http://ipython.org).

Scenarios that are generally more difficult to handle with the current generation of literate programming tools and interactive notebooks are (1) where
computations take considerable time (hours or days) to run; (2) where
computations are distributed on parallel hardware or are queued for later
execution; and (3) where code is split among many modules, so that the
code included in the literate document or notebook is only a small part of
the whole.

Visual workflow management or pipeline tools, such as Kepler [8],
Taverna [10], and VisTrails [3,4], are aimed at scientists with limited coding
experience or who prefer visual programming environments. They are particularly straightforward to use in domains where there some standardiza-
tions of data formats and analysis methods—for example, in bioinformatics
and in fields that make extensive use of image processing. The main disadvantage is that where there are no preexisting components for a given need,
creating a new component can require considerable effort and a detailed
knowledge of the workflow system architecture. Most widely used systems include provenance tracking either as an integral part or as an optional
module.

Environment capture systems, such as Sumatra, are generally the easiest
to adopt for an existing workflow. The simplest approach is to capture the
entire operating system as a virtual machine (VM) image—see the chapter
by Howe [7] in the current volume. A more lightweight alternative to this
is CDE [5,6], which archives only those executables and libraries actually
used by the computation. The main disadvantages with such approaches are
(1) your results risk being highly sensitive to the particular configuration of
your computer and (2) it is difficult or impossible to index, search, or analyse
the provenance information. Sumatra aims to overcome both of these disadvantages by capturing the information needed to recreate the experimental
context rather than the context itself in binary form. Some combination of
Sumatra and CDE would perhaps give the best of both worlds. Integration
of CDE is planned in a future version of Sumatra.

In summary, at the time of writing, Sumatra is most suitable for scientists who prefer to write their own code and run it from the command line,
especially when factors such as computation time, parallelism, or remote
execution make it difficult to work interactively, or where code is highly
modular so that literate programming tools capture only the tip of the code
iceberg. In any case, Sumatra is fast to set up, easy to use, and requires no
changes to existing code, so there is little to be lost in trying it out.

We have seen that Sumatra makes it much easier to replicate computational research in capturing the details of the software and hardware
environment that was used. In particular, Sumatra makes it much easier to
identify, in the case of failure to reproduce a result, what are the differences
between the original and current environments. However, Sumatra cannot
guarantee reproducibility, for two reasons. First, there are some details that
are not captured. For example, in Figure 3.1, you can see that for some of the

dependencies the version is unknown, either because the version information is genuinely not present or because Sumatra does not yet have a heuristic for finding it. Similarly, the compilation procedure and software library versions used to compile third-party programs, such as the Python interpreter, are not currently captured, and it may sometimes be impossible to capture this information. Second, with the passage of time, even if you know the particular versions of the libraries used, these versions may no longer be available, or the particular hardware architecture needed may not even be available. This problem is not restricted to Sumatra, of course. The use of VMs and careful archiving of old hardware is one partial solution, while for code that continues to be useful, a program of maintenance and ongoing updates can avoid obsolescence.

In the future, we plan to add support for using Sumatra with interactive notebooks (i.e., supporting a more granular unit of computation than an entire script), automated re-creation of software environments using the captured information, support for pipelines (where the output in one Sumatra record is the input in another), better support for compiled languages and software build systems, and interoperability with other provenance tracking tools, probably using the Open Provenance Model [9].

Sumatra is open-source software and is developed as an open community—if you have ideas or wish to contribute in any way, please join us at http://neuralensemble.org/sumatra.

Acknowledgments

We thank Eilif Muller, Konrad Hinsen, Stephan Gabler, Takafumi Arakaki, Yoav Ram, Tristan Webb, and Maximilian Albert for their contributions to Sumatra, as well as everyone who has reported bugs on the issue tracker. The code examples of SEM image analysis were based on the SciPy tutorial at http://scipy-lectures.github.com/.

References

1. AP Davison. Automated capture of experiment context for easier reproducibility in computational research. *Computing in Science and Engineering*, 14:48–56, 2012.
2. M Delescluse, R Franconville, S Joucla, T Lieury, and C Pouzat. Making neurophysiological data analysis reproducible: Why and how? *Journal of Physiology Paris*, 106:159–170, 2012.

3. J Freire. Making computations and publications reproducible with VisTrails. *Computing in Science and Engineering*, 14(4):18–25, 2012.

4. J Freire, D Koop, F Seabra Chirigati, and CT Silva. Reproducibility using VisTrails. In V Stodden, F Leisch, and R Peng, editors, *Implementing Reproducible Computational Research*. CRC Press/Taylor & Francis, Boca Raton, FL, 2013.

5. PJ Guo. CDE: A tool for creating portable experimental software packages. *Computing in Science and Engineering*, 14:32–35, 2012.

6. PJ Guo. CDE: Automatically package and reproduce computational experiments. In V Stodden, F Leisch, and R Peng, editors, *Implementing Reproducible Computational Research*. CRC Press/Taylor & Francis, Boca Raton, FL, 2013.

7. B Howe. Reproducibility, virtual appliances and cloud computing. In V Stodden, F Leisch, and R Peng, editors, *Implementing Reproducible Computational Research*. CRC Press/Taylor & Francis, Boca Raton, FL, 2013.

8. B Ludäscher, I Altintas, C Berkley, D Higgins, E Jaeger, M Jones, EA Lee, J Tao, and Y Zhao. Scientific workflow management and the Kepler system. *Concurrency and Computation: Practice and Experience*, 18(10):1039–1065, 2006.

9. L Moreau, B Clifford, J Freire, J Futrelle, Y Gil, P Groth, N Kwasnikowska, S Miles, P Missier, J Myers, B Plale, Y Simmhan, E Stephan, and J Van den Bussche. The Open Provenance Model core specification (v1.1). *Future Generation Computer Systems*, 27:743–756, June 2011.

10. T Oinn, M Greenwood, M Addis, M Nedim Alpdemir, J Ferris, K Glover, C Goble, A Goderis, D Hull, D Marvin, P Li, P Lord, MR Pocock, M Senger, R Stevens, A Wipat, and C Wroe. Taverna: Lessons in creating a workflow environment for the Life Sciences. *Concurrency and Computation: Practice and Experience*, 18(10):1067–1100, 2006.

11. Y Xie. knitr: A comprehensive tool for reproducible research in R. In V Stodden, F Leisch, and R Peng, editors, *Implementing Reproducible Computational Research*. CRC Press/Taylor & Francis, Boca Raton, FL, 2013.

4

CDE: *Automatically Package and Reproduce Computational Experiments*

Philip J. Guo

CONTENTS

4.1 Motivation

The simple-sounding task of taking software that runs on one person's machine and getting it to run on another machine can be painfully difficult in practice, even if both machines have the same operating system. Since no two machines are identically configured, it is hard for developers to predict the exact versions of software and libraries already installed on potential users' machines and whether those conflict with the requirements of their own software. Thus, software companies devote considerable resources to creating and testing one-click installers for products such as Microsoft Office, Adobe Photoshop, and Google Chrome. Similarly, open-source developers must carefully specify the proper dependencies in order to integrate their software into package management systems [2] (e.g., RPM on Linux, MacPorts on Mac OS X). Despite these efforts, online forums and mailing lists are filled with discussions of users' troubles in compiling, installing, and configuring software and dependencies.

Researchers are unlikely to invest the effort to create one-click installers or wrestle with package managers since their job is not to release production-quality software. Instead, they usually "release" their software by uploading their source code and data files to a server and writing some informal installation instructions. There is a slim chance that their colleagues will be able to run their research code "out-of-the-box" without some technical support.

4.2 CDE System Overview

In this chapter, we present a tool called CDE [1] that makes it easy for people to get their software running on other machines without the hassle of manually creating a robust installer or dealing with user

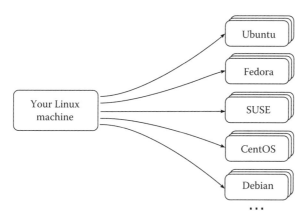

FIGURE 4.1
CDE enables users to package up any Linux program and deploy it to all modern Linux distros.

complaints about dependencies. CDE automatically packages up the Code, Data, and Environment required to run a set of x86-Linux programs on other x86-Linux machines without any installation (see Figure 4.1). To use CDE, the user simply:

1. Prepends any set of Linux commands with the `cde` executable. `cde` executes the commands and uses `ptrace` system call interposition to collect all code, data, and environment variables used during execution into a self-contained package.
2. Copies the resulting CDE package to an x86-Linux machine running any distribution (*distro*) from the past ~5 years.
3. Prepends the original packaged commands with the `cde-exec` executable to run them on the target machine. `cde-exec` uses `ptrace` to redirect file-related system calls so that executables can load the required dependencies from within the package. Execution can range from ~0% to ~30% slower.

The main benefits of CDE are that creating a package is as easy as executing the target program under its supervision, and that running a program within a package requires no installation, configuration, or root permissions.

In addition, CDE offers an *application streaming mode*, described in Section 4.5.3. Figure 4.2 shows its high-level architecture: The system administrator first installs multiple versions of many popular Linux distros in a "distro farm" in the cloud (or an internal compute cluster). The user connects to that distro farm via an ssh-based protocol from any x86-Linux machine. The user can now run *any* application available within the package managers

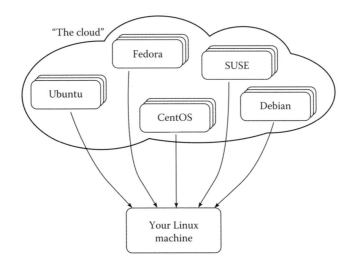

FIGURE 4.2
CDE's streaming mode enables users to run any Linux application on demand by fetching the required files from a farm of preinstalled distros in the cloud.

of any of the distros in the farm. CDE's streaming mode fetches the required files on demand, caches them locally on the user's machine, and creates a portable distro-independent execution environment. Thus, Linux users can instantly run the hundreds of thousands of applications already available in the package managers of all distros without being forced to use one specific release of one specific distro.*

We will use an example to introduce the core features of CDE. Suppose that Alice is a climate scientist whose experiment involves running a Python weather simulation script on a Tokyo dataset using this Linux command:

```
python weather_sim.py tokyo.dat
```

Alice's script (`weather_sim.py`) imports some third-party Python extension modules, which consist of optimized C++ numerical analysis code compiled into shared libraries. If Alice wants her colleague Bob to run and build upon her experiment, then it is not sufficient to just send her script and `tokyo.dat` data file to him. Even if Bob has a compatible version of Python on his machine, he will not be able to run her script until he compiles, installs, and configures the extension modules that she used (and all of their transitive dependencies).

---

* The package managers included in different releases of the same Linux distro often contain incompatible versions of many applications!

4.2.1 Creating a New Package with cde

To create a self-contained package with all dependencies required to run her experiment on another machine, Alice prepends her command with the cde executable:

```
cde python weather_sim.py tokyo.dat
```

cde runs her command normally and uses the Linux ptrace mechanism to monitor all files it accesses throughout execution. cde creates a new subdirectory called cde-package/cde-root/ and copies all of those accessed files into there, mirroring the original directory structure. For example, if her script dynamically loads an extension module (shared library) named /usr/lib/weather.so, then cde will copy it to cde-package/cde-root/usr/lib/weather.so (see Figure 4.3). cde also saves the values of environment variables in a file within cde-package/.

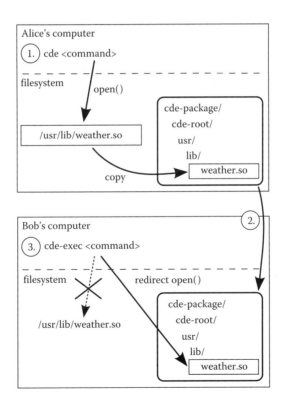

FIGURE 4.3
Example use of CDE: 1. Alice runs her command with cde to create a package, 2. Alice sends her package to Bob's computer, and 3. Bob runs that same command with cde-exec, which redirects file accesses into the package.

When execution terminates, the `cde-package/` subdirectory (which we call a "CDE package") contains all of the files required to run Alice's original command.

4.2.2 Executing a Package with `cde-exec`

Alice zips up the `cde-package/` directory and transfers it to Bob's Linux machine. Now Bob can run Alice's experiment without installing anything on his machine. He unzips the package, changes into the subdirectory containing the script, and prepends the original command with the `cde-exec` executable*:

```
cde-exec python weather_sim.py tokyo.dat
```

`cde-exec` sets up the environment variables saved from Alice's machine and executes the version of `python` and its extension modules from within the package. `cde-exec` uses `ptrace` to monitor all system calls that access files and rewrites their path arguments to the corresponding paths within the `cde-package/cde-root/` subdirectory. For example, when her script requests to load the `/usr/lib/weather.so` extension library using an `open` system call, `cde-exec` rewrites the path argument of the `open` call to `cde-package/cde-root/usr/lib/weather.so` (see Figure 4.3). This path redirection is essential because `/usr/lib/weather.so` probably does not exist on Bob's machine.

Not only can Bob reproduce Alice's exact experiment, but he can also edit her script and dataset and then rerun to explore variations and alternative hypotheses, as long as his edits do not cause the script to import new Python extension modules that are not in the package. Also, since a CDE package is a directory tree, Bob can add additional dataset files into the package to run related experiments.

4.2.3 CDE Package Portability

Alice's CDE package can execute on any Linux machine with an architecture and kernel version that are compatible with its constituent binaries. CDE currently works on 32- and 64-bit variants of the x86 architecture (i386 and x86-64, respectively). In general, a 32-bit `cde-exec` can execute 32-bit packaged applications on 32- and 64-bit machines. A 64-bit `cde-exec` can execute both 32- and 64-bit packaged applications on a 64-bit machine. Extending CDE to other architectures (e.g., ARM) is straightforward because the `strace` tool that CDE is built upon already works on many architectures.

---

* The package contains a copy of `cde-exec`.

However, CDE packages cannot be transported *across* architectures without using a CPU emulator.

In practice, CDE packages are portable across Linux distros released within approximately 5 years of the distro where the package originated [13]. Besides sharing with colleagues such as Bob, Alice can also deploy her package to run on a cluster for more computational power or to a public server for real-time weather simulation reporting. Since she does not need to install any software or libraries as the root user, she does not risk perturbing existing software on those machines. Finally, having her script and all of its dependencies (including the Python interpreter and extension modules) encapsulated within a CDE package makes it somewhat "future-proof" and likely to still run on her machine even when its version of Python and associated extensions are upgraded in the future.

Users can combine CDE with a virtual machine (VM) to achieve greater portability. For example, if Alice wants her colleagues who run Windows, Mac OS, or an antiquated Linux to reproduce her experiments, she can put her CDE package within a Linux VM and distribute the entire VM image. However, the price to pay for such portability is increased file size: A VM image file can be 10–100 times larger than a CDE package because it contains the entire operating system.

Finally, unlike language-based portability technologies (such as Java or Python `virtualenv`), CDE works on Linux programs written in any language or mix of languages.

4.2.4 Ignoring Files and Environment Variables

By convention, Linux directories such as `/dev`, `/proc`, and `/sys` contain pseudo-files (e.g., device files) that do not make sense to include in a CDE package. Also, environment variables such as `$XAUTHORITY` and the corresponding `.Xauthority` file (for X Window authorization) are machine specific. Informed by our debugging experiences and user feedback, we have manually created a blacklist of directories, files, and environment variables for CDE to ignore so that packages can be portable across machines. By "ignore" we mean that `cde` will not copy those files (or variables) into a package, and `cde-exec` will not redirect their paths and instead access the real versions on the machine. This user-customizable blacklist is implemented as a plain-text options file. Figure 4.4 shows this file's default contents.

CDE also allows users to customize which paths it should ignore (leave alone) and which it should redirect into the package, thereby making its sandbox "semi-permeable." For example, one computational scientist chose to have CDE ignore a directory that mounts an NFS share containing huge data files because he knew that the machine on which he was going to execute the package also mounts that NFS share at the same path. Therefore, there was no point in bloating up the package with those data files.

```
# These directories often contain pseudo-files that shouldn't be tracked
ignore_prefix=/dev/
ignore_exact=/dev
ignore_prefix=/proc/
ignore_exact=/proc
ignore_prefix=/sys/
ignore_exact=/sys
ignore_prefix=/var/cache/
ignore_prefix=/var/lock/
ignore_prefix=/var/log/
ignore_prefix=/var/run/
ignore_prefix=/var/tmp/
ignore_prefix=/tmp/
ignore_exact=/tmp

ignore_substr=.Xauthority      # Ignore to allow X Window programs to work

ignore_exact=/etc/resolv.conf # Ignore so networking can work properly

# Access the target machine's password files:
# (some programs like texmacs need these lines to be commented-out,
#  since they try to use home directory paths within the passwd file,
#  and those paths might not exist within the package.)
ignore_prefix=/etc/passwd
ignore_prefix=/etc/shadow

# These environment vars might lead to 'overfitting' and hinder portability
ignore_environment_var=DBUS_SESSION_BUS_ADDRESS
ignore_environment_var=ORBIT_SOCKETDIR
ignore_environment_var=SESSION_MANAGER
ignore_environment_var=XAUTHORITY
ignore_environment_var=DISPLAY
```

FIGURE 4.4
The default CDE options file, which specifies the file paths and environment variables that CDE should ignore. `ignore_exact` matches an exact file path, `ignore_prefix` matches a path's prefix string (e.g., directory name), and `ignore_substr` matches a substring within a path. Users can customize this file to tune CDE's sandboxing policies (see Section 4.2.4).

4.2.5 Nongoals

Our philosophy in designing CDE was to create the simplest possible tool that would allow a large class of real-world Linux programs to be portable across a range of contemporary distros. One way we have kept CDE's design simple was to limit its scope. Here are some tasks that CDE is *not* designed to perform:

- *Deterministic replay*: CDE does not try to replay exact execution paths like record-replay tools [8,17,20] do. Thus, CDE does not need to capture sources of randomness, thread scheduling, and other nondeterminism. It also does not need to create snapshots of filesystem state for rollback/recovery.

- *OS/hardware emulation*: CDE does not spoof the OS or hardware. Thus, programs that require specialized hardware or device drivers will not be portable across machines. Also, CDE cannot capture remote network dependencies.
- *Security*: Although CDE isolates target programs in a chroot-like sandbox, it does not guard against attacks to circumvent such sandboxes [11]. Users should only run CDE packages from trusted sources. (Of course, the same warning applies to *all* downloaded software.)
- *Licensing*: CDE does not attempt to "crack" software licenses, nor does it enforce licensing or distribution restrictions. It is ultimately the package creator's responsibility to make sure that he/she is both willing and able to distribute the files within a package, abiding by the proper software and dataset licenses.

4.3 Use Case Categories

Since we released the first version of CDE on November 9, 2010, it has been downloaded at least 10,000 times as of November 2012 [1]. We cannot track how many people have directly checked out its source code from GitHub [1], though.

We have exchanged hundreds of e-mails with CDE users and discovered five salient real-world use cases as a result of these discussions:

4.3.1 Creating Reproducible Computational Experiments

The results of many computational science experiments can be reproduced within CDE packages because their code is output-deterministic [8], always producing the same outputs (e.g., statistics, tables, graphs) for a given input. We have received several e-mails describing how researchers have used CDE to make their experiments reproducible, including

- Robotics motion planning experiments using C++ and OpenGL code [22]
- Genetic algorithms for social networking using C++ and R code [18]
- Biological fingerprint identification using the LibSVM machine learning library and the Open Babel computational chemistry toolbox [15]

4.3.2 Distributing Research Software

The creators of several research tools found CDE online and used it to create portable packages that they uploaded to their websites:

The website for Graph-Tool, a Python/C++ module for analyzing graphs, lists these (direct) dependencies: "GCC 4.2 or above, Boost libraries, Python 2.5 or above, expat library, NumPy and SciPy Python modules, GCAL C++ geometry library, and Graphviz with Python bindings enabled" [6]. Unsurprisingly, lots of people had trouble compiling it: 47% of all messages on its mailing list (137 out of 289) were questions related to compilation problems. The author of Graph-Tool used CDE to automatically create a portable package (containing 149 shared libraries and 1909 total files) and uploaded it to his website so that users no longer needed to suffer through the pain of manually compiling it.

Arachni, a Ruby-based tool that audits web application security [5], requires six hard-to-compile Ruby extension modules, some of which depend on versions of Ruby and libraries that are not available in the package managers of most modern Linux distributions. Its creator, a security researcher, created and uploaded CDE packages and then sent us a grateful e-mail describing how much effort CDE saved him: "My guess is that it would take me half the time of the development process to create a self-contained package by hand; which would be an unacceptable and truly scary scenario."

A British research programming team used CDE to make portable packages for their protein crystallography software.* Similarly, a team at Johns Hopkins University made CDE packages for CAWorks,† medical visualization software for computational anatomy. The following e-mail snippet from a conversation with its lead developer provides a sense of the complex dependencies that CDE automatically encapsulated: "My program which is called CAWorks is huge with a massive dependency list. ParaView the base program uses vtk, python, zlib and of course Qt and all of their dependent libraries. While my program CAWorks adds to this libcurl, openssl, ITK and vxl, blitz, dicom, ivcon, clapack, getopt, gts, md5, quazip, and glib and all of their dependent libraries, none of which ParaView uses."

Finally, we used CDE to create portable binary packages for two of our Stanford‡ colleagues' research tools, which were originally distributed as tarballs of source code: PADS [10] and Saturn [7]. Forty-four percent of the messages on the PADS mailing list (38/87) were questions related to troubles with compiling it (22% for Saturn). Once we successfully compiled these projects (after a few hours of improvising our own hacks since the documentation was grossly outdated), we created CDE packages by running their

---

* http://www.ccp4.ac.uk/.
† http://cis.jhu.edu/software/caworks/.
‡ CDE originated as a research project in the Computer Science Department at Stanford University.

regression test suites. Now our fellow researchers no longer need to suffer through the compilation process.

To see the benefits of CDE here, note that the Saturn team leader admitted in a public e-mail, "As it stands the current release likely has problems running on newer systems because of bit rot—some libraries and interfaces have evolved over the past couple of years in ways incompatible with the release" [3]. In contrast, CDE packages are largely immune to "bit rot" (until the user-kernel ABI changes) because they contain all dependencies.

4.3.3 Deploying Computations to Cluster or Cloud

People working on computational experiments on their desktop machines often want to run them on a cluster for greater performance and parallelism. However, before they can deploy their computations to a cluster or cloud computing (e.g., Amazon EC2), they must first install all of the required executables and dependent libraries on the cluster machines. At best, this process is tedious and time consuming, since cluster/cloud machines run older versions of software and libraries due to both slow upgrade cycles and concerns about security and stability. At worst, installation can be impossible since regular users often do not have root access on cluster machines.

Using CDE, a user can create a self-contained package on their desktop machine and then execute that package on the cluster or cloud (possibly many instances in parallel), without needing to install any dependencies or to get root access on the remote machines.

For example, our Stanford colleague Peter wanted to use a department-administered 100-CPU cluster to run a parallel image processing job on topological maps. However, since he did not have root access on those older machines, it was nearly impossible for him to install all of the dependencies required to run his computation, especially the image processing libraries. Peter used CDE to create a package by running his job on a small dataset on his desktop, transferred the package and the complete dataset to the cluster, and then ran 100 instances of it in parallel there.

Similarly, we worked with labmates to use CDE to deploy the CPU-intensive Klee [9] automated bug finding tool from the desktop to Amazon's EC2 cloud computing service without needing to compile Klee on the cloud machines. Klee can be hard to compile since it depends on LLVM, which is very picky about specific versions of GCC and other build tools being present on the machine before it will compile.

Researchers have also used CDE to deploy computational experiments to internal compute clusters within several software companies, to the European Grid distributed computing infrastructure, and to the iPlant* cloud infrastructure (NSF-funded cyberinfrastructure for plant biologists).

---

* http://www.iplantcollaborative.org/.

On a related note, several researchers have used CDE to deploy their research software not to a cluster but rather to a webserver; that way, users can interact with their code via a web interface. Since researchers often do not have root access on shared web hosting machines, it can be impossible to install all of the required dependencies on there.

4.3.4 Running Production Software on Incompatible Distros

Even production-quality software might be hard to install on Linux distros with older kernel or library versions, especially when system upgrades are infeasible. A user can run software under CDE supervision on a modern distro to create a package and then run that package on an older distro, regardless of what libraries are present on there.

For example, an engineer at Cisco wanted to run some new open-source tools on his work machines, but the IT department mandated that those machines run an older, more secure enterprise Linux distro. He could not install the tools on those machines because that older distro did not have up-to-date libraries, and he was not allowed to upgrade. Therefore, he installed a modern distro at home, ran CDE on there to create packages for the tools he wanted to port, and then ran the tools from within the packages on his work machines.

Hobbyists applied CDE in a similar way: A game enthusiast could only run classic games within a DOS emulator on one of his Linux machines, so he used CDE to create a package and can now play the games on his other machines. We also helped a user create a portable package for the Google Earth 3D map application, so he can now run it on older distros whose libraries are incompatible with Google Earth.

4.3.5 Class Programming Projects

A teaching assistant for Stanford's Parallel Computing course (CS 149) used CDE to package up the toolchain required to compile and run class programming projects; thus, students can focus on the actual programming rather than on the drudgery of installation and configuration.

In addition, two users sent us CDE packages they created for collaborating on class assignments: Rahul, a Stanford grad student, was using NLTK [19], a Python module for natural language processing, to build a semantic e-mail search engine for a machine learning class. Despite much struggle, Rahul's two teammates were unable to install NLTK on their Linux machines due to conflicting library versions. This meant that they could run only one instance of the project at a time on Rahul's laptop for query testing and debugging. When Rahul discovered CDE, he created a package for their project and was able to run it on his two teammates' machines so that all three of them could test and debug in parallel. Joshua, an undergrad from Mexico, e-mailed us a similar story about how he used CDE to collaborate on and demo his virtual reality class project.

4.4 Implementation Details

This section describes the implementation of CDE in some detail, so it is relevant only for readers who are either curious about those details or who want to implement a similar tool.

CDE uses the Linux `ptrace` system call to monitor the target program's processes and threads, read/write to its memory, and modify its system call arguments, all without requiring root permission. System call interposition using `ptrace` is a well-known technique that computer systems researchers have used for implementing tools such as secure sandboxes [12,16], record-replay systems [17], and user-level filesystems [21].

We implemented CDE by adding 3000 lines of C code to the `strace` system call monitoring tool. CDE works only on x86-based Linux machines (32-bit and 64-bit) but should be easy to extend to other hardware architectures. Although implementation details are Linux-specific, these same ideas could be used to implement CDE for another OS such as Mac OS X or Windows.

4.4.1 Creating a New Package with `cde`

4.4.1.1 Primary Action

The main job of `cde` is to use `ptrace` to monitor the target program's system calls and copy all of its accessed files into a self-contained package. The only relevant syscalls here are those that take a file path string as an argument, which are listed in the "File path access" category in Table 4.1 (and also `execve`). After the kernel finishes executing one of these syscalls and is about to return to the target program, `cde` wakes and observes the return value. If the return value signifies that the indicated file exists, then `cde` copies that file into the package (see Figure 4.5).

Note that many syscalls operate on files but take a file descriptor as an argument rather than a file path (e.g., `mmap`); `cde` does not need to track those since it already tracks the preceding syscalls (e.g., `open`) that create file descriptors from file paths.

4.4.1.2 Copying Files into Package

Prior to copying a file into the package, `cde` creates all necessary subdirectories and symbolic links to mirror the original file's location. In our example from Figure 4.3, `cde` copies `/usr/lib/weather.so` into the package as `cde-package/cde-root/usr/lib/weather.so`. For efficiency, copies are done via Linux hard links if possible.

If a file is a symlink, then both it and its target must be copied into the package. Multiple levels of symlinks, to both files and directories, must be

TABLE 4.1

The 48 (out of 338 Total) Linux 2.6 System Calls Intercepted by `cde` and `cde-exec`, and Actions Taken for Each Category of Syscalls

Category	Linux Syscalls	cde Action	cde-exec Action
File path access	`open[at]`, `mknod[at]`, `fstatat64`, `access`, `faccessat`,`readlink[at]`, `truncate[64]`, `stat[64]`, `creat`,`lstat[64]`, `oldstat`, `oldlstat`, `chown[32]`,`lchown[32]`, `fchownat`, `chmod`, `fchmodat`, `utime`, `utimes`, `futimesat`	Copy file into package	Redirect path into package
Local sockets	`bind`, `connect`	None	Redirect path into package[a]
Mutate filesystem	`link[at]`, `symlink[at]`, `rename[at]`, `unlink[at]`, `mkdir[at]`, `rmdir`	Repeat in package	Redirect path into package
Get current dir.	`getcwd`	Update current directory	Spoof current directory
Change directory	`chdir`, `fchdir`	Update current directory	
Spawn child	`fork`, `vfork`, `clone`	Track child process or thread	
Execute program	`execve`	Copy binary into package	Maybe run dynamic linker

Syscalls with suffixes in [brackets] include variants both with and without the suffix: for example, `open[at]` means `open` and `openat`.

[a] For `bind` and `connect`, `cde-exec` only redirects the path if it is used to access a file-based socket for local IPC.

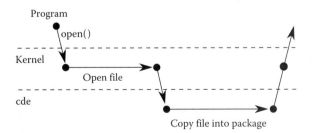

FIGURE 4.5
Timeline of control flow between the target program, kernel, and `cde` process during an `open` syscall.

properly handled. More subtly, *any component* of a path may be a symlink to a directory, so the exact directory structure must be replicated within the package for cde-exec to work. For example, we once encountered a path /usr/lib/gcc/4.1.2/libgcc.a, where 4.1.2 is a symlink to a directory named 4.1.1. We observed that some programs are sensitive to exact filesystem layout, so cde must faithfully replicate symlinks within the package, or else those programs will fail with cryptic errors when run from within the package.

Finally, if the file being copied is an ELF binary (executable or library code), then cde searches through the binary's contents for constant strings that are filenames and then copies those files into the package. Although this hack is simplistic, it works well in practice to partially overcome CDE's limitation of only being able to gather dependencies on executed paths (see Section 4.5.1 for more details). It works because many binaries dynamically load libraries whose filenames are constant strings. For example, we encountered a Python extension library that dynamically loads one of a few versions of the Intel Math Kernel Library based on the current CPU's capabilities. Without this hack, any given execution will copy only *one* version of the Intel library into the package, so packaged execution will fail when running on another machine with different CPU capabilities. Finding and copying all versions of the Intel library into the package makes the program more likely to run on machines with different hardware.

Here is how cde handles the other syscalls in Table 4.1:

Mutate filesystem: After each call that mutates the filesystem, cde repeats the same action on the corresponding copies of files in the package. For example, if a program renames a file from foo to bar, then cde also renames the copy of foo in the package to bar. This way, at the end of execution, the package's contents mirror the "poststate" of the original filesystem's contents, not the "prestate" before execution.

Updating current working directory: At the completion of getcwd, chdir, and fchdir, cde updates its record of the monitored process's current working directory, which is necessary for resolving relative paths.

Tracking subprocesses and threads: If the target program spawns subprocesses, cde also attaches onto those children with ptrace (it attaches onto spawned threads in the same way). cde keeps track of each monitored process's current working directory and shared memory segment address (needed for Section 4.4.2). cde remains single-threaded and responds to events queued by ptrace.

This feature is useful for packaging up workflows consisting of multiple program invocations, such as a compilation job. Running "cde make" will track all subprocesses that the Makefile spawns and package up the source files and compiler toolchain. Now you can edit and compile the given project on another Linux machine by simply running "cde-exec make" without needing to install any compilation tools or header files on that machine.

execve syscall: cde copies the executable's binary into the package. For a script, cde finds the name of its interpreter binary from the she-bang (#!) line. If the binary is dynamically linked, cde also finds its dynamic linker (e.g., ld-linux.so.2) and copies it into the package. The dynamic linker is responsible for loading the shared libraries that a program needs at start-up time.

4.4.2 Executing Package with cde-exec

4.4.2.1 Primary Action

The main job of cde-exec is to use ptrace to redirect file paths that the target program requests into the package. Before the kernel executes most syscalls listed in Table 4.1, cde-exec rewrites their path argument(s) to refer to the corresponding path within cde-package/cde-root/ (Figure 4.6). By doing so, cde-exec creates a chroot-like sandbox that fools the target program into "believing" that it is executing on the original machine. Unlike chroot, this sandbox does not require root access to set up, and it is user-customizable (see Section 4.2.4).

In our running example, suppose that Alice runs her experiment within the /expt directory on her computer:

```
cd /expt
cde python weather_sim.py tokyo.dat
```

She then sends the package to Bob's computer. If Bob unzips it into his home directory (/home/bob), then he can run these commands to execute her Python script:

```
cd /home/bob/cde-package/cde-root/expt
cde-exec python weather_sim.py tokyo.dat
```

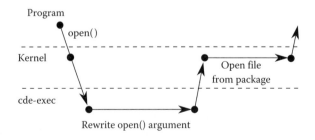

FIGURE 4.6
Timeline of control flow between the target program, kernel, and cde-exec process during an open syscall.

Note that Bob needs to first change into the /expt subdirectory within the package, since that is where Alice's scripts and data files reside. When cde-exec starts, it finds Alice's python executable within the package (with the help of $PATH) and launches it. Now if her program requests to open, say, /usr/lib/weather.so, cde-exec rewrites the path argument of the open call to /home/bob/cde-package/cde-root/usr/lib/ weather.so, so that the kernel opens the library version from within the package.

4.4.2.2 Implementing Syscall Rewriting

Since ptrace allows cde-exec to directly read and write into the target program's memory, the easiest way to rewrite a syscall's argument is to simply override its buffer with a new string. However, this approach does not work because the new path string is always longer than the original, so it might overflow the buffer. Also, if the program makes a system call with a constant string, the buffer would be read-only.

Instead, what cde-exec does is redirect the *pointer* to the buffer. When the target program (or one of its subprocesses) first makes a syscall, cde-exec forces it to make another syscall to attach a 16kB shared memory segment (a trick from Spillane et al. [21]). Now cde-exec can write data into that shared segment and have it be visible in the target program's address space. The two large rectangles in Figure 4.7 show the address spaces of the target program and cde-exec, respectively. Figure 4.7 illustrates the three steps involved in syscall argument rewriting:

1. cde-exec uses ptrace to read the original argument from the traced program's address space.
2. cde-exec creates a new string representing the path redirected inside of the package and writes it into the shared memory buffer.

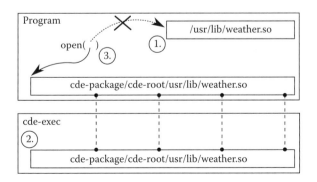

FIGURE 4.7
Example address spaces of target program and cde-exec when rewriting path argument of open. The two boxes connected by dotted lines are shared memory.

This value is immediately visible in the target program's address space.

3. cde-exec uses ptrace to mutate the syscall's filename char*
   argument(s) to point to the start of the shared memory buffer (in the
   target program's address space). x86-Linux syscall arguments are
   stored in registers, so ptrace mutates the target program's registers prior to executing the call. Most syscalls take only one filename
   argument, which is stored in %ebx on i386 and %rdi on x86-64.
   Syscalls such as link, symlink, and rename take two filename
   arguments; their second argument is stored in %ecx on i386 and
   %rsi on x86-64.

4.4.2.3 Spoofing Current Working Directory

At the completion of the getcwd syscall, cde-exec mutates the return value
string to eliminate all path components up to cde-root/. For example,
when Bob runs Alice's script:

```
cd /home/bob/cde-package/cde-root/expt
cde-exec python weather_sim.py tokyo.dat
```

If her Python script requests its current working directory using getcwd,
the kernel will return the true full path: /home/bob/cde-package/
cde-root/expt. Then cde-exec will truncate that string so that it
becomes /expt, which is the value it would have returned if it were running on Alice's machine. We have encountered many programs that break
when getcwd is not spoofed.

There is no danger of buffer overflow here since the new string is always
shorter, and the char* buffer passed into getcwd cannot be read-only,
since the kernel must be able to update its contents. Some programs call
readlink("/proc/self/cwd") to get the current working directory, so
cde-exec also spoofs the return value for that particular syscall instance.

4.4.2.4 Execve Syscall

When the target program executes a dynamically linked binary, cde-exec
rewrites the execve syscall arguments to execute the dynamic linker stored
in the package (with the binary as its first argument) rather than directly
executing the binary. For example, if Bob invokes "cde-exec python
weather_sim.py tokyo.dat", cde-exec will prepend the dynamic
linker filename to the argv array argument of the execve syscall:

```
argv[0]: cde-package/cde-root/lib/ld-linux.so.2
argv[1]: /usr/bin/python
argv[2]: weather_sim.py
argv[3]: tokyo.dat
```

(Also note that although `argv[1]` is `/usr/bin/python`, that path will get redirected into the version of the binary file within the CDE package during the open syscall.)

Here is why `cde-exec` needs to explicitly execute the dynamic linker: When a user executes a dynamically linked binary, Linux first executes the system's default dynamic linker to resolve and load its shared libraries. However, we have found that the dynamic linker on one Linux distro might not be compatible with binaries created on another distro, due to minor differences in ELF binary formats. Therefore, to maximize portability across machines, `cde` copies the dynamic linker into the package, and `cde-exec` executes the dynamic linker from the package rather than the target machine's builtin dynamic linker. Without this hack, we have noticed that even a trivial "hello world" binary compiled on one distro (e.g., Ubuntu with Linux 2.6.35) will not run on an older distro (e.g., Knoppix with Linux 2.6.17).*

A side effect of rewriting `execve` to call the dynamic linker is that when a target program inspects its own executable name, the kernel will return the name of the dynamic linker, which is incorrect. Thus, `cde-exec` spoofs the return values of calls to `readlink("/proc/self/exe")` and `readlink("/proc/<$PID>/exe")` to return the original executable's name. This spoofing is necessary because some narcissistic programs crash with cryptic errors if their own names are not properly identified!

4.5 Advanced Features

We now describe three advanced CDE features that are relevant for power users: semiautomated package completion, seamless execution mode, and application streaming mode.

4.5.1 Semiautomated Package Completion

CDE's main limitation is that it packages only the files accessed on executed program paths. Thus, programs run from within a CDE package will fail when executing paths that access new files (e.g., libraries, configuration files) that the original execution(s) did not.

Unfortunately, *no automatic tool* (static or dynamic) can find and package up all of the files required to successfully execute all possible program paths since that problem is undecidable in general. Similarly, it is also impossible to automatically quantify how "complete" a CDE package is or determine what files are missing since every file-related system call instruction could

---

* It actually crashes with a cryptic "Floating point exception" error message.

be invoked with complex or nondeterministic arguments. For example, the Python interpreter executable has only one `dlopen` call site for dynamically loading extension modules, but that `dlopen` could be called many times with different dynamically generated string arguments derived from script variables or configuration files.

There are two ways to cope with this package incompleteness problem. First, if the user executes additional program paths, then CDE will add new files into the same `cde-package/` directory. However, making repeated executions can get tedious, and it is unclear how many or which paths are necessary to complete the package.*

Another way to make CDE packages more complete is by manually copying additional files and subdirectories into `cde-package/cde-root/` (see Section 4.5.1.3 for more details). For example, while executing a Python script, CDE might automatically copy the few Python standard library files it accesses into, say, `cde-package/cde-root/usr/lib/python/`. To complete the package, the user could copy the entire `/usr/lib/python/` directory into `cde-package/cde-root/` so that *all* Python libraries are present.

However, programs also depend on shared libraries that reside in system-wide directories such as `/lib` and `/usr/lib`. Copying the entire contents of those directories into a package results in lots of wasted disk space. In Section 4.5.1.2, we present an automatic heuristic technique that finds nearly all shared libraries that a program requires and copies them into the package.

4.5.1.1 OKAPI: Deep File Copying

Before describing our heuristics for completing CDE packages, we first introduce a utility library we built called OKAPI (pronounced *oh-copy*), which performs detailed copying of files, directories, and symlinks. OKAPI does one seemingly simple task that turns out to be tricky in practice: copying a filesystem entity (i.e., a file, directory, or symlink) from one directory to another while fully preserving its original subdirectory and symlink structure (a process that we call *deep-copying*). CDE uses OKAPI to copy files into the `cde-root/` subdirectory when creating a new package, and the support scripts of Sections 4.5.1.2 and 4.5.1.3 also use OKAPI.

For example, suppose that CDE needs to copy the `/usr/bin/java` executable file into `cde-root/` when it is packaging a Java application. The straightforward way to do this is to use the standard `mkdir` and `cp` utilities. Figure 4.8 shows the resulting subdirectory structure within `cde-root/`, with the boxes representing directories and the bold ellipse representing the copy of the `java` executable file located at `cde-root/usr/bin/java`. However, it turns out that if CDE were to use this straightforward copying

---

* Similar to trying to achieve 100% coverage during software testing.

FIGURE 4.8
The result of copying a file named /usr/bin/java into cde-root/.

method, the Java application would *fail to run* from within the CDE package! This failure occurs because the java executable introspects its own path and uses it as the search path for finding the Java standard libraries. On our Fedora Core 9 machine, the Java standard libraries are actually installed in /usr/lib/jvm/java-1.6.0-openjdk-1.6.0.0, so when java reads its own path as /usr/bin/java, it cannot possibly use that path to find its standard libraries.

For Java applications to run from within CDE packages, all of their constituent files must be "deep-copied" into the package while replicating their original subdirectory and symlink structures. Figure 4.9 illustrates the complexity of deep-copying a single file, /usr/bin/java, into cde-root/. The diamond-shaped nodes represent symlinks, and the dashed arrows point to their targets. Notice how /usr/bin/java is a symlink to /etc/alternatives/java, which is itself a symlink to /usr/lib/jvm/jre-1.6.0-openjdk/bin/java. Another complicating factor is that /usr/lib/jvm/jre-1.6.0-openjdk is itself a symlink to the /usr/lib/jvm/java-1.6.0-openjdk-1.6.0.0/jre/ directory, so the actual java executable resides in /usr/lib/jvm/java-1.6.0-openjdk-1.6.0.0/jre/bin/. Java can only find its standard libraries when these paths are all faithfully replicated within the CDE package.

The OKAPI utility library automatically performs the deep-copying required to generate the filesystem structure of Figure 4.9. Its interface is as simple as ordinary cp: The caller simply requests for a path to be copied into a target directory, and OKAPI faithfully replicates the subdirectory and symlink structure.

OKAPI performs one additional task: rewriting the contents of symlinks to transform absolute path targets into relative path targets within the destination directory (e.g., cde-root/). In our example, /usr/bin/java

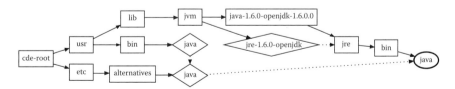

FIGURE 4.9
The result of using OKAPI to deep-copy a single /usr/bin/java file into cde-root/, preserving the exact symlink structure from the original directory tree. Boxes are directories (solid arrows point to their contents), diamonds are symlinks (dashed arrows point to their targets), and the bold ellipse is the real java executable.

is a symlink to `/etc/alternatives/java`. However, OKAPI cannot simply create the `cde-root/usr/bin/java` symlink to also point to `/etc/alternatives/java` since that target path is outside of `cde-root/`. Instead, OKAPI must rewrite the symlink target so that it actually refers to `../../etc/alternatives/java`, which is a relative path that points to `cde-root/etc/alternatives/java`.

The details of this particular example are not important, but the high-level message that Figure 4.9 conveys is that deep-copying even a single file can lead to the creation of over a dozen subdirectories and (possibly rewritten) symlinks. The problem that OKAPI solves is not Java-specific; we have observed that many real-world Linux applications fail to run from within CDE packages unless their files are deep-copied in this intricate way.

Aside from being an integral part of CDE, OKAPI is also available as a free standalone command-line tool [1]. To our knowledge, no other Linux file copying tool (e.g., `cp`, `rsync`) can perform the deep-copying and symlink rewriting that OKAPI does.

4.5.1.2 Heuristics for Copying Shared Libraries

When Linux starts executing a dynamically linked executable, the dynamic linker (e.g., `ld-linux*.so*`) finds and loads all shared libraries that are listed in a special `.dynamic` section within the executable file. Running the `ldd` command on the executable shows these start-up library dependencies. When CDE is executing a target program to create a package, CDE finds all of these dependencies as well because they are loaded at start-up time via `open` system calls.

However, programs sometimes load shared libraries in the middle of execution using, say, the `dlopen` function. This run-time loading occurs mostly in GUI programs with a plug-in or extension architecture. For example, when the user instructs Firefox to visit a web page with a Flash animation, Firefox will use `dlopen` to load the Adobe Flash Player shared library. `ldd` will not find that dependency since it is not hard-coded in the `.dynamic` section of the Firefox executable, and CDE will only find that dependency if the user actually visits a Flash-enabled web page while creating a package for Firefox.

We have created a simple heuristic-based script that finds most or all shared libraries that a program requires.* The user first creates a base CDE package by executing the target program once (or a few times) and then runs our script, which works as follows:

---

* Always a superset of the shared libraries that `ldd` finds.

1. Find all ELF binaries (executables and shared libraries) within the package using the Linux `find` and `file` utilities.

2. For each binary, find all constant strings using the `strings` utility and look for strings containing ".so" since those are likely to be shared libraries.

3. Call the `locate` utility on each candidate shared library string, which returns the *full absolute paths* of all installed shared libraries that match each string.

4. Use OKAPI to copy each library into the package.

5. Repeat this process until no new libraries are found.

This heuristic technique works well in practice because programs often list all of their dependent shared libraries in string *constants* within their binaries. The main exception occurs in dynamic languages such as Python or MATLAB®, whose programs often dynamically generate shared library paths based on the contents of scripts and configuration files. Of course, our technique provides no completeness guarantees since the package completeness problem is undecidable in general.

4.5.1.3 OKAPI-Based Directory Copying

In general, running an application once under CDE monitoring only packages up a subset of all required files. In our experience, the easiest way to make CDE packages complete is to copy entire subdirectories into the package. To facilitate this process, we created a script that repeatedly calls OKAPI to copy an entire user-specified directory into `cde-root/`, automatically following symlinks to other directories and recursively copying as needed. (Note that simply running "cp -aR" is not sufficient since that does not follow and preserve symlinks.)

Although this approach might seem primitive, it is effective in practice because applications often store all of their files in a few top-level directories. When a user inspects the directory structure within `cde-root/`, it is usually obvious where the application's files reside. Thus, the user can run our script to copy those directories into the package.

4.5.2 Seamless Execution Mode

When executing a program from within a CDE package, `cde-exec` redirects all file accesses into the package by default, thereby creating a chroot-like sandbox with `cde-package/cde-root/` as the pseudo-root directory (see Figure 4.3, Step 3).

This default chroot-like execution mode is fine for running self-contained GUI applications such as games or web browsers, but it is a somewhat awkward way to run most types of UNIX-style command-line programs that researchers often prefer. If users are running, say, a compiler or command-line image processing utility from within a CDE package, they would need to first move their input data files into the package, run the target program using `cde-exec`, and then move the resulting output data files back out of the package, which is a cumbersome process.

Let us consider a modified version of the Alice-and-Bob example from Section 4.2. Suppose Alice is a researcher who is developing a Python script to detect anomalies in network log files. She normally runs her script using this Linux command:

```
python detect_anomalies.py net.log
```

Let us say she packages up her command with CDE and sends that package to Bob, who can now rerun her original analysis on the `net.log` file from within the package. However, if Bob wants to run Alice's script on his own log data (e.g., `bob.log`), then he needs to first move his data file inside of `cde-package/cde-root/`, change into the appropriate subdirectory deep within the package, and run:

```
cde-exec python detect_anomalies.py bob.log
```

In contrast, if Bob had actually installed the proper version of Python and its required extension modules on his machine, then he could run Alice's script from *anywhere* on his filesystem with no restrictions.

Some CDE users wanted CDE-packaged programs to behave just like regularly installed programs rather than requiring input files to be moved inside of a `cde-package/cde-root/` sandbox, so we implemented a *seamless execution mode* that largely achieves this goal.

Seamless execution mode works using a simple heuristic: If `cde-exec` is being invoked from a directory *not* in the CDE package (i.e., from somewhere else on the user's filesystem), then only redirect a path into `cde-package/cde-root/` if the file that the path refers to actually exists within the package. Otherwise simply leave the path unmodified so that the program can access the file normally. No user intervention is needed in the common case.

The intuition behind why this heuristic works is that when programs request to load libraries and other mandatory components, those files must exist within the package, so their paths are redirected. On the other hand, when programs request to load an input file passed via, say, a command-line argument, that file does not exist within the package, so the original path is used to load it from the native filesystem.

In the example shown in Figure 4.10, if Bob ran Alice's script to analyze an arbitrary log file on his machine (e.g., his web server log, `/var/log/httpd/access_log`), then `cde-exec` will redirect Python's

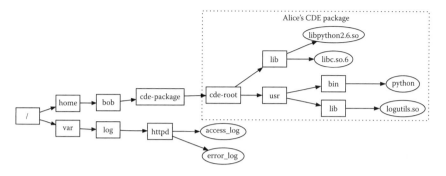

FIGURE 4.10
Example filesystem layout on Bob's machine after he receives a CDE package from Alice (boxes are directories, ellipses are files). CDE's seamless execution mode enables Bob to run Alice's packaged script on the log files in /var/log/httpd/ without first moving those files inside of cde-root/.

request for libraries (e.g., /lib/libpython2.6.so and /usr/lib/logutils.so) inside of cde-root/ since those files exist within the package, but cde-exec will *not* redirect /var/log/httpd/access_log and instead load the real file from its original location.

Seamless execution mode fails when the user wants the packaged program to access a file from the native filesystem, but an identically named file actually exists within the package. In the aforementioned example, if cde-package/cde-root/var/log/httpd/access_log existed, then that file would be processed by the Python script instead of /var/log/httpd/access_log. There is no automated way to resolve such name conflicts, but cde-exec provides a "verbose mode" where it prints out a log of what paths were redirected into the package. The user can inspect that log and then manually write redirection/ignore rules in a configuration file (see Figure 4.4) to control which paths cde-exec redirects into cde-root/. For instance, the user could tell cde-exec to *not* redirect any paths starting with /var/log/httpd/*.

4.5.3 On-Demand Application Streaming

With CDE's streaming mode, users can instantly run any Linux application on demand without having to create, transfer, or install any packages. Figure 4.2 shows a high-level architectural overview. The basic idea is that a system administrator first installs multiple versions of many popular Linux distros in a "distro farm" in the cloud (or an internal compute cluster). When a user wants to run some application that is available on a particular distro, they use sshfs (an ssh-based network filesystem [4]) to mount the root directory of that distro into a special cde-remote-root/ mount point. Then, the user can use CDE's streaming mode to run any application from that distro locally on their own machine.

4.5.3.1 Implementation and Example

Figure 4.11 shows an example of streaming mode. Let us say that Alice wants to run the Eclipse 3.6 IDE on her Linux machine, but the particular distro she is using makes it difficult to obtain all the dependencies required to install Eclipse 3.6. Rather than suffering through finding, installing, and configuring all dependent libraries and software, Alice can simply connect to a distro in the farm that contains Eclipse 3.6 and then use CDE's streaming mode to "harvest" the required dependencies on demand.

Alice first mounts the root directory of the remote distro at `cde-remote-root/`. Then, she runs "`cde-exec -s eclipse`" (-s activates streaming mode). `cde-exec` finds and executes `cde-remote-root/bin/eclipse`. When that executable requests shared libraries, plugins, or any other files, `cde-exec` will redirect the respective paths into `cde-remote-root/`, thereby executing the version of Eclipse 3.6 that resides in the cloud distro. However, note that the application is running locally on Alice's machine, not in the cloud.

Astute readers will recognize that running applications in this manner can be slow since files are being accessed from a remote server. While sshfs performs some caching, we have found that it does not work well enough in practice. Thus, we have implemented our own caching layer within CDE: When a remote file is accessed from `cde-remote-root/`, `cde-exec` uses

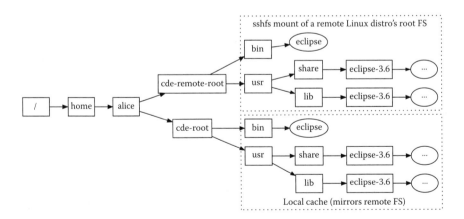

FIGURE 4.11

An example use of CDE's streaming mode to run Eclipse 3.6 on any Linux machine without installation. `cde-exec` fetches all dependencies on demand from a remote Linux distro and stores them in a local cache.

OKAPI to make a deep-copy into a local `cde-root/` directory and then redirects that file's path into `cde-root/`. In streaming mode, `cde-root/` initially starts out empty and then fills up with a subset of files from `cde-remote-root/` that the target program has accessed.

To avoid unnecessary filesystem accesses, CDE's cache also keeps a list of file paths that the target program tried to access from the remote server, even keeping paths for *nonexistent files*. On subsequent runs, when the program tries to access one of those paths, `cde-exec` will redirect the path into the local `cde-root/` cache. It is vital to track nonexistent files since programs often try to access nonexistent files at start-up while, say, searching for shared libraries by probing a list of directories in a search path. If CDE did not track nonexistent files, then the program would still access the directory entries on the remote server before discovering that those files *still* do not exist, thus slowing down execution.

With this cache in place, the first time an application runs, all of its dependencies must be downloaded, which could take several seconds to minutes. This one-time delay is unavoidable. However, subsequent runs simply use the files already in the local cache, so they execute at regular `cde-exec` speeds. Even running a *different* application for the first time might still result in some cache hits for, say, generic libraries such as `libc`, so the entire application does not need to be downloaded.

Finally, the package incompleteness problem faced by regular CDE (see Section 4.5.1) no longer exists in streaming mode. When the target application needs to access new files that do not yet exist in the local cache (e.g., Alice loads a new Eclipse plug-in for the first time), those files are transparently fetched from the remote server and cached.

4.5.3.2 Synergy with Package Managers

Nearly all Linux users are currently running one particular distro with one default package manager that they use to install software. For instance, Ubuntu users must use APT, Fedora users must use YUM, SUSE users must use Zypper, and Gentoo users must use Portage. Moreover, different releases of the *same* distro contain different software package versions since distro maintainers add, upgrade, and delete packages in each new release.*

As long as a piece of software and all of its dependencies are present within the package manager of the exact distro release that a user happens to be using, then installation is trivial. However, as soon as even one

---

* We once tried installing a machine learning application that depended on the `libcv` computer vision library. The required `libcv` version was found in the APT repository on Ubuntu 10.04, but it was not found in the repositories on the two neighboring Ubuntu releases: 9.10 and 10.10.

dependency cannot be found within the package manager, then users must revert to the arduous task of compiling from source (or configuring a custom package manager).

CDE's streaming mode frees Linux users from this single-distro restriction and allows them to run software that is available within the package manager of any distro in the cloud distro farm. The system administrator is responsible for setting up the farm and provisioning access rights (e.g., ssh keys) to users. Then users can directly install packages in any cloud distro and stream the desired applications to run locally on their own machines.

Philosophically, CDE's streaming mode maximizes user freedom since users are now free to run any application in any package manager from the comfort of their own machines, regardless of which distro they choose to use. CDE complements traditional package managers by leveraging all of the work that the maintainers of each distro have already done and opening up access to users of all other distros. This synergy can potentially eliminate quasi-religious squabbles and flame-wars over the virtues of competing distros or package management systems. Such fighting is unnecessary since CDE allows users to freely choose from among all of them.

4.6 Discussion

Our design philosophy underlying CDE is that people should be able to package up their Linux software and deploy it to run on other Linux machines with as little effort as possible. However, we are not proposing CDE as a replacement for traditional software installation. CDE packages have a number of limitations. Most notably,

- They are not guaranteed to be complete.
- Their constituent shared libraries are "frozen" and do not receive regular security updates. (Static linking also shares this limitation.)
- They run slower than native applications due to `ptrace` overhead. We measured slowdowns of up to 28% in our experiments [13], but slowdowns can be worse for I/O-heavy programs.

Software engineers who are releasing production-quality software should obviously take the time to create and test one-click installers or integrate with package managers. But for the millions of research scientists, prototype designers, system administrators, programming course students and teachers, and hobby hackers who just want to deploy their ad hoc software as quickly as possible, CDE can emulate many of the benefits of traditional software distribution with much less required labor: In just minutes, users can create a base CDE package by running their program under CDE supervision, use our *semiautomated heuristic tools* (Section 4.5.1) to make the package complete, deploy to the target Linux machine, and then execute it in *seamless execution mode* (Section 4.5.2) to make the target program behave like it was installed normally.

4.6.1 Practical Lessons Learned

Here are some generalizable lessons that we have learned in the past two years of developing CDE and supporting thousands of diverse users.

- First and foremost, start with a conceptually clear core idea, make it work for basic nontrivial cases, document the still-unimplemented tricky cases, launch your tool, and then get feedback from real users. User feedback is by far the easiest way for you to discover what bugs are important to fix and what new features to add next.
- A simple and appealing quick-start web page guide and screencast video demo are essential for attracting new users. No potential user is going to read through dozens of pages of an academic research paper before deciding to try your tool. In short, even hackers need to learn to be great salespeople.
- To maximize your tool's usefulness, you must design it to be easy to use for beginners but also to give advanced users the ability to customize it to their liking. One way to accomplish this goal is to have well-designed default settings, which can be adjusted via command-line options or configuration files. The defaults must work effectively "out-of-the-box" without any tuning, or else new users will get frustrated.
- Resist the urge to add new features just because they are "cool," interesting, or potentially useful. Only add new features when there are compelling real users who demand it. Instead, focus your development efforts on fixing bugs, writing more test cases, improving your documentation, and, most importantly, attracting new users.
- Users are by far the best sources of bug reports since they often stress your tool in ways that you could have never imagined. Whenever a user reports a bug, send them a sincere *thank you* note, try to create a representative minimal test case, and add it to your regression test suite.
- If a user has a conceptual misunderstanding of how your tool works, then think hard about how you can improve your documentation or default settings to eliminate this misunderstanding.

Reflecting on the past two years of serving CDE's users, we believe that its success thus far is largely due to it being a conceptually simple tool that has been meticulously engineered to do one thing well—eliminating Linux software dependency problems.

4.7 Future Vision: PhD-In-A-Box

Despite the fact that so much modern scientific research is being done on computers, research papers are still the primary means of disseminating new knowledge. Dead trees are an impoverished communications medium, though. The ideas in this chapter could be extended to build a richer electronic medium where one's colleagues, apprentices, and intellectual adversaries can interactively explore the results of one's experiments. Imagine enhancing CDE packages with fine-grained versioning history and notes, perhaps hosted on a cloud service where readers can visit a website to rerun and tweak those experiments. If we could capture and present a rich history of a project's progression over time, then readers can learn from the *entire research process*, not merely digest the final products.

To convey the potential benefits of learning from research *processes* rather than just end results, we will make an analogy to mathematics. Mathematics research papers are written in a concise manner presenting a minimal set of proofs of lemmas and theorems. Readers unfamiliar with the process of discovery in mathematics might mistakenly assume that some lofty genius must have dreamt up the perfect proof fully formed and scribbled it down on paper. The truth is far messier: Much like computational researchers, mathematicians explore numerous hypotheses, go down dead ends, backtrack repeatedly, talk through ideas with colleagues, and gradually cobble together a finished proof. Then they painstakingly whittle down that proof until it can be presented as elegantly as possible, and only then do they write up the final paper. The vast majority of intellectual wisdom lies in the *process* of working through the problems, and such knowledge is not represented anywhere in the published paper. Mathematicians learn their craft not just by reading papers, but by actually watching their colleagues and mentors at work. Imagine if there was a way to capture and present the entire months-long process of a famous mathematician coming up with a particular proof. Aspiring mathematicians could learn far more from such an interactive presentation than from simply reading the final polished paper.

We believe that such a goal of "total recall" is easier to accomplish in the context of computational research. Since most of the work is being done on the computer, it is easier to trace the entire workflow history and provide user interfaces for in-context annotations and notetaking [14]. First-class support for branching and backtracking are vital needs in such a system since much of the wisdom gained from a research apprenticeship is learning from what did not work and what dead ends to avoid. In this vision, all PhD students would maintain a hard disk image containing the complete trials and tribulations of their five to seven (or more) years' worth of computational experiments. This "PhD-In-A-Box" could be used to train new students and to pass down all of the implicit knowledge, experiences, tricks, and wisdom that are often lost in a dead-tree paper dissertation.

Extending this analogy further, imagine an online library filled with the collected electronic histories of all research projects, not just their final results in published form. It now becomes possible to perform pattern recognition and aggregation across multiple projects to discover common "tricks of the trade." Someone new to a field, say machine learning, can now immersively learn from the collective wisdom of thousands of expert machine learning researchers rather than simply reading their papers. One could argue that, in the limit, such a system would be like "indexing" all of those researchers' brains and making that knowledge accessible. We speculate that such a system can be more effective than "brain indexing," since people subconsciously apply tricks from their intuitions and often forget the details of what they were working on (especially failed trials). In this vision of the future, a paper is merely a facade for the real contributions of the full research process.

Such a dream "PhD-In-A-Box" system can approach the holy grail of universally reproducible research in three main ways:

1. Researchers can revisit and reflect upon their old experiments, perhaps uncovering hidden biases that might have skewed results (e.g., multiple comparison problems in statistics).

2. Colleagues and students can reproduce, learn from, and scrutinize the entire workflow history of fellow researchers, not just the final products in the form of published papers. If properly authenticated, such an audit trail can be used to detect evidence of "cherry picking" and inappropriate "slicing and dicing" of data and code parameters to achieve statistically significant results.

3. "Meta-researchers" can perform meta-analyses of research processes, methodologies, trends, and findings within a field by data mining the computational archives of all researchers in that field.

The good news is that the technology to realize this dream already exists. A properly motivated team can use much of the ideas and tools described in this book to create a "PhD-In-A-Box" system that makes the still-elusive ideal of reproducible research into a pervasive reality. Just like how we now take for granted that e-mails and documents are stored and synchronized in the cloud, perhaps in a decade or two, computational researchers will take for granted that all of their experiments are versioned, annotated, archived, curated, and fully reproducible.

References

1. CDE public source code repository, https://github.com/pgbovine/CDE (Accessed October, 2012.)
2. List of software package management systems, http://en.wikipedia.org/wiki/List_of_software_package_management_systems (Accessed October, 2012.)
3. Saturn online discussion thread, https://mailman.stanford.edu/pipermail/saturn-discuss/2009-August/000174.html (Accessed October, 2012.)
4. SSH Filesystem, http://fuse.sourceforge.net/sshfs.html (Accessed October, 2012.)
5. arachni project home page, https://github.com/Zapotek/arachni (Accessed October, 2012.)
6. graph-tool project home page, http://projects.skewed.de/graph-tool/ (Accessed October, 2012.)
7. A Aiken, S Bugrara, I Dillig, T Dillig, B Hackett, and P Hawkins. An overview of the Saturn project. In *PASTE '07*, pp. 43–48. ACM, New York, 2007.
8. G Altekar and I Stoica. ODR: Output-deterministic replay for multicore debugging. In *SOSP '09*, pp. 193–206. ACM, New York, 2009.
9. C Cadar, D Dunbar, and D Engler. KLEE: Unassisted and automatic generation of high-coverage tests for complex systems programs. In *OSDI '08*, pp. 209–224. USENIX Association, San Diego, CA, 2008.
10. K Fisher and R Gruber. PADS: A domain-specific language for processing ad hoc data. In *PLDI '05*, pp. 295–304. ACM, New York, 2005.
11. T Garfinkel. Traps and pitfalls: Practical problems in system call interposition based security tools. In *NDSS '03*, San Diego, CA, 2003.
12. T Garfinkel, B Pfaff, and M Rosenblum. Ostia: A delegating architecture for secure system call interposition. In *NDSS '04*, San Diego, CA, 2004.
13. PJ Guo. CDE: Run any Linux application on-demand without installation. In *Proceedings of the 2011 USENIX Large Installation System Administration Conference, LISA '11*. USENIX Association, Boston, MA, 2011.
14. PJ Guo and M Seltzer. Burrito: Wrapping your lab notebook in computational infrastructure. In *TaPP '12: Proceedings of the 4th USENIX Workshop on the Theory and Practice of Provenance*, Boston, MA, 2012.
15. M Heinonen, H Shen, N Zamboni, and J Rousu. Metabolite identification and molecular fingerprint prediction via machine learning. *Bioinformatics*, 28(18):2333–2341, 2012.
16. K Jain and R Sekar. User-level infrastructure for system call interposition: A platform for intrusion detection and confinement. In *NDSS '00*, San Diego, CA, 2000.

17. O Laadan, N Viennot, and J Nieh. Transparent, lightweight application execution replay on commodity multiprocessor operating systems. In *SIGMETRICS '10*, New York, pp. 155–166, 2010.
18. M Lahiri and M Cebrian. The genetic algorithm as a general diffusion model for social networks. In *Proceedings of the 24th AAAI Conference on Artificial Intelligence*. AAAI Press, Atlanta, GA, 2010.
19. E Loper and S Bird. NLTK: The Natural Language Toolkit. In *In ACL Workshop on Effective Tools and Methodologies for Teaching NLP and Computational Linguistics*, Philadelphia, PA, 2002.
20. Y Saito. Jockey: A user-space library for record-replay debugging. In *AADEBUG*, pp. 69–76. ACM Press, New York, 2005.
21. RP Spillane, CP Wright, G Sivathanu, and E Zadok. Rapid file system development using ptrace. In *Experimental Computer Science*. USENIX Association, San Diego, CA, 2007.
22. IA Sucan and LE Kavraki. Kinodynamic motion planning by interior-exterior cell exploration. In *International Workshop on the Algorithmic Foundations of Robotics*, pp. 449–464, 2008.

5

Reproducible Physical Science and the Declaratron

Peter Murray-Rust and Dave Murray-Rust

CONTENTS

5.1 Introduction

In this chapter, we address the issue of reproducibility of computation in the physical sciences. We focus on disciplines concerned with chemistry, crystallography, and materials science, although the strategies, and the Declaratron software we describe, have much greater applicability. Here, we will concentrate on the computation of the properties of materials, where there is a long tradition of reproducibility. Much of this is based on the almost Platonic identity of materials: to a first practical approximation, sodium chloride crystals have the same properties wherever and however they are produced. This consistency of properties means that suppliers can offer materials for sale with reproducible physical and chemical properties.

Over the last 50 years, it has become possible to measure, and now calculate, the properties of substances to a high degree of consistency. Substances are key to both research and technological exploitation (semiconductors, optical materials, piezoelectrics, second-harmonic generators, etc.). We estimate that over one billion USD is spent annually on the computation of materials properties (cf. the Materials Genomics Project [15]). There is a dynamic interplay between the measurement of *observables* (properties that are discovered through measurement of physical substances) and *computables* (properties that can be computed through simulation of physical laws, machine learning, or heuristics). As knowledge and techniques are improved, sometimes the observable is more accessible or accurate, and sometimes the computable, as observations feed back into more accurate computation and vice versa. There is also a fundamental reliance on instrumentation as a mechanism for reproducibility: given instruments of similar quality, there is an expectation that if an experiment is rerun, the results will be *noncontradictory* to the previous measurements.

Instrumental measurements are fundamental to materials science, and all modern instruments produce digital output. Many instrumental techniques have now developed both automation (e.g., robot sample feeders) and high throughput (hundreds of samples per day or more), such as single-crystal x-ray crystallography, powder x-ray diffraction, IR/UV/VIS [8], nuclear magnetic resonance spectroscopy [4], and mass spectrometry [7]. In principle, it should be possible to compare the output from two instruments using automated methods, but this requires agreed communal semantics, which are unfortunately rare. While some standards exist (e.g., JCAMP-DX* and AnIML† [17]), most output is proprietary, so there are very few effective communal dictionaries and semantics. We shall see the same problem later for computation.

---

* http://www.jcamp-dx.org/.
† http://animl.sourceforge.net/.

As well as self-consistency of results, reproducibility increasingly requires agreement between experiment and theory. Computational experiments, including calculation and simulation, are part of the architecture of an agreement between theory and experiment. There is a need for mechanisms and expectations of reproducibility to be applied equally to scientific computation as instrumentation. In this chapter, we describe how existing technologies can be combined with a novel approach to semantic calculation to carry out reproducible scientific computation of materials properties.

5.1.1 What Do We Mean by Reproducible Computation?

5.1.1.1 Archetypal Example of the Problem

There are two main methods for computing the properties of matter based on physical principles. Quantum mechanics (QM) involves solving Schroedinger's equation for a multi-nucleus, multielectron system. There is no analytical solution, and the approximations can be very expensive often rising with N^3 (N is the number of atoms) or greater. Increasing the accuracy requires additional expense. The more tractable alternative is "Forcefields" (FFs), which use empirical parameterization of Newton's laws. We use FFs as the main example in this chapter.

In the FF approach, the energy of a molecule can be approximated by a number of empirical terms, which are added together to form a "forcefield."* A typical and widely used example is the AMBER program[†] for computing molecular energy. It contains five terms (see Figure 5.1). The energy of the molecule is described (empirically) by bonds, angles, torsions (dihedrals), nonbonded contacts ("bumping atoms"), and electrostatic

$$V(\mathbf{r}) = \sum_{bonds} K_b (b - b_0)^2 + \sum_{angles} K_\theta (\theta - \theta_0)^2$$

$$+ \sum_{dihedrals} \left(\frac{V_n}{2} \right) (1 + \cos[n\phi - \delta])$$

$$+ \sum_{nonbij} \left(\frac{A_{ij}}{r_{ij}^{12}} \right) - \left(\frac{B_{ij}}{r_{ij}^6} \right) + \left(\frac{q_i q_j}{r_{ij}} \right)$$

FIGURE 5.1
Functional form of the AMBER forcefield. (Taken from http://ambermd.org/doc11/Amber11. pdf, p. 1.9 reproduced verbatim including typographic errors) The last summation should be split into separate sums, the first with A and B terms and the last with the electrostatics ($q_i q_j / r_{ij}$) (Coulomb's inverse power law).

---

* Technically this is a slight misnomer, as force is the derivative of energy w.r.t. distance.
[†] http://ambermd.org.

interactions. A molecule has many such interactions, and the energy of each is computed and summed to give the total energy. The scale rises rapidly. Water has 2 bonds and 1 angle; acetic acid, with 8 atoms, has 7 bonded terms, 10 angle terms, 8 torsional terms, and 11 nonbonded terms (the precise number varies with the FF formulation). A typical protein molecule might have several thousand bonds and angles and over a million nonbonded interactions.

The central problem of reproducibility is that each program/FF may have a slightly different formulation. The number of nonbonded terms in AMBER may not be the same as in MM2 [1]. Some programs will include atoms separated by three or more bonds as nonbonded; for others, it must be four or more. These differences are often not formalized in the documentation and may only be described in human language. The actual "algorithm" is usually hidden deep in FORTRAN code and can only be determined by reading the program source. Additionally, the code may have been modified by later developers, and these changes may not be reflected in the human documentation. In some cases, the parameterization may change during the progress of a calculation.

The current example also shows imprecision in the mathematical representation. The dihedral function for a given bond is often not a single cosine but a multiple sum (over Fourier terms), so formally, a double sum ($\Sigma_{\text{dihedrals}}\Sigma_{\text{Fourier terms}}$) is required. Also, the electrostatic function has omitted a factor of ($1/4\pi\epsilon_0$), where ϵ_0 represents the permittivity of free space.*

One of our goals is to provide a precise, unambiguous formulation that can be implemented by a competent programmer with no previous knowledge of physics or chemistry. This can potentially be extended by using machines to generate code from semantic specifications, and we see no fundamental reason why molecular energy calculations cannot be specified in this way. By contrast, it would currently be almost impossible for a nonphysicist/chemist to recreate programs such as AMBER correctly from the documentation.

At this point, it is necessary to discuss what we mean by reproducibility in scientific computation. When dealing with *observables*, the definition is relatively clear: by following the same experimental procedure, one should obtain the same results, allowing for known and unknown causes of experimental variance. However, with computation, there are alternative expectations and possibilities for what the same "experimental procedure" and "results" should mean. Taking these separately, the "same experimental procedure" could mean:

- Download the original software and data and run it.
- Download the original software, compile it for a different machine, and run it with the original data.

---

* http://en.wikipedia.org/wiki/Vacuum_permittivity.

- Download software that carries out the same operations as originally described, and apply it to the original data.
- Read a paper, produce a new implementation of the algorithms described, and run it on the original data.
- Run any of the aforementioned programs on a refined or updated data set.

And the "same results" could mean:

- Identical output at the bit level
- Exactly the same numbers
- Exactly the same numbers (when run on a similar machine)
- Numbers that are within some bound of error
- Ensembles of outputs that share certain characteristics (again, within the bound of error)

This discussion of what the "same results" means is partially motivated by the chaotic nature of some physical calculations. Many algorithms in physical sciences are completely deterministic—for example, calculating the length of a bond should give and identical result no matter what code is used (within the bounds of floating-point imprecision). However, some algorithms contain branch points that are sensitive to precision and instabilities. For example, the QM calculation of molecular energy requires two independent optimizations—the self-consistent field (SCF) of the molecular orbitals and the optimization of energy against geometry to get the minimum energy structure. For many molecules, these are well-constrained, and the calculation proceeds essentially identically on different machines and often with different programs that use the same basic physical model. However, calculations for some systems are unstable (e.g., near a transition state in a reaction), and the behavior is effectively unpredictable at a detailed level. Similarly, the dynamics of molecules (e.g., when simulated by Newton's laws) is inherently unpredictable. Although formally deterministic, small imprecisions cause bifurcations in trajectories, which rapidly diverge. Hence, when discussing reproducibility for calculations of this type, it is problematic to compare results from individual runs, or exact results, and the focus must be moved to ensemble or aggregate properties. Additionally, since there is no experimental validation—for example, trajectories of individual molecules are not usually observable—special human care is required to validate code and parameters, as mistakes will be very difficult to detect later.

For the purpose of this chapter, we take "semantically defined reproducible science" to be defined as follows:

> Can a computational scientist (or machine) with no intrinsic domain knowledge, when given the specification, build a system which can be

guaranteed to compute problems in a scientific domain and produce results which are semantically consistent, and in some sense similar.*

5.1.2 What's Wrong with Business as Usual?

Currently, computable semantics are not commonplace within scientific practice. Indeed, very few scientific domains have fully addressed computable semantics. While computational chemistry (CompChem) is more advanced than some areas, it is still far from (1) having complete computational semantics and (2) integrating the use of computational semantics into the daily lives of computational chemists. We will illustrate the problem of missing semantics using examples from widely used programs; these have been chosen as typical examples in widespread use. We are not attempting to single out egregious offences, and many other examples—most commonly used programs—display the same issues.

This is one line of input for MOPAC, a widely used CompChem program, taken from http://openmopac.net/manual/index.html:

```
1    H 1.092 0 120.615 1 179.979 1 10 9 11
```

In this single line, there are no *explicit* semantics at all. Taking each field (separated by groups of spaces) in turn, the *implicit* semantics are as follows:

1. H is the element symbol for hydrogen. Although this seems like a precise, commonly accepted designation, many other programs use arbitrary, nonstandard abbreviations for elements, with integers or floats for nuclear charge, for example, "W1" 8 for oxygen in water, which could also be mistaken for tungsten.
2. 1.092 is the distance in angstrom units to 10th atom. The number 10 in field 8 is what specifies it as the 10th atom.
3. 0 is an integer flag: should this distance be allowed to vary during the computation? 0 means yes, it should be allowed to vary.
4. 120.615 is the angle in degrees between this atom, atom 10, and atom 9. Again, the 10th atom is specified by the 10 in field number 8 and 9th by the 9 in field 9.
5. 1 integer flag: do not allow this angle to change.
6. 179.979 is the dihedral angle in degrees between this atom, atom 10, atom 9, and atom 11.
7. 1 integer flag: do not allow this dihedral angle to change.
8. 10 means that bond length is between this atom and atom 10.

---

* We will leave open the question of exactly what "similar" means here, but specify that numbers should be approximately the same.

9. 9 means that angle is between this atom, atom 10 and atom 9.

10. 11 means that dihedral angles are between this atom and atoms 10, 9, and 11.

This is just a *typical* example, and similar issues can be found in the input specifications of many programs. This type of ad hoc, unmarked up yet implicitly meaningful data format has enormous scope for catastrophic errors. Common causes of error include fields mistyped when the file is edited by hand; users have an old copy of the documentation, so column ordering or meaning can change without causing obvious errors; field boundaries can be misplaced: Is it one space between each field or any number of spaces? Are tabs or spaces used as delimiters? Do fields need to be justified to exact column positions? The input modules of the programs usually have no validation. "User-friendly" GUI editors are usually program specific and proprietary, although sometimes there is an ecology of ad hoc converters; both of these situations bring a different set of issues.

In addition to the possibility for error when managing data, a huge burden is placed on the programmers who maintain applications that read these files. They are forced to maintain parsers for poorly defined specifications and may have to deal with different dialects of the data language as alternative interpretations come into fashion. It leads to (1) brittle code with poor error handling; (2) a high barrier to entry for new programmers wanting to join projects; and (3) an excessive proportion of programming effort being given over to reading.

Output is similarly problematic. This example is taken from http://www.cup.uni-muenchen.de/ch/compchem/energy/MOPAC_output.html:

1		ATOM NO.	TYPE	CHARGE	ATOM ELECTRON DENSITY
2		1	O	-.3827	6.3827
3		2	H	.1914	.8086
4		3	H	.1914	.8086
5	DIPOLE	X	Y	Z	TOTAL
6	POINT-CHG.	.677	.859	.000	1.094
7	HYBRID	.475	.602	.000	.767
8	SUM	1.151	1.461	.000	1.860

This cannot be understood without being a practitioner and/or having a manual (often out of date) and/or asking questions to humans. You must know or guess that charges are in units of *electrons* and that dipoles are in Debyes—neither are SI units. It is not clear what a HYBRID is or how it is calculated. It appears that SUM = POINT-CHG + HYBRID, and so we might infer that it is probably the predicted quantity. Without complete understanding of a quantity, it is by definition irreproducible—although this particular calculation could be run again to give the same numbers, it would be impossible to construct an alternative, clean room implementation, which computed the same result.

In certain cases, computations are not reproducible due to *licensing* restrictions on distribution of the *output* of proprietary programs. One major program manufacturer legally forbids the publication of complete output files; in order to have any chance of creating reproducible science, it is fundamentally necessary to publish exactly these computational details.

We believe that, in addition to inhibiting reproducibility, the issues outlined earlier are responsible for many millions of hours of wasted work each year, by allowing errors to go unvalidated and unnoticed, propagating through chains of experiments; through time spent understanding unclear semantics and editing brittle config files; and by forcing chemists to learn to parse semistructured text, and programmers to maintain code that has to be compatible with a fuzzy, moving-target data specification.

In an example from our own experience, we autogenerated input for the GAMESS program. GAMESS has a limit of 80 characters per line (cf. Hollerith cards), and some of out-generated lines exceeded this. Although the program noted this in the (voluminous) output, it did not halt but quietly discarded the offending atom records. The result was that erroneous calculations were carried out, without a strongly visible warning. These errors were only discovered when the output was reused in further calculations, where it caused crashes. This could have been avoided by carrying out syntactic and semantic validation in the input stage and refusing to produce output from invalid input. In general, not much trust can be placed in legacy CompChem programs to carry out sufficient validation on input or output; even when such validation is carried out, it is not clear how to verify that it has happened.

5.2 Constructing Chemical Semantics

We have been inspired by the practice of crystallography in developing a completely semantic approach to physical science. For half a century, IUCr* and the community have insisted that crystallography is reproducible by such means as comparing experimental data, testing programs against experiments, and most critically the creation of a computable ontology: Crystallographic Information Framework (CIF). CIF has provided a model for Chemical Markup Language (CML) [11], which is now being adopted in computational and other chemistry communities. Since fundamental chemistry concepts were probably solidified 80 years ago, the ontologies used are simpler than those in, for example, bioscience or high energy physics.

---

* International Union of Crystallography.

To promote awareness of the need for and value of semantics we ran two meetings at Cambridge [10,14].* These brought together a group of scientists who cared about reproducibility and interpretability through developing shared semantics (dictionaries and code). These have led to further meetings (e.g., at Pacific Northwest National Laboratory in 2011) and the determination to make key tools such as NWChem [18] and Avogadro available. We believe that if there are enough components available the world will come to see the value of semantics, leading to a gradual change in practice.

5.2.1 Note about Our Software Status and Availability

Some parts of the software described here have been developed over two decades with a large focus on reproducibility. The main parts (JUMBO, CML) have been distributed and are widely used, but only *implicitly* for reproducibility. JUMBOConverters (templates) provides semantic conversion for legacy files during the transition to completely reproducible computation. The Declaratron itself is novel and provides complete reproducibility. All software is in public repositories. In order to give an indication to the reader of the status of any given software component, in the spirit of the five-star open data nomenclature,[†] we use the following symbols: ∅ = vaporware; *— = prototype (has worked for us); ** = "alpha" (hackable by others); *** = usable by others; and **** = in widespread use.

Our software is written in Java and Scala, using XML and XPath libraries also in Java. Other CML libraries have been written in C#, C++, and Python—an object-oriented approach is almost essential. However, many of the main third-party legacy computational programs are written in FORTRAN and are too expensive to change. To interface them into this framework therefore requires an XML/CML wrapper; the FoX **** library has been developed by Toby White and Andrew Walker for this purpose. It deals with a subset of languages and concentrates on program output [6].

5.2.2 CIF and CML as Semantic Languages

CIF **** [2] uses a lightweight set of primitives (items and tables) with datatypes (char and numb) and a very extensive set of dictionaries compiled by the community. A wide range of crystallographers—experimentalists, instrument manufacturers, and computational—use it for interchange. CIF is also used for journal submission and supports computable semantic articles.

CML was launched in 1994 [11] to support semantic chemistry. Since there are few other semantic tools in physical science, this required the

---

* "Visions of a Semantic Molecular Future," and "Semantic Physical Science," sponsored by the EPSRC "Pathways to Impact" program, which supports the dissemination of research done under their auspices.
[†] http://5stardata.info/.

creation of a basic infrastructure for STM computation, scientific technical medical markup language (STMML) [12].* STMML supports basic quantities, error estimations, datatypes, and scientific units of measurement, and we believe it is very widely applicable, certainly to any discipline where typed quantities with units can be understood as stand-alone objects. For instance, temperature is not a specifically chemical quantity, and using STMML, we can write

```
1   <html:p>It was a nice day, 21 degrees
2     (<cml:scalar dataType="xsd:double" dictRef="iupac:T06321"
3         min="19.1" max="23.1" units="nist:sp811.08.8.5"/>)
4           </html:p>
```

The `dictRef` points to the IUPAC GoldBook's [9] temperature[†] and the `units` points to the National Institute of Standards and Technology.[‡] This illustrates some key virtues of CML/STMML: (1) It can be mixed with text (our "datument" approach [13]) and other markup languages (here HTML) through namespaces; and (2) it builds on W3C work (XSD datatypes). However, even though STMML was published 11 years ago, there has been very little adoption of *any* markup languages in physical science. There is full support in CML software , including FoX.[§]

For this chapter, we will make use of CML [****] and introduce a few self-explanatory terms: <molecule> with <atom> and <bond> and a <propertyList> (which may be measured or computed). A <property> has a structural type (<scalar>, <array>, or <matrix>) annotated with @dataType (xsd:string, xsd:integer, or xsd:double) and annotated with a reference to a dictionary (@dictRef). This covers the vast majority of CompChem data needs.[¶]

Here is how a molecule with atoms and coordinates can be completely described:

```
1   <molecule
2     xmlns="http://www.xml-cml.org/schema"
3     xmlns:xsd="http://www.w3.org/2001/XMLSchema"
4     xmlns:units="http://www.xml-cml.org/schema/units"
5     xmlns:compchem="http://www.xml-cml.org/dict/compchem"
6   >
7   <atomArray>
8     <atom id="a1" elementType="O"
9         x3="0.0" y3="0.0" z3="0.0"/>
```

---

* http://www.ch.ic.ac.uk/rzepa/codata2/.
† http://goldbook.iupac.org/T06321.html.
‡ http://physics.nist.gov/Pubs/SP811/sec08.html#8.5.
§ Although markup languages map well onto object languages, FORTRAN needs special support, and we thank Toby White and Andrew Walker for writing a FORTRAN library for CML and XML.
¶ Where necessary, larger dimensions are supported by using CML pointers into (say) HDF or NETCDF, but these extensions are not necessary for the examples at hand.

```
10    <atom id="a2" elementType="H" x3="0.96" y3="0.0" z3="0.0
         "/>
11    <atom id="a3" elementType="H" x3="-0.23" y3="0.93" z3="
         0.0"/>
12   </atomArray>
13   <propertyList>
14     <property dictRef="compchem:dipole">
15       <scalar dataType="units:debye" dataType="xsd:double">
           1.85</scalar>
16     </property>
17   </propertyList>
18  </molecule>
```

5.2.3 Dictionaries

Dictionaries are fundamental to semantic, and therefore reproducible, computing. There is a hierarchy of power:

- Give every semantic object or concept a unique ID. Where possible, we reuse authorities, so
  - A float is defined as xsd:double (defined by W3C)
  - Temperature by IUPAC (http://goldbook.iupac.org/T06321.html)
  - Kelvin (units) NIST (http://physics.nist.gov/cuu/Units/kelvin.html)
  These fit well into RDF/URI and could be written as `nist:kelvin` and `iupac:T06321` using standard prefix notations.
- Create a dictionary entry with an id and type and, if possible, definition and description.

```
1  <cml:entry id="electricdipole" dataType="cml:vector3">
2    <cml:definition>The electric dipole a molecule</
        cml:definition>
3    <cml:description>Dipole moments in molecules are
          responsible for the behavior of a substance in
          the presence of external electric fields.
4      See http://en.wikipedia.org/wiki/
            Electric_dipole_moment
5    </cml:description>
6  </cml:entry>
```

- Add semantic validation of transformation to the entry. This might be done through OWL ontologies or alternatively by adding CML/Declaratron snippets.

The IUCr dictionaries are an excellent example of community-created dictionaries. There is a core dictionary applicable to most crystallography and many subdomain dictionaries for areas such as proteins, diffraction, and powder. We recommend the use of multiple dictionaries as this gives each community a chance to create well-developed subcomponents, which are then rationalize later. For example, we propose one dictionary per computational code (e.g., for NWChem) and then rationalizing parts of these at a higher communal level where possible. Within CML, as well as defining objects such as molecule and atom, in an XML schema, there are thousands of unit tests in JUMBO* that act to resolve possible ambiguities in the textual descriptions.

The use of namespaces and semantic dictionary-based annotation is fundamental to CML (and to the MathML in the Declaratron and the Declaratron itself). Here, we use the following namespaces:

- http://www.xml-cml.org/schema. CML with its domain semantics hard coded. We can rely on a consistent interpretation of the chemistry.
- http://www.w3.org/2001/XMLSchema. W3C XSD datatypes ****. Complete semantic description of xsd:double, for example, giving min/max values and representations. It is possible to use an XSD toolkit as a black box to manage these with complete confidence. xsd:date could be normalized with (say) JodaTime.
- http://www.xml-cml.org/schema/units. ** Units are fundamental. Although NIST started a UnitsML nearly 20 years ago, it was aimed at a database of units ****. There are no agreed computable semantics for units, and we use the ones we proposed in STMML. Note the link later to a CML dictionary of units, which is somewhat ad hoc.
- http://www.xml-cml.org/dict/compchem. There are many hundreds of essential concepts and property definitions required in CompChem. Despite 40+ years of CompChem programs, there is no communal dictionary of terms, let alone a structured ontology. At the Cambridge SPS meeting, we started a call for CompChem dictionaries, and this is being taken forward in Cambridge, CSIRO, PNNL, and Kitware, and we hope it will spread to a wider community of materials science informatics *—.
- atom, elementType, x3.... CML elements (e.g., atom) represent structured objects and attributes (elementType, and x3 represent properties with hard-coded semantics). Thus, @elementType must be found in a standard periodic table (avoiding the problem of, say, "W1" authored for water, misread as tungsten). Similarly, x3 is the Cartesian x-coordinate of an atom in angstrom; this avoids confusion with x2 (chemical formula) and xFract (crystallography). Legacy formats very commonly ambiguate these concepts causing massive wasted work.

---

* "JUMBO" is a library for processing CML, which is based on XOM and subsumes CMLXOM and XML DOM.

- `property`. This is a property of the molecule (atoms and bonds can also have properties). It uses STMML syntax to define a scalar quantity with defined units, defined datatype, and defined semantics (through a linked dictionary). The dictionary can, in principle, contain computable semantics for validation and transformation.

5.3 Components for Defining Computation

In this chapter, we take the view that scientific computation can be broken down into three components:

1. **Data** to use in computation. In a chemical calculation, this might consist of structures representing atoms, bonds, and molecules; their relationships; and any parameterization.
2. **Formulae** to be applied to the data—for example, the functional form of a molecular FF.
3. **Computational specifications** detailing how the formulae should be applied, and to which bits of data; for example, which objects are we computing over, are we computing a single value for a given formula, or is it being used as input to an optimizer.

We demonstrate techniques for representing data and formulae in a semantically well-defined manner, validated with unit tests. We then introduce a tool for specifying computation that preserves semantics, supports data transclusion, and allows a separation of domain knowledge and programming, allowing the combination of well-tested "black-box" domain objects with declaratively specified computation.

5.3.1 Black-Box Libraries

Because chemistry is stable, we have been able to create a black-box library (JUMBO ****). JUMBO was developed to act as a reference implementation for CML, with full unit testing. However, that also means it is deployable as a reliable component for the Declaratron—we give an example:

```
public double getDistanceTo(CMLAtom atom2);
```

This returns the Euclidean distance between two atoms (or throws an exception), a result which is as well-defined in chemistry as a square root is in mathematics. The method is tested with a common set of `fixtures` and tolerances (EPS). Note the use of `assert` that fails with an error if the condition is not true—we adopt a similar strategy in the following for the Declaratron:

```
1    public final void testGetDistanceTo() {
2      double d = fixture.atom[0].getDistanceTo(fixture.atom
           [1]);
3      Assert.assertEquals("distance", Math.sqrt(3.), d, EPS);
4      d = fixture.atom[0].getDistanceTo(fixture.atom[0]);
5      Assert.assertEquals("distance", 0.0, d, EPS);
6    }
```

The test not only confirms the correct operation but gives guidance to a human developer about the intention behind the method. As well as JUMBO, there are other open libraries (e.g., the Chemistry Development Kit, CDK) that can also be reliably used as black boxes.

In order to compare XML documents, JUMBO must go beyond syntactic equivalence, and consider issues such as character encoding, whitespace, and line endings. A typical JUMBO test is

```
1    JumboTestUtils.assertEqualsIncludingFloat(
2      "MOPAC", referenceXML, textXML, ignoreWhitespace, 1.0E-6)
```

5.3.2 JUMBOConverters and FoX

Because almost all physical science is in nonsemantic form, there have been many X-to-Y converters written to get the output of one program into another; in chemistry, an excellent example is Open Babel.* Obviously, conversion can only be provided for those concepts that exist in both programs or can be generated algorithmically or looked up. The converters are rarely complete and generally do not expose any semantics. We strongly recommend conversion to a validatable semantic form and, for chemistry, provide the JUMBOConverter *** framework, which currently converts about 60 to CML.

Converters are often written where the source code is not visible, so intentions and meanings must be inferred. The traditional approach is to create procedural programs, using Python, Java, C++, etc. However, this does not always lead to visible semantics that can be validated against dictionaries. JUMBOConverter templates *** provide a different approach. Here, legacy output is mapped semantically onto validatable results. This forces the translator to define the dictionary entries used and leads to a robust implementation of unit and regression tests. It is also suited to analyzing large corpora.

5.3.3 MathML

MathML **** is a W3C math working group recommendation for representing mathematics on the web. This is an established web standard, with widepsread support in popular browsers, and an ecosystem of supporting

---

* http://openbabel.org/.

tools such as MathJax* to aid in-line display. On the authoring side, there are many tools available: the W3C lists more than 30 editors and viewers.†

There are two distinct dialects of MathML, with different goals:

1. **Presentation MathML** represents the visual layout of mathematical equations. It concerns itself with how symbols are displayed and arranged on the page, but not what they *mean*.

2. **Content MathML** [3] is oriented toward representing the semantics of mathematics "to provide an explicit encoding of the underlying mathematical meaning of an expression"[19].

Content MathML *** (CoMML) is highly suited to representing computation in a semantically aware manner—it has been designed to add an extra layer of formalism to the communication of mathematical equations, removing several sources of potential ambiguity.

5.3.4 Executable MathML

In order to use CoMML in an executable document, it must be linked to a tool that can run the computations that it encodes. Here, we use ScMathML,‡ a Scala engine for running computations specified using CoMML. Scala is a functional language that runs on the Java VM, combining the power of functional programming with easy interoperability with Java code. It was used here because

- It is very concise—most of the MathML entities are defined in about a hundred lines of code
- Inbuilt XML support makes starting to work with XML easy—in the listings as follows, XML blocks define real Scala objects, rather than strings to be parsed
- Support for building domain-specific languages, with flexible parsing

ScMathML "parses" CoMML into a tree of Scala objects, which can carry out computation. We will illustrate, using unit tests from the framework, how this works, and how it can be used for computation in the physical sciences. Unit tests are written using the ScalaTest framework, which—along with some wrapper functions—leads to clean, self-documenting test code: all of the code examples in the following are taken directly§ from the unit

---

* http://www.mathjax.org/.
† http://www.w3.org/Math/Software/mathml_software_cat_editors.html.
‡ www.mo-seph.com/projects/SCMathML.
§ Formatting has been changed, and some variables have been renamed for clarity out of context.

test files. For example, to check that a MathML `<cn>` element is parsed into a SCMathML constant, we can write

```
1   parsing(<cn>5.3</cn>) should equal( DoubleConstant(5.3))
```

Two of the basic elements of MathML are constant numbers (`<cn>` for content numeric) and variables (`<ci>` for content identifier). Variables need to have values provided if a function is to be computed—for example, when given $y = x^2 + c$, if we want to get a number out, we need to provide values for both x and c. In ScMathML, this is done through a *context*, where objects can be passed in:

```
1   evaluating(<ci>x</ci>, "x"->5) should equal( 5 )
```

In this example, "evaluating" is a function defined to take a MathML expression, and some mappings of strings to objects, and evaluate the expression. `should` and `equal` (and later `be`, `plusOrMinus`) are Scala Test functions that allow a natural reading of unit tests. This example can be read that if we take the expression `<ci>x</ci>`, parse it, and then evaluate it in a context where x has been set to 5, we should get 5 out. This is an illustration of how objects are *bound* to variables and used to evaluate abstract mathematical expressions and obtain concrete results.

CoMML is strongly influenced by Scheme and related languages: it uses `<apply>` tags to denote function application, with the first argument being the function to apply. For example, `<apply><plus/><cn>2</cn><cn>2</cn></apply>` is roughly equivalent to (plus 2 2) in Scheme, or

```
1   evaluating( <apply><sin/><ci>x</ci></apply>, "x"->3 )
2       should equal( Math.sin(3))
```

For a slightly larger example, we can implement Leibniz's method of approximating π:

$$\sum_{k=0}^{n} \frac{(-1)^k}{2k+1} \approx \frac{\pi}{4}$$

```
1   evaluating(
2   <apply><times/><cn>4</cn>
3     <apply><sum/>                    <!--carry out a summation -->
4       <bvar><ci>k</ci></bvar>        <!-- for k in ... -->
5       <lowlimit><cn>0</cn></lowlimit> <!-- start at 0 -->
6       <uplimit><ci>n</ci></uplimit> <!-- go up to the value
                 bound to n -->
7       <apply><divide/>
8         <apply><power/><cn>-1</cn><ci>k</ci></apply>
9         <apply><plus/>
10          <apply><times/><cn>2</cn><ci>k</ci></apply>
11          <cn>1</cn>
12        </apply>
```

```
13      </apply>
14  </apply></apply>,  "n"->4000)  should  be  (  3.1415  plusOrMinus
            0.01  )
```

Finally, there is often a need to work with values obtained from domain entities. In order to do this, we have defined a small set of extensions to the MathML specification to interface with existing objects, using the `<csymbol>` tag. Figure 5.2 gives a worked example where values are extracted from domain objects. It is based on Hooke's law, which would typically be written as

$$E = \sum_{\text{bonds}} \frac{1}{2}kx^2$$

However, since it has been translated into MathML, with bindings added, it is clear that

1. x actually refers to displacement from equilibrium length, and so has to be split into l and l_0
2. All of the values are specific to a given spring, including the spring coefficient k

Contrast this with the first term in the AMBER FF equation (Figure 5.1) where it is up to the reader to interpret that (1) b is a subscript for the current bond, (2) b on its own means the length of the current bond, and (3) b_0 is the equilibrium length *of the current bond*.

It should be noted, however, that this example only defines the *mathematical* semantics—the operations to be carried out, and the bits of data to lay them to. It does not deal with any *domain* semantics, as the Springs class is not semantically explicit.

5.4 Semantic Physical Computation

We have discussed the representation of scientific entities using domain-specific markup languages, and formalization of computation using domain-independent markup. Now, we introduce a system—"the Declaratron"[**]— which brings these together to carry out reproducible scientific calculations.

The Declaratron consists of

1. An XML dialect for specifying declarative computation (Figure 5.3). Our current vocabulary is
   - `<sem:computationalDocument>`, the overall container and organizer.

```
1   var sum = //Define Hookes law in Content MathML
2   <apply>
3     <sum/>
4     <bvar><ci>spring</ci></bvar><!-- this is the variable to bind -->
5     <condition>                  <!-- and this is the set to bind over -->
6       <apply><in/><ci>spring</ci><ci type='set'>springs</ci></apply>
7     </condition>
8     <apply><times/>
9       <cn>0.5</cn>
10      <apply><times/>
11        <apply><csymbol function='elasticity'>k</csymbol><ci>spring</
             ci></apply>
12        <apply><power/>
13          <apply><minus/>
14            <apply><csymbol function='length'>l</csymbol><ci>bond</ci><
                /apply>
15            <apply><csymbol function='equilibrium'>l0</csymbol><ci>bond
                </ci></apply>
16          </apply>
17          <cn>2</cn>
18        </apply>
19      </apply>
20    </apply>
21  </apply>
22  //We have a spring class which takes 3 arguments:
23  //length, equilibrium length, and elasticity
24  //Create two Springs to test:
25  var bonds = List(new Spring("A",8,7,3),new Spring("B",10,9,4))
26
27  // Target equation is: sum of
28  // 0.5 * elasticity(spring)*(length(spring)-equilibrium(spring))^2
29  // With the test springs, the expected value is:
30  val exp =
31      0.5 * 3 * Math.pow(8-7, 2) + //Spring A
32      0.5 * 4 * Math.pow(10-9, 2); //Spring B
33
34  //Now run the test:
35  evaluating(<math>{sum}</math>, "bonds"->bonds) should equal( exp )
```

FIGURE 5.2
Example MathML equation, showing a test summation over a set of domain-specific objects. Target equation is $E = (k(l - l_0)^2/2)$. Note that this example is not semantically bound.

- <sem:editor> that allows the document to modify itself using copy, transform, move, and delete operations.
- <sem:assert> allows components to be tested against scalar values or complete (XML) files.
- @href allows input of files (transclusion and copy).

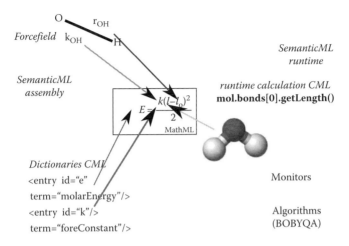

FIGURE 5.3

Declarative creation and execution of `computationalDocument`. The mathematics is linked to domain semantics (e.g., for each element of the summation l links (arrows) to reference equilibrium bond length l_0 in FF (e.g., for an O-H bond) and k to the reference O-H force constant. Dictionaries are used to ensure semantic mapping (e.g., that energy can be related to force-constants and lengths) and also to provide human prose. The document is assembled and modified by repeated XPath queries to create nodeSets (parts of the tree) and then application of editor commands (copy, create, move, delete) to modify it. In this way, a human-friendly (minimal) document is automatically expanded to constancy and machine-interpretability. In the second phase, the document is executed using visitors to traverse the (now constant) tree and process nodes. Because the tree is constant, it is possible to attach monitors to the nodes (e.g., recording which terms had significant values or which took longest to compute). Single functions and nodes are declarative, it becomes possible to change algorithms (e.g., for optimization, such as the single-point optimizer BOBYQA).

- `<sem:writer>` allows output of sections of the document.
- {`<sem:functionalForm>` specification of a MathML expression that can be bound to other domain semantics.
- `<sem:computation>` evaluation of a `<sem:functionalForm>` either once or in an algorithm.

2. Core libraries that support the operations in this dialect: transclusion, copying, merging, and validation.

3. Support for replacing XML nodes with domain objects that bring useful code with them (decoration). This includes decorating data structures (e.g., replacing a plain `<cml:atom/>` element with a Java object and creating executable objects, such as unit converters).

4. Links to domain libraries to bring in necessary semantics and computational elements, including:
   - ScMathML for evaluating mathematical formulae in the context of a scientific computation

- General STM information, such as units and their conversions
- CML/JUMBO for representing chemical data and computing common properties

It is entirely possible for users to add their own domain libraries.

5.4.1 XML and XPath Design

Much of the power of the Declaratron comes through the XML data structures used, in particular, the ease with which it can be navigated and transformed. XML is generally represented as a tree, and all XML libraries support concepts of parent/child—`cml:atoms` are children of `cml:atomArray`, which itself is a child of `cml:molecule`. The XPath language* allows easy navigation of the document tree, and it is central to the Declaratron. Xpath can reference any set of nodes in the tree (nodeSet) with a natural and powerful syntax (based on tree structures). To give a feel for XPath, we provide some examples in the following[†]; the syntax can be likened to a directory structure, with / representing direct children, and [...] a condition:

1. `//cml:atom`—find all atoms in the document (`//` addresses any level in the hierarchy).
2. `//cml:molecule[@id='acetic']/cml:atomArray/cml:atom`—find all atoms in the `cml:molecule` with id attribute `acetic`.
3. `//cml:molecule[@id='acetic']/cml:atomArray/cml:atom[not(@elementType='H')]`—find all nonhydrogen atoms in the earlier set.
4. `//cml:molecule[count(.//cml:atom[@elementType='H'])=0`—find all molecules without any hydrogen atoms.

A series of Declaratron editor commands allows documents to be modified. Since Declaratron documents are in XML, they can be self-modifying. XML acts as both input and output and so can provide a full record of computations. Snippets from files such as schemas and dictionaries can be included in the output so that it is clear exactly what versions were used and what was done.

The Declaratron works as follows, illustrated in Figure 5.4:

1. Read in a computational document
2. Manipulation: transclusion and substitution (see Section 5.4.2)

---

* http://www.w3.org/TR/xpath/.
[†] For a tutorial, see http://www.w3schools.com/xpath/.

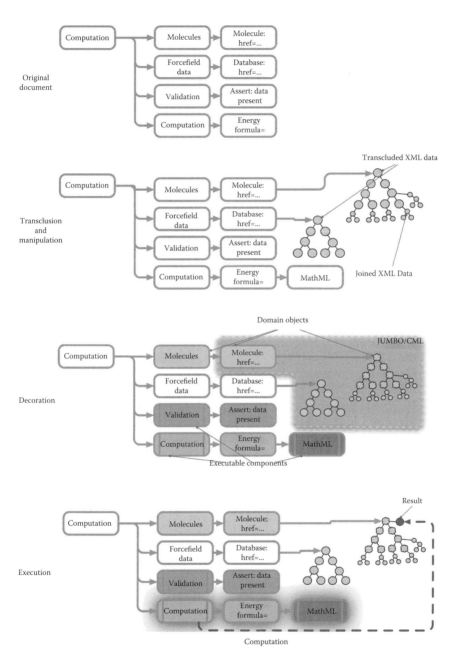

FIGURE 5.4
Overview of Declaratron operation: (1) original document; (2) manipulated XML document, (3) decorated document with executable domain objects, and (4) executing a computation.

3. Decoration: replacing standard XML elements with domain objects

4. Computation

Validation is threaded through the entire process, to check that incoming data are in the correct form, manipulated data have the right properties, and the results of computations are correct.

We will use a case study—calculating the energy of water using a very simple FF—to demonstrate this.

5.4.2 Bridging the Gap between Human- and Machine-Readable Semantics

Humans work with implicit semantics and require documents to be small and nonrepetitive in order to extract meaning. Machines require explicit, formal semantics and can work with large (typically 10–100 times larger), repetitive documents that are deeply human-unfriendly. In order to address this division, the Declaratron can use documents that are written in a relatively concise, human-readable form, and automatically expanded to the complete, explicit computable form. The human form uses

- Key-value syntactic substitution, which both reduces repetition, and allows for human-readable names to be attached to complex structures. For example, replacing occurrences of `xpath="//cml:molecule[@id='molecule']"` with `xpath="${molpath}"` makes the document more readable as the intent of the XPath expression is clear and makes it more robust as complex expressions can be defined and tested once and then reused.
- Transclusion of files; there is a `<sem:editor>` that expands `href` attributes recursively. Again, this enhances readability, especially where large files are brought in and reused, as common elements can be put into files and shared (e.g., `<maths href="$mathsPath/hookesLaw.xml"/>`). It also allows the use of data from nonlocal sources, as any URI that provides an XML stream can be used.

Using this system of transclusion, we have successfully computed the energy of acetic acid (8 atoms) using the current AMBER (parm94) FF, by expanding a human-readable input, to machine form and evaluating it. However, the human form hides many important details, and for this chapter, we have therefore chosen a very simple complete example (water—3 atoms), using a highly simplified functional form (only bonds), with a later indication of how this would be expanded to a more complete example.

5.4.3 Example: Water–Energy Calculation

To illustrate the Declaratron's operation, we will walk through a calculation file step by step[*]. This file carries out single-point energy computation and atomic optimization on water, using the Hooke's law term from the AMBER FF ($\sum_{\text{bonds}} K_b(b - b_0)^2$) and associated parameters.[*] This is described in four stages:

- Setting up the document: namespaces and named variables
- Identifying the data to be used: the molecule, atoms, and bonds
- Specifying the FF
- Specifying the computations to be carried out

In a real-world file, the definitions of data and computations might be given in a different order—for example, putting the computation first in files makes it easy for a human to find out what a file does. The order of elements is fairly flexible, so in this explanation, we start with the data and then later specify what we intend to compute. As noted previously, we also go through some elements that would typically be expanded from databases or refactored into individual files to make a humane document.

5.4.3.1 Set Up the Document

The `<computationalDocument>` node is a container for the entire computation. We also define XML namespaces that can be used throughout the file:

```
1  <computation xmlns="http://www.xml-cml.org/
       semanticcomputation"
2      xmlns:m="http://www.w3.org/1998/Math/MathML"
3      xmlns:cml="http://www.xml-cml.org/schema"
4      xmlns:amber='http://www.xml-cml.org/dict/amber:gaffType'
       >
```

In order to carry out XPath queries over namespaced documents, it is necessary to set up namespace prefixes that can be used by the XPath engine:

```
5      <!-- setup XML namespaces for use in XPath queries -->
6      <queryNS prefix="semc" uri="http://www.xml-cml.org/
           semanticcomputation"/>
7      <queryNS prefix="semf" uri="http://www.xml-cml.org/
           semanticforcefields"/>
8      <queryNS prefix="cml" uri="http://www.xml-cml.org/schema
           "/>
```

---

[*] The full file can be found at: https://bitbucket.org/petermr/semantic-forcefield/src/cf7ef9b03020/src/main/resources/org/xmlcml/cml/examples/amberNew.xml?at=default.

```
9    <queryNS prefix="m" uri="http://www.w3.org/1998/Math/
         MathML"/>
10   <queryNS prefix="amber" uri="http://www.xml-cml.org/dict
         /amber:gaffType" />
11   <queryNS prefix="cmlx" uri="http://www.xml-cml.org/
         schema/cmlx" />
```

Throughout the file, key/value pairs can be used to reduce repetition. Here, we set up variables for XPath queries for (1) the molecule to analyze, using an id to ensure only the desired molecule is used; (2) finding all the bonds belonging to that molecule; and (3) finding all the atoms:

```
12   <!-- XPath references to molecule, atoms, bonds -->
13   <keyValue name="molpath" value="//cml:molecule[@id='
         molecule']"/>
14   <keyValue name="bondpath" value="${molpath}/
         cml:bondArray/cml:bond"/>
15   <keyValue name="atompath" value="${molpath}/
         cml:atomArray/cml:atom"/>
```

5.4.3.2 *Identify and Verify Chemical Data*

Now we define the chemical data to be used. First, the molecule. As this is a container for chemistry, it defines some extra namespaces for chemical entities:

```
16   <cml:molecule
17     xmlns:cml="http://www.xml-cml.org/schema"
18   xmlns:xsd="http://www.w3.org/2001/XMLSchema"
19   xmlns:units="http://www.xml-cml.org/schema/units"
20   xmlns:compchem="http://www.xml-cml.org/dict/compchem"
21     >
```

Next, the atoms. Note that each atom has an *atomType* defined by AMBER. This is *not* the same as the elementType—AMBER uses atomTypes as well as elementTypes for atoms, the distinction being that atomTypes change depending on the surrounding atoms. For example, an oxygen bonded to a hydrogen has atomType "OH". Bond parameters are looked up using atomTypes rather than elementTypes, and bonds are often annotated with different atomTypes in different communities of practice (i.e., different laboratories):

```
22   <cml:atomArray>
23     <cml:atom id="a1" elementType="O" x3="0.0" y3="0.0" z3
         ="0.0">
24       <atomType dictRef="amber:parm94Type">OH</atomType>
25     </cml:atom>
26     <cml:atom id="a2" elementType="H" x3="0.96" y3="0.0"
         z3="0.0">
```

```
27        <atomType dictRef="amber:parm94Type">H</atomType>
28      </cml:atom>
29    <cml:atom id="a3" elementType="H" x3="-0.23" y3="0.93"
             z3="0.0">
30        <atomType dictRef="amber:parm94Type">H</atomType>
31      </cml:atom>
32    </cml:atomArray>
```

The id structure is very important and is used to link components of the document (e.g., the bonds reference the atom ids) or even to aggregate them (through the editor).

The final component of the chemical data is a set of bonds, with parameters looked up by atomType. For each bond, k is the spring constant, req is the equilibrium bond length, and desc is the formal description (e.g., a literature reference):

```
33    <cml:bondArray>
34      <cml:bond atomRefs2="a1 a2">
35          <cml:property>
36              <cml:list id="c_ct" cmlx:atomTypesId="OH__H">
37                  <cml:atomType dictRef="amber:parm94Type">OH<
                        /atomType>
38                  <cml:atomType dictRef="amber:parm94Type">H</
                        atomType>
39                  <cml:scalar dictRef="ff:k" dataType="
                        xsd:double">317.0</scalar>
40                  <cml:scalar dictRef="ff:req" dataType="
                        xsd:double">1.522</scalar>
41                  <cml:scalar dictRef="ff:desc" dataType="
                        xsd:string">JCC,7,(1986),230;AA</scalar>
42              </cml:list>
43          </cml:property>
44      </cml:bond>
45      <cml:bond atomRefs2="a1 a3"> <!-- contents omitted for
             brevity --> </cml:bond>
46    </cml:bondArray>
```

For the purposes of this example, we have declared the bond properties inline. In general, these would be pulled in from a knowledgebase automatically, but this is too complex for this chapter.

Once all of the data are in place, we can verify it. To ensure that all atoms can be annotated with the AMBER parm94 dictionary, an <assert> element checks that (1) all atoms have a valid id (expressed as the inverse: there are 0 atoms without a valid id) and (2) there are 0 atoms without a valid atomType:

```
47    <!-- all atoms in the document must have ids and
             atomTypes (expressed as negation) -->
48    <assert count="0" xpath="${atompath}[not(@id)]"/>
```

```
49    <assert count="0" xpath="${atompath}[not(cml:atomType[
          @dictRef='amber:parm94Type'])]"/>
```

5.4.3.3 Specify the FF to Be Used

After specifying the data, we specify the functional form of the FF ($E = k(l - l_0)^2/2$). It starts with the summation over bonds:

```
50    <functionalForm id="hookes"
51      hrefSource="src/main/resources/org/xmlcml/cml/forcefield
          /functional/harmonicBond.xml"
52      xmlns="http://www.xml-cml.org/semanticforcefields">
53      <math xmlns="http://www.w3.org/1998/Math/MathML">
54        <apply><sum/>                     <!-- Sum -->
55          <bvar><ci>bond</ci></bvar> <!-- For bond -->
56          <condition>                     <!-- in bonds -->
57            <apply><in/><ci>bond</ci><ci type="set">bonds</ci><
                /apply>
58          </condition>
59          <!-- This is what is inside the sum -->
60          <apply><times/><cn>0.5</cn> <!-- divide by 2 -->
```

We need to bind the value of k to the actual bond (`property` is child of bond). Again, note the use of `dictRef`—this uses a defined id reference, which means that the semantics of the value to be used can be looked up in the dictionary:

```
61              <apply><times/>               <!-- get k for the current
                  bond -->
62                <apply>
63                  <csymbol xpath="./cml:property/cml:list/
                    cml:scalar[@dictRef='ff:k']">k</csymbol>
64                  <ci>bond</ci>
65                </apply>
66              </apply>
67              <apply><power/>               <!-- start the squared
                  term -->
68                <apply><minus/>             <!-- start l-l_0 -->
```

The value of l is bound to the result of calling the JUMBO function `cml:bond.getBondLength()` for each bond:

```
69                <apply>
70                  <csymbol function="getBondLength">l</csymbol>
71                  <ci>bond</ci>
72                </apply>
```

and the reference (equilibrium) length is looked up for each bond with an XPath expression that selects the relevant property from its descendants:

```
73        <apply>
74          <csymbol xpath="./cml:property/cml:list/
              cml:scalar[@dictRef='ff:req']">10</csymbol>
75          <ci>bond</ci>
76        </apply>
77      </apply>                    <!-- end 1-1_0 -->
78      <cn>2</cn>                  <!-- end of the squared
            term -->
79    </apply>
80   </apply>
81  </apply>
82  </math>
83  </functionalForm>
84 </computation>
```

At this point, we have defined a molecule, and the FF which is to be applied to it.

5.4.3.4 Specify the Computation to Be Carried Out

We now specify a computation to carry out with these entities (do we want simply to evaluate the energy, or adjust the geometry to optimize the structure against its energy?). First, let's create a node (child of molecule to hold the result of a single-point energy calculation):

```
85    <editor method="createChild" xpath=".//molecule" element
         ="cml:scalar" targetId="singlePoint"/>
```

Next, specify that an evaluation of the functional form with the molecule in its initial configuration should be carried out. This will locate the functional form and ask it to evaluate itself, using the molecule as input. When the molecule is passed in, the set of bonds will be bound to the variable bonds. The functional given earlier will iterate over the set of bonds, and for each bond, call the JUMBO function to find the current bond length, subtract the equilibrium length, etc., as detailed earlier. After the calculation, we ensure that the output has the correct value and has the correct units:

```
86    <computation method="singleEvaluation"
87      formula="//functionalForm[@id='hookes']" input="${
         molpath}">
88      <variable name="bonds" xpath="${bondPath}"/>
89    </computation>
90    <assert value="1.234" xpath=".//scalar[@id='singlePoint
         ']"/>
91    <assert value="units:joule" xpath=".//scalar[@id='
         singlePoint']@units"/>
```

The optimum geometry of a molecule is that of lowest energy, and many calculations attempt to find this using a variety of algorithms. We have chosen the recent BOBYQA method [16], which does not require analytical derivatives (or second derivatives).

The same functional form can also be used in the optimization of geometry to find the minimum energy. Here, we use the nonderivative optimizer BOBYQA by giving it (1) a target function to evaluate (the functional form); (2) the data to work over (the molecule); and (3) an XPath expression to find the free variables in the optimization. The optimization happens *in place*, so the modified atom positions are now part of the document:

```
92    <computation method="optimise" algorithm="BOBYQA"
93        formula="//functionalForm[@id='hookes']" input="${
            molpath}"
94        freeVariables=".//@x3 or .//@y3 or .//@z3"/>
```

Finally, we can compare the output geometry and energy with a previous computation stored in another file. This is a complete nodewise comparison of XML, which will ensure both semantic identity and numerical identity, including a tolerance (eps) for floating-point variations:

```
95    <assert href="expected.xml" ref="${molpath}" eps="1.0E
            -06"/>
96  </computation>
```

5.4.4 Moving beyond Toy Examples

The Declaratron has a wider range of features than we have illustrated here. There is a one-off cost to transforming legacy files to CML (some of which can be done with JUMBOConverter templates). Most problems then require a complex process of locating transcludable information (see Section 5.4.2), extracting the desired nodes, and inserting into the growing `semanticDocument.` Although this is a complex operation, once constructed, the semanticDocument subtrees can be reused without change for future computations. This means that computations can be stated very simply in terms of the major free variables and the operations to be performed on them.

As an example, to use the parm94 database in a semantic calculation, we would carry out the following steps:

First, the JUMBO atomTypeTool can be used to add the required ids:

```
97    <!-- list is the default type for general data -->
98    <cml:list id="parm94Test" href="${forcefield}/amber/
            parm94test.xml"/>
99    <!-- transform (add id) to atomType children -->
100   <atomTypeTool method="addAtomTypesId" using="./
            cml:atomType"
101       xpath="//cml:list[@id='parm94Test']/cml:list/cml:list
            [count(cml:atomType)>0]" />
```

Then we can merge functional form and parameters for each bond from the database, by copying the relevant information into each `cml:bond` element:

```
102
103     <moleculeTool method="getOrCreateBonds" xpath="${molpath
            }" setId="createdBonds"/>
104     <editor method="copyChild"
105        xpath="//cml:molecule[@id='molecule']/cml:bondArray"
106           from="//cml:list[@id='parm94Test']/cml:list[@title
                 ='bonds']/semf:functionalForm" />
```

We can also ensure that the bond angles are in the correct units:

```
107     <unitsVisitor xpath="//cml:list[@id='parm94Test']//
           cml:scalar[@dictRef='ff:angeq'
108        or @dictRef='ff:phase']" method="degrees2Radians"/>
```

And finally, we can save this annotated molecule into its own file—coarse-grain memoization—so we do not have to do the conversion again in the future:

```
109     <writer xpath="//cml:list[@id='parm94Test']" file="
           output/parm94testNew.xml" />
```

All this can be packaged into standard operations—a file can be created for any given conversion and transcluded where necessary—so that the final calculation mirrors the human-readable form and is expanded into the detailed semantic form at runtime.

The program output can contain as detailed a list as we like of the operations and their outputs/results. It could contain fine-grained information for debugging or simple summary data. It will have a complete record of the input—not just the values but the semantic parameters, the dictionaries, and the functional forms. This means the output is immediately rerunnable. Note that XML has a very wide range of open document manipulation tools, so we can build high-quality print or semantic indexes.

The output is directly transformable into the inputs of other programs that share some or all of the semantics (e.g., chemistry and mathematics). In many cases, these can be understood without the wider context—a molecule optimized by an FF could then be read into a QM program or posted in an online CML repository for use in chemical informatics.

5.4.5 Integrating Semantic Physical Computation with Emerging Architectures and Automation

One of the most immediate and powerful benefits of declarative computational science is the *parameter sweep*. In many computations, we have a number of independent parameters that the experimenter may wish to vary. Examples are as follows:

- Compute the optimum geometry for a series of molecules.
- Use different methods ("functionals") on a well-understood test molecule and compare results to find the best method.
- Vary the temperature at which a zeolite is modeled and thence compute the coefficient of thermal expansion.

Figure 5.5 shows an example of CompChem architecture. The input (LHS) consists of about six orthogonal axes: (1) molecules; (2) commands—the scientific problem to be solved; (3) basis set—the Qm parameterization; (4) method of solving QM equations; (5) physical parameters, for example, temperature and pressure; (6) computer environment, for example, CPU limits, memory, number of processors. While this is a particular example, and would be used for experiments such as "run 1000 molecules with 3 basis sets," many experiments have a similar structure to this, that is, "explore a defined subset of the parameter space."

The sweep has similarities to the "MapReduce" approach—a large number of job inputs are created, farmed out to processors, and then collected and analyzed. The example in Figure 5.5 has several independent input axes, and it is clear that these must be semantically defined if the experiment is to be reproducible—ambiguity or the possibility for disagreement about the meanings or intents of specifications will result in different or unpredictable results. More generally, no science can be reproducible without agreed semantics.

The Declaratron is very well suited to the flexible generation of input—it allows parameter axes to be declared semantically and combined through domain-specific commands (expanding the scope and precision of "Convolution"). It is also a critical part of marshaling and validating output, especially transforming documents to have different structures and components.

5.5 Conclusions

The adoption of semantics by long-tail physical science has been extremely slow, and its absence causes millions of lost hours and costs hundreds of millions of dollars. We do not believe science is reproducible without a committed community (as in crystallography or astronomy or much of bioscience).

Instrumental and sensor output is now massive, but there are very few semantic implementations or dictionaries; this is an essential task for the communities to tackle.

In this chapter, we have demonstrated a system that addresses several of these concerns. It integrates existing semantically aware components with

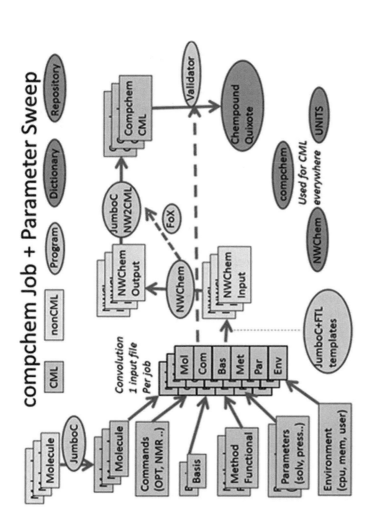

FIGURE 5.5

A multiaxis parameter sweep for CompChem using NWChem ***** as a semantic framework *_ — for computing properties of matter. The experiments often involve several axes and could involve many thousands of jobs. Our architecture allows the automatic creation of jobs with combinations of settings ("parameter sweeps"), which can be filtered semantically. NWChem outputs CML through FoX calls; alternatively, the legacy output is converted by an NWChem JUMBO converter to CML. The output is semantically validated against (a) NWChem, (b) general CompChem, and (c) units dictionaries and normalized, before archive and redistribution in a Quixote repository [5]. *Note:* We thank PNNL for making NWChem open and agressively CML-izing it.

legacy data and provides a strongly semantically grounded computation environment, with a high level of reproducibility.

The usefulness of any semantically aware, reproducible system is dependent on its context: the further the semantic frontier is pushed back, the more use we can make of each component in the system. To make this happen, we have a set of recommendations:

1. Build a community of practice around semantic computation and reproducibility. Ideally, this should be through learned societies or international scientific unions—without community semantics, there is no interoperability and hence no effective reproducibility.

2. In the absence of any general approach from the scientific community, adopt STMML semantics where possible: using strongly typed quantities with dictionaries. Dictionaries can be created in a semiformal manner, being ratified by formal groups when they are proved to work.

3. Create black-box libraries for fundamental domain-specific operations and algorithms. These should be tested using declarative approaches, to allow for integration with semantic systems.

4. Build declarative validators for legacy code bases (e.g., AMBER) so that their correctness can be verified on a wide range of archetypal problems.

5. Make all "documentation" semantic and computable: write examples in manuals as executable code so that the documentation is always in sync with the code.

References

1. N.L. Allinger. Conformational analysis. 130. MM2. A hydrocarbon force field utilizing V1 and V2 torsional terms. *Journal of the American Chemical Society*, 99(25):8127–8134, 1977.
2. I.D. Brown and B. McMahon. CIF: The computer language of crystallography. *Acta Crystallographica Section B: Structural Science*, 58(3):317–324, 2002.
3. D. Carlisle. OpenMath, MathML, and XSL. *ACM SIGSAM Bulletin*, 34(2):6–11, 2000.
4. A.N. Davies and P. Lampen. JCAMP-DX for NMR. *Applied Spectroscopy*, 47(8):1093–1099, 1993.
5. P. de Castro, P. Echenique, J. Estrada, M.D. Hanwell, P. Murray-Rust, and J. Thomas. The quixote project: Collaborative and open quantum

chemistry data management in the internet age. *Journal of Cheminformatics*, 3(1):1–27, 2011.

6. MT Dove, AM Walker, TOH White, RP Bruin, KF Austen, I Frame, GT Chiang, P Murray-Rust, RP Tyer, PA Couch, et al. Usable grid infrastructures: Practical experiences from the eMinerals project. In *Proceedings of the UK e-Science All Hands Meeting 2007*, Nottingham, UK, pp. 48–55, 2007.

7. P Lampen, H Hillig, AN Davies, and M Linscheid. JCAMP-DX for mass spectrometry. *Applied Spectroscopy*, 48(12):1545–1552, 1994.

8. RS McDonald and PA Wilks. JCAMP-DX: A standard form for exchange of infrared spectra in computer readable form. *Applied Spectroscopy*, 42(1):151–162, 1988.

9. AD McNaught and A Wilkinson. *Compendium of Chemical Terminology*, vol. 1669. Blackwell Science, Oxford, U.K., 1997.

10. P Murray-Rust. Semantic science and its communication—A personal view. *Journal of Cheminformatics*, 3(1):1–7, 2011.

11. P Murray-Rust and HS Rzepa. Chemical markup, XML, and the worldwide web. 1. Basic principles. *Journal of Chemical Information and Computer Sciences*, 39(6):928–942, 1999.

12. P Murray-Rust and HS Rzepa. STMML. A markup language for scientific, technical and medical publishing. *Data Science Journal*, 1:128–192, 2002.

13. P Murray-Rust and HS Rzepa. The next big thing: From hypermedia to datuments. *Journal of Digital Information*, 5(1), 2006.

14. P Murray-Rust and HS Rzepa. Semantic physical science. *Journal of Cheminformatics*, 4(1):1–7, 2012.

15. S Ping Ong, A Jain, G Hautier, M Kocher, S Cholia, D Gunter, D Bailey, D Skinner, KA Persson, and G Ceder. The Materials Project, 2011.

16. MJD Powell. The BOBYQA algorithm for bound constrained optimization without derivatives. Cambridge NA Report NA2009/06, University of Cambridge, Cambridge, U.K., 2009.

17. A Roth, R Jopp, R Schäfer, and GW Kramer. Automated generation of AnIML documents by analytical instruments. *Journal of the Association for Laboratory Automation*, 11(4):247–253, 2006.

18. M Valiev, EJ Bylaska, N Govind, K Kowalski, TP Straatsma, HJJ Van Dam, D Wang, J Nieplocha, E Apra, TL Windus et al. NWChem: A comprehensive and scalable open-source solution for large scale molecular simulations. *Computer Physics Communications*, 181(9):1477–1489, 2010.

19. W3C Math Working Group. Content MathML. http://www.w3.org/TR/MathML3/chapter4.html

Part II

Practices and Guidelines

6

Developing Open-Source Scientific Practice*

K. Jarrod Millman and Fernando Pérez

CONTENTS

---

* Dedicated to the memory of John D. Hunter III, 1968–2012.

149

6.1 Introduction

Computational tools are at the core of modern research. In addition to experiment and theory, the notions of simulation and data-intensive discovery are often referred to as "third and fourth pillars" of science [12]. It is probably more accurate to simply accept that computing is now inextricably woven into the DNA of science, as today, even theory and experiment are computational. Experimental work requires computing (whether in data collection, preprocessing, or analysis), and theoretical work requires symbolic manipulation and numerical exploration to develop and refine models. Scanning the pages of any recent scientific journal, one is hard-pressed to find an article that does not depend on computing for its findings.

Yet, for all its importance, computing receives perfunctory attention in the training of new scientists and in the conduct of everyday research. It is treated as an inconsequential task that students and researchers learn "on the go" with little consideration for ensuring computational results are trustworthy, comprehensible, and ultimately a secure foundation for reproducible outcomes. Software and data are stored with poor organization, little documentation, and few tests. A haphazard patchwork of software tools is used with limited attention paid to capturing the complex workflows that emerge. The evolution of code is not tracked over time, making it difficult to understand what iteration of the code was used to obtain any specific result. Finally, many of the software packages used by scientists in research are proprietary and closed source, preventing complete understanding and control of the final scientific results.

We argue that these considerations must play a more central role in how scientists are trained and conduct their research. Our approach grows out of our experience as part of both the research and the open-source scientific Python communities. We begin (Section 6.2) by outlining our vision for the scientific software development in everyday research. In the remaining sections, we provide specific recommendations for computational work. First, we describe the routine practices (Section 6.3) that should be part of the daily conduct of computational work. We next discuss tools and practices developed by open-source communities to enable and streamline collaboration (Section 6.4). Finally, we present an approach to developing and communicating computational work that we call *literate computing* in contrast to the traditional approach of literate programming (Section 6.5).

6.2 Computational Research

Consider a researcher using MATLAB® for prototyping a new analysis method, developing high-performance code in C, postprocessing by twiddling controls in a graphical user interface, importing data back into MATLAB for generating plots, polishing the resulting plots by hand in Adobe Illustrator, and finally pasting the plots into a publication manuscript or PowerPoint presentation. What if months later they realize there is a problem with the results? Will they be able to remember what buttons they clicked to reproduce the workflow to generate updated plots, manuscript, and presentation? Can they validate that their programs and overall workflow are free of errors? Will other researchers or students be able to reproduce these steps to learn how a new method works or understand how the presented results were obtained?

The pressure to publish encourages us to charge forward chasing the goal of an accepted manuscript, but the term "reproducibility" implies repetition and thus a requirement to also move *back*—to retrace one's steps, question or change assumptions, and move forward again. Unfortunately, the all-too-common way scientists conduct computational work makes this necessary part of the research process difficult at best, often impossible.

The open-source software development community* has cultivated tools and practices that, if embraced and adapted by the scientific community, will greatly enhance our ability to achieve reproducible outcomes. Open-source software development uses public forums for most discussion and systems for sharing code and data. There is a strong culture of public disclosure, tracking and fixing of bugs, and development often includes exhaustive validation tests that are executed automatically whenever changes are made to the software and whose output is publicly available on the Internet. This detects problems early, mitigates their recurrence, and ensures that the state and quality of the software is known under a wide variety of situations (operating systems, inputs, parameter ranges, etc.). The same systems used for sharing code also track the authorship of contributions. All of this ensures an open collaboration that recognizes the work of individual developers and allows for a meritocracy to emerge.

As we learn from the open-source process how to improve our scientific practice, we recognize that the ideal of scientific reproducibility is by necessity a reality of shades. We see a gradation from a pure mathematical result whose proof should be accessible to any person skilled in the necessary specialty to one-of-a-kind experiments such as the Large Hadron Collider or the Hubble Space Telescope, which cannot be reproduced in any realistic sense. However, it is always possible to improve our confidence in the results: whether we reexamine the same unique datasets with independently developed packages run by separate groups or we reacquire partial sampling of critical data multiple times.

---

*We take it as a forgone conclusion (see [16]) that to share our research code with one another, we must use open-source tools. Instead of discussing the need for using open-source software, we focus on adopting development practices used by open-source communities.

Similarly, in computational research, we also have certain areas where complete reproducibility is more challenging than others. Some projects require computations carried on the largest supercomputers, and these are expensive resources that cannot be arbitrarily allocated for repeated executions of the same problem. Others may require access to enormous datasets that cannot easily be transferred to the desktop of any researcher wishing to reexecute an analysis. But again, alternatives exist: it is possible to partially validate scaled versions of the largest problems against smaller runs created on the same supercomputing environments. Similarly, coarse resolution datasets can be used to conduct an analysis that may provide insights into the reliability of the full analysis. While not every quantity can be studied in this manner and there are deep research questions embedded in this problem, we should not consider this to be a paralyzing impediment to the quest for better computational reproducibility. Fortunately, the vast majority of research is conducted in smaller, simpler environments where full replication is feasible.

6.2.1 Computational Research Life Cycle

We advocate an integrated approach to computing where the entire life cycle of scientific research is considered, from the initial exploration of ideas and data to the presentation of final results. Schematically, this life cycle can be broken down into the following phases:

- *Individual exploration*: a single investigator tests an idea, algorithm, or question, likely with a small-scale test, dataset, or simulation.
- *Collaboration*: if the initial exploration appears promising, more often than not some kind of collaborative effort ensues to bring together complementary expertise from colleagues.
- *Production-scale execution*: large datasets and complex simulations often require the use of clusters, supercomputers, or cloud resources in parallel.
- *Publication*: whether as a paper or an internal report for discussion with colleagues, results need to be presented to others in a coherent form.
- *Education*: ultimately, research results become part of the corpus of a discipline that is shared with students and colleagues, thus seeding the next iteration in the cycle of research.

Before presenting our approach, we examine the typical patchwork of tools and approaches that researchers use to navigate these phases and discuss how the standard approach makes the goal of reproducibility nearly unattainable.

For **individual work**, researchers use various interactive computing environments: Microsoft Excel, MATLAB, Mathematica, Sage, and more specialized systems like R, SPSS, SAS, and STATA for statistics. These environments combine interactive, high-level programming languages with a rich set of numerical and visualization libraries. The impact of these environments cannot be overstated; researchers use them for rapid prototyping, interactive exploration, and data analysis, as well as visualization. However, they have limitations: (a) some of them are proprietary and/or expensive (Excel, MATLAB, Mathematica); (b) most (except for Sage) are focused on coding in a single, relatively slow, programming language; and (c) most (except for Sage and Mathematica) do not have a document format that is rich, that is, that can include text, equations, images, and video in addition to source code. While the use of proprietary tools is not a problem *per se* and may be a good solution in industry, it is a barrier to scientific collaboration and to the construction of a common scientific heritage where anyone can validate the work of others and build upon it. Scientists cannot share work unless all colleagues can purchase the same package; students are forced to work with black boxes they are legally prevented from inspecting. Furthermore, because of their limitations in performance and handling large, complex codebases, these tools are mostly used for prototyping: researchers eventually have to switch tools for building production systems.

For **collaboration**, researchers tend to use a mix of e-mail, version control systems (VCSs) and shared network folders (Dropbox, etc.). VCSs (see Section 6.3.1) are critically important in making research collaborative and reproducible. They allow groups to work collaboratively on documents and track how they evolve over time. Ideally, all aspects of computational research would be hosted on publicly available version control repositories, such as GitHub or Google Code. Unfortunately, the common approach is for researchers to e-mail documents to each other with *ad hoc* naming conventions that provide a poor man's version control (and are the source of endless confusion and frequent mistakes). This form of collaboration makes it nearly impossible to track the development of a large project and establish reproducible and testable workflows. While a small group can make it work, this approach most certainly does not scale beyond a few collaborators, as painfully experienced by anyone who has participated in the madness of a flurry of e-mail attachments with oddly named files such as `paper-final-v2-REALLY-FINAL-john-OCT9.doc`.

For **production-scale execution**, researchers typically turn away from the convenience of interactive computing environments to compiled code (C, C++, Fortran) and libraries for distributed and parallel processing (Hadoop, MPI). These tools are specialized enough that their mastery requires a substantial investment of time. We emphasize, that before production-scale computations begin, the researchers already have a working prototype in an interactive computing environment. Therefore, turning to new parallel tools means starting over and maintaining at least two versions of the code moving forward. Furthermore, data produced by the compiled version are often imported back into the interactive environment for visualization and analysis. The resulting back-and-forth workflow is nearly impossible to capture and put into VCSs, making the computational research difficult to reproduce. Obviously the alternative, taken by many, is simply to run the slow serial code for as long as it takes. This is hardly a solution to the reproducibility problem, as runtimes in the weeks or months become in practice single-shot efforts that no one will replicate.

For **publications** and **education**, researchers use tools such as LATEX, Google Docs, or Microsoft Word and PowerPoint. The most important attribute of these tools in this context is that, LATEX excepted, they integrate poorly with VCSs and are ill-suited for workflow automation. Digital artifacts (code, data, and visualizations) are often manually pasted into these documents, which easily leads to a divergence between the computational outcomes and the publication. The lack of automated integration requires manual updating, something that is error-prone and easy to forget.

From this perspective, we now draw a few lessons:

1. The common approaches and tools used today introduce discontinuities between the different stages of the scientific workflow. Forcing researchers to switch tools at each stage, which in turn makes it difficult to move fluidly back and forth.

2. A key element of the problem is the gap that exists between what we view as "final outcomes" of the scientific effort (papers and presentations that contain artifacts such as figures, tables, and other outcomes of the computation) and the pipeline that feeds these outcomes. Because most workflows involve a manual transfer of information (often with unrecorded changes along the way), the chances that these final outcomes match what the computational pipeline actually produces at any given time are low.

3. The problems listed earlier are *both* technical and social. While we largely focus on the tools aspect in this chapter, it is critical to understand that at the end of the day, only when researchers make a conscious decision to adopt improved work habits will we see substantial improvements on this problem. Obviously, higher-quality tools will make it easier and more appealing to adopt such changes; but other factors—from the inertia of ingrained habits to the pressure applied by the incentive models of modern research—are also at play.

Asking about reproducibility by the time a manuscript is ready for submission to a journal is simply too late: this problem must be tackled from the start, not as an afterthought tacked on at publication time. We must therefore look for approaches that allow researchers to fluidly move back and forth between the previous stages and that integrate naturally into their everyday practices of research, collaboration, and publishing, so that we can simultaneously address the technical and social aspects of this issue.

6.2.2 Open-Source Ecosystem

With the previous discussions in mind, our approach focuses on the need for tools and practices that enable researchers to naturally consider the entire cycle of research as a continuum and where "doing the right thing" is the easy and natural path rather than an awkward and cumbersome one. Rather than the haphazard patchwork of tools and processes described previously, we promote the development and adoption of a robust, open-source ecosystem that makes reproducible research a central aim.

To illustrate our point, we briefly describe the scientific Python ecosystem [23,26,30] and introduce a few core projects, which serve as examples throughout this chapter. While strong proponents of the Python programming language, we understand Python is not the only choice for scientific computing or reproducible research. Rather we consider the scientific Python ecosystem as a case study for the type of community-developed software stack that we believe necessary for improving the reliability and reproducibility of our computational results.

Initially written for teaching, the Python language has a simple, expressive, and accessible syntax that emphasizes code readability (see Section 6.3.4). Rather than imposing a single programming paradigm, it allows one to code at many levels of sophistication, including the procedural programming style familiar to many scientists. Python is available in an easily installable form for almost every platform and, therefore, ideal for a heterogeneous computing environment. It is also powerful enough to manage the complexity of large applications, supporting functional programming, object-oriented programming, generic programming, and metaprogramming. Due to excellent support for scripting tools written in other languages (including C, C++, Fortran, and R), Python is often used as an *integration language* for calling routines from a wide array of high-quality scientific libraries. Finally, it has an extensive standard library that provides built-in functionality for many tasks including database access, Internet protocols, data compression, and operating system services.

Importantly, from our perspective, Python is not specifically designed for scientific computing. So it is extremely capable at a diverse set of general purpose tasks. This benefits the scientific community, by providing an assortment of useful functionality and features while we focus on extending them with the specific features necessary for our research. While there are numerous libraries and extensions for scientific computing in Python, the three most widely used are NumPy,* SciPy,† and matplotlib.‡ NumPy [34] provides a high-level multidimensional array object and basic operations to manipulate them. SciPy is a collection of common numerical operations used in scientific computing. Matplotlib [14,15] is the standard 2D plotting library. In addition to these tools, there are even more specialized packages to provide advanced support and algorithms for machine learning, image processing, graph theory, symbolic mathematics, etc. On top of these general scientific libraries, there are even more domain-specific projects developed by those scientific communities. For instance, we are both members of the Neuroimaging in Python [24] community in addition to participating in the more general parts of the scientific Python software stack. The ability to participate and contribute at multiple levels of the toolchain is possible because of the adoption of common tools, standards, and procedures—many of which will be discussed in this chapter.

In addition to this stack of scientific software packages, we briefly introduce IPython,§ a system for interactive and parallel computing that has become the *de facto* standard environment for scientific computing and data analysis in the Python community. It was created by one of us (FP) in 2001 as an interactive command-line shell for Python and has evolved into a large collaborative open-source project with contributions from a broad team of scientists [29]. We call special attention to it as the natural focus of our integrated approach to the computational life cycle. As such, it will serve as a primary example throughout this chapter and will be discussed in detail in Section 6.5.3.

6.2.3 Communities of Practice

While the case can be made for the use of open-source software in science, even more important is the benefit that comes with open-source community-driven development practices. In community-developed projects, the distinction between users and developers is more fluid than it is in proprietary software projects where this distinction is not just expected but often rigorously enforced by legal mechanisms. This does not mean that everyone must become a *core developer*. There are still differing levels of contribution, which includes reporting issues, suggesting functionality, contributing enhancements, discussing use cases, and answering questions.

Communities of practice must drive the development of our scientific software [32]. A participatory community of active researchers using and contributing to the development of the code we depend on for our scientific output is necessary for robust software ecosystems where we can share and

---

* http://numpy.org.
† http://scipy.org.
‡ http://matplotlib.org.
§ http://ipython.org.

verify our work. As this work becomes more reliant on computational tools and techniques, the questions we can ask will be constrained by what our software can do and how easy it is to extend. Hence, moving a field forward will increasingly require scientists to be computationally literate, part of which includes embracing the tools and practices widely adopted by the open-source community.

There are real concerns that arise when attempting to transplant the practices of open-source development directly to computational research. The open-source development model is one where, in practice, the copyright and authorship of any large collaborative project is spread among many authors, possibly thousands. While the source control tools in use allow for a precise provenance analysis to be performed, this is rarely done and its success is contingent on the community having followed certain steps rigorously to ensure that attribution was correctly recorded during development.

This is not a major issue in open-source, as the rewards mechanisms tend to be more informal and based on the overall recognition of any one contributor in the community. Sometimes people contribute to open-source projects as part of their official work responsibilities, and in that case, a company can enact whatever policies it deems necessary; often contributions are made by volunteers for whom acknowledgment in the project's credits is sufficient recognition.

In the academic world, the authorship of scholarly articles in scientific journals and conference proceedings is currently the main driver of professional advancement and reward. In this system, the order of authorship matters enormously (with the many unpleasant consequences familiar to all of us), and so does the total number of authors in a publication. While in certain communities papers with thousands of authors do exist (experimental high-energy physics being the classic example), most scientists need the prominent visibility they can achieve in a short author list. The dilution of authorship resulting from a largely open collaborative development model is an important issue that must be addressed.

Furthermore, the notion of a fully open development model typical of open-source projects is at odds with another aspect of the scientific publication and reward system: the "first to publish" race. Many scientists are, understandably, leery of exposing their projects on an openly accessible website when in their embryonic stages. The fear of being scooped by others is real, and again we must properly address it as we consider how to apply the lessons of open-source development to the scientific context.

6.3 Routine Practice

The practices recommended in this section are distilled from writing and maintaining software, teaching programming courses to students and scientists, as well as extensive interaction and discussion with a diverse group of scientists and engineers. Whole books have been dedicated to best practices in software development with highly specialized tools and habits for individual programming languages and methodologies. In this short section, we highlight the practices and tools essential to any computational work. For a more detailed discussion, we recommend [1,13,17].

We begin by discussing practices and tools that should be applied to even exploratory, individual research. These practices are so essential to efficient and productive use of computational resources that we routinely use them whenever we use a computer. In Section 6.4, we discuss how these practices and tools extend to collaborative work.

6.3.1 Version Control

When collecting data, running analyses, or writing papers, you inevitably need to keep track of the various versions of your work: data is augmented and curated, code is adapted and improved, and writing is revised and expanded. While only keeping the most recent version of your work is possible, this is seldom sufficient. There are tentative new directions, detours, and dead ends.

We have witnessed numerous researchers attempting to manage different versions of their work using manual and laborious kludges. The most common patterns include using *ad hoc* naming schemes (e.g., `file.txt.bak` and `file.txt.1st`), e-mailing different versions to yourself, or using the application specific functionality such as Microsoft Word's "Track Changes" feature. While these approaches are partial solutions to the problem, they are also cumbersome, prone to failure, or limited to specific applications. More importantly, they are unsustainable beyond simple scenarios with only one or two files and do not scale to any kind of sensible collaboration workflow.

Because tracking and managing how work evolves over time is so fundamental to the workflow of software development, programmers have created specialized software tools to do exactly this. These tools are called version control systems or VCSs. Several open-source VCSs have been developed over the years, the most well known being CVS, SVN, Git, and Mercurial.

While there are notable differences among these tools, they all share some basic concepts. All project files (code, text, figures, etc.) are stored in a *repository* (often represented on disk in a directory hierarchy). There are commands to add to and remove files from a repository. To track changes to a file, it must be *committed* to the repository, ideally with a meaningful *commit message*. The repository and commit mechanism provide a complete historical log of the project from inception to current state, including every change made along with time stamps, author, comments, and other metadata for each modification.

Code changes may follow a linear progression of commits. However, it is more common for projects to include alternate development paths.* Given the exploratory nature of research, several approaches to a problem are often pursued simultaneously. In such cases, commits will resemble a tree with several *branches* diverging from a common base or trunk. When exploring these alternative approaches on different branches, several branches may eventually converge and need to be *merged* back together. If the changes in each of these branches do not overlap with one another, the VCS can merge them together in a completely automated fashion. When there are *conflicting* changes in different branches (e.g., edits to the same line of code), then manual intervention is required. But in all cases, a VCS is the only reasonable solution for managing the evolution of multiple branches of parallel development in a set of files (whether written documents, computer code, or data).

In the design of more modern VCS such as Git,[†] an important consideration is woven into the core of the system: built-in *data integrity verification* via cryptographically robust fingerprinting of all content. The basic idea is that at every commit, the VCS computes a "fingerprint" of the content being committed as well as the data it depended on.[‡] This makes it possible to establish the integrity of the entire history of a repository at any point, by computing these fingerprints and comparing them against the stored one. By the nature of hash functions, even small changes will result in new hashes. This key design idea is used by Git for all kinds of internal operations; but it also means that when a scientist gets a copy of a repository, he or she can be confident the content (including every recorded change) has not been tampered with in any way.

Strong guarantees on data integrity are a necessary condition of any reproducible workflow and one of the reasons why we emphasize so much the pervasive use of modern VCSs as the foundation of a reproducible research environment.

It is important to note that VCSs were developed for the management of human-generated content such as computer source code or text documents, not for the handling of large binary data that is common in science. By virtue of their design, they tend to be somewhat inefficient if you attempt to store all the changes in a project with many frequently changing large binary files, which somewhat limits their use for the tracking of all assets in a research

---

* This tendency becomes more pronounced in collaborative projects (see Section 6.4).

[†] From this point on, we will mainly focus on Git, which is our preferred VCS. It is also the one that is mostly widely used in the scientific Python community.

[‡] More precisely, a hash function is evaluated on the content of the commit and the hash of all commits it depends on, which creates a directed acyclic graph of hash values that signs the entire repository. Today, these systems employ the SHA1 hash function, but other hashes could be equally used if necessary.

project. But new efforts exist to mitigate these limitations, such as the git-annex* project, which uses Git for storing all metadata about large binary assets, along with a static (configurable) storage resource external to Git for the assets themselves. This approach makes it possible to smoothly integrate the management of binary data within a VCS workflow, without creating an explosion in the size of the VCS storage area.

The use of version control should become second nature; we routinely use it for everything—including the writing of this document.[†] We suggest researchers adopt a practice of pervasive version control: research codes, teaching materials, manuscripts, and data analysis projects should be developed, from the beginning, *always* using VCSs that track the actual history of everyone's contributions.

6.3.2 Execution Automation

Just as it is impossible to reproduce old results if you don't have access to the code and data that created them (hence the need for version control), it is equally impossible if you did not record somewhere how the code and data were used. You could write everything down and manually follow these instructions again later on, but a more sensible approach is to record them in a machine-readable way so that the computer can execute them. Furthermore, since most computational processes are a chain of executions where each step depends on the previous or on inputs that may have been modified, ideally you should be able to understand the structure of these dependencies and only run things when necessary.

Since building complex software with many source files is repetitive, full of detail, and time consuming, this is another task for which the software development world has developed powerful, automated solutions. The venerable `make` system is the workhorse of process automation [21]. It has a declarative syntax for expressing dependencies between sources and targets and a simple (timestamp-based) mechanism for resolving when dependencies need to be rebuilt. To get an idea how this works, consider the situation where a plot is created by a script, which reads a data file. In the parlance of `make`, the output `plot` is a *target* that depends on two *sources*—the data file and the script. If you type `make plot`, for example, `make` checks whether the script or data has been modified after the current plot was generated; if so, it calls the script on the data to generate an updated plot. In this simple scenario, using `make` does not offer much more than just running the script by hand. However, if the data this script consumes are generated by a chain of other scripts and data files, then the benefit of `make` becomes apparent.

---

* http://git-annex.branchable.com.
† http://github.com/fperez/repro-chapter-oss.

More modern systems also exist, and a detailed review of the options is beyond our scope. But whether running a sequence of scripts to produce some figures, compiling your software, or creating the final PDFs for a grant proposal, you should be able to do so by typing `make results` or the equivalent syntax in your system of choice. Once things are automated in this way, it becomes possible for others (humans or machines and even yourself on a new system or months later) to reliably repeat the process.

6.3.3 Testing

Computing is error-prone. While there is no foolproof way to rid computing of error, there are ways to limit and reduce it. One of the most successful and widely used techniques involves comprehensive testing, so that bugs (i.e., errors) are found quickly. Finding bugs as soon as possible in the development process is extremely valuable. Depending on the nature of the bug, it may reveal a fundamental problem with the overall design of your code requiring months more of coding. Even small errors that are easily fixed may require rerunning months of analysis. To reduce the amount of time it takes to uncover bugs and to ease the pain of debugging your code, it is essential to adopt a rigorous testing practice up front.*

Testing should be performed on multiple levels and begun as early as possible in the development process. For programs that accept input either from a user or file, it is important that the code validates the input is what it expects to receive. Tests that ensure individual code elements (e.g., functions, classes, and class methods) behave correctly are called *unit tests*. Writing unit tests early in the process of implementing new functionality helps you think about what you want a piece of code to do, rather than just how it does it. This practice improves code quality by focusing your attention on use cases rather than getting lost in implementation details. By thinking about the test at the outset, you can avoid finding that the code you just wrote is a huge, untestable mess. It also improves documentation because an example (i.e., the test case) is often better than an explanation. And if you regularly run the test, you will quickly know when your code no longer works for the example (something you may never notice in the case of explanatory text). Finally, unit testing leads to more robust code as you will more quickly isolate bugs, which makes them easier to fix [27].

Testing is mainly a language-specific pursuit (as it must be implemented in the programming language of a given project to be most effective). The authors are most familiar with the Python-based world, and [4] is a good hands-on starting point for the tool most widely used in scientific Python projects, namely, the nose[†] testing framework.

---

* While testing is an extremely useful practice, we should also point out that it is often more interesting work than debugging.
[†] http://nose.readthedocs.org.

6.3.4 Readability

While writing code that is well tested and systematically managed by a modern VCS is important, code that is not easy to read will be difficult to understand, correct, and modify. Readable code is written with explanatory names, clear logical structure, and comprehensive documentation where necessary. There is an extensive and growing literature on stylistic aspects of good programming [3,9,13,17,22]. Because scientific papers and grant submissions have become the currency of the scientific realm, many scientists have read classics such as Strunk and White's *Elements of Style*. Yet, even as an increasing amount of our work is produced in lines of code, there is a paucity of scientists paying the same attention to the elements of good programming style. The emphasis on readability is included in this section because even when you are the only one using or working on your code, the chance that you will need to read your own code is high. Even when your code is widely used and shared, you will still often be the one most frequently reading it.

Self-documenting code, as the name implies, reduces the need for external documentation by placing an emphasis on clear, well-written code that is easy to read and understand. In mathematics, it is an accepted practice to follow established naming conventions (e.g., capital letters for sets and lowercase letter for set elements). It is equally expected that when making a mathematical argument, one shouldn't arbitrarily switch from functional to relational notation. Similarly, using consistent and uniform naming conventions when programming should be a standard practice. Brevity in naming should be balanced against explicit and descriptive words. For example, you might use the term `download` rather than `get` in a function call to download a specific dataset from the Internet. Expressions are the next block to readability. While mathematical manipulation (e.g., De Morgan's laws) can be used to great effect in making your expressions more easily understood, it is often important to use the right level of abstraction. Higher-level programming languages (e.g., Python and R) provide data structures (such as n-dimensional vectors or statistical formulas) that enable the code to be more readily understood at the level of the mathematical ideas they implement. Finally, the overall control flow of your code must be clear and easy to follow. Finding the best control flow requires a deep understanding of your problem and an in-depth knowledge of programming methodology and the specifics of the language you are using. Like good writing, good coding is achieved through deliberate practice.

Inspired by the idea of self-documenting code, some argue that good code does not need comments. Indeed, liberally commenting your program to compensate for poorly written, obscure code is counterproductive. Comments that merely explain how a piece of code works add limited benefit. If code is so obscure to need explanation, it is better to revise or rewrite it. Another limitation of comments (as with many types of documentation) is that it is often uncoupled from the actual code. This means that there is no way to ensure that the two do not diverge. And, if they diverge, it may not be obvious which is correct.

To illustrate how comments and documentation *can* enhance readability of your code, we discuss the commenting and documentation system that has been developed by NumPy and is used by other scientific Python projects. While this section is specific to the tools and processes put in place in the scientific Python community, the general ideas are more broadly applicable.

In 2007, NumPy lacked good reference information for the various functions, classes, and modules it provided. Users and developers had access to the source code, a nearly 400-page "Guide to NumPy" and an active mailing list. Yet, it was clear that this level of documentation was not enough. To address this, the community began a yearlong effort to develop a *documentation string standard*.* In Python, a documentation string (or docstring) is any string in the first line in an object's (e.g., function and class) definition. Since docstrings are embedded in the source code, they are readily available to anyone directly viewing the source. When the code is executed, this string is associated with the object and can be programmatically accessed and used by introspection tools such as IPython. Docstrings can also be accessed for autogenerating documentation. Similar functionality exists in other programming languages such as R. Even in languages that don't include this functionality, it is common practice to include comments at the beginning of object definitions that are used similarly.

Given our desire for better documentation and wanting to leverage Python docstrings, the discussion focused on what they should include. Besides the information such as a brief statement of purpose as well as input and output parameters, we identified several issues with particular relevance for scientific applications. For instance, many algorithms had simple mathematical expressions that were not immediately obvious from their implementation in Python, but which a few equations written in LaTeX make clear. Often, our code implemented functionality described in peer-reviewed academic journals that could be referenced. Finally, we could provide short mini-examples of how to use the code. Since Python makes it easy to include examples in the docstrings as part of the test suite, this also improves test coverage and helps ensure that the documentation doesn't get out of sync with the actual code. These types of standards encourage contributors to explicitly think about input, output, equations, examples, and references. This, in turn, helps promote more deliberate and rigorous coding practices. And when reviewing code already written, having these details recorded aids in understanding whether the code is performing as expected.

---

* http://github.com/numpy/numpy/blob/master/doc/HOWTO_DOCUMENT.rst.txt.

6.3.5 Infrastructure

For small projects, managing everything by hand may be straightforward. But as your research project (code, data, and text) evolves, the burden of running your tests, building your project, and generating reports will become overwhelming. Eventually you will need tools and procedures in place to take care of these details for you. Even when the project is small enough that you can manually manage things, automating these tasks can be extremely beneficial [7].

We have often seen colleagues shy away from adopting certain practices related to the infrastructure that supports their computational research with claims of not having enough time or energy to invest in learning how to use them. This is a good example of being penny wise and pound foolish: a small initial investment in learning best practices pays off manyfold over time in increased productivity and smoother workflows that can support collaboration and scale to complex scenarios. The manual execution and repetition of common computational tasks may appear like an easy solution, but it is error-prone and impossible to apply reliably in collaborative settings beyond two or three people.

In the next section, we will discuss collaborative scenarios, but we want to address first how certain tools and practices have enormous value even for the individual researcher. And these are precisely the foundation that will then make it possible to naturally evolve a project from a single-person effort into a collaboration without a breakdown of complexity.

6.3.5.1 *Hosted Version Control*

While Git can be used purely locally, there are many advantages to having your repositories replicated on a server that is externally accessible. Git's design allows it to simultaneously keep track of multiple repositories tied to a single project, and it can synchronize and merge work between these multiple sources. Each of these sources is denoted a *remote* in Git lingo, and while a remote can be simply another location in your hard drive, the most useful kind of remotes are those that are physically in other computers. By synchronizing your local repository with an external remote, you simultaneously have an automatic backup of your entire project's history. But more importantly, this external remote is now available to synchronize with *other* computers, so you can cleanly and robustly synchronize multiple machines, even if you do independent development on each of them at some point.

There are many services online that host repositories from Git and other VCSs; in recent years, GitHub* has gained wide adoption among the community of scientists who write open-source software in Python and R. As discussed in Section 6.4.2, these systems truly shine once you use them to

---

* http://github.com.

collaborate with others, since collaboration hinges on the ability of multiple parties to synchronize their work.

6.3.5.2 Continuous Integration

Once your computational code is stored in version control repositories and, has tests and scripts that automate the execution of these tests, then it becomes possible to have a machine do this for you, all the time, bugging you only when something goes wrong. This is known as *continuous integration* (CI). CI systems are servers that grab the most recent version of a project from version control, execute the test suite, and gather statistics of this process. They are typically configured to log and summarize these results and to only produce alerts when something goes wrong (typically by e-mail, but more aggressive options such as SMS are possible). The amount of data collected during the test execution can be configured, so it is possible to have setups that range from a basic summary of success and failure to a detailed collection of metrics on the performance evolution of a codebase.

These systems are called *continuous* because they are meant to be used all the time: as the codebase evolves (typically when changes are committed to an official version control repository), the system fires automatically and collects its data. Therefore, these systems can also fulfill an important role: over time, they accumulate a *historical retrospective* of a project's evolution. And this is where the importance of collecting detailed metrics is realized: a CI system configured to do a fairly detailed analysis of a project when it runs becomes an invaluable tool to analyze what is happening over time. Is performance degrading in subtle ways that are not evident from day to day? Is the fraction of code that is tested (known as the *test coverage*) going down over time, indicating that new contributions are not being tested as thoroughly as the older code? Questions like these are impossible to answer in a manually managed workflow, yet they come *for free* once a few tools are set up, and can be an extremely important part of managing a healthy computational pipeline. A more detailed discussion of CI can be found in [5].

While a number of these tools exist, one of the most widely used by projects from many different programming languages and communities is called *Jenkins*.* Jenkins is a highly configurable CI system that can be run on a personal laptop or internal server and that is available hosted in the cloud as a service from a variety of sources. *Travis CI*† is a purely hosted CI system that, while not as configurable for fine-grained statistics as Jenkins, requires minimal setup, is free for open-source projects, and is tightly integrated with the version control hosting service GitHub.

---

* http://jenkins-ci.org.
† http://travis-ci.org.

6.3.5.3 Documentation Generation Systems

We have already discussed the importance of documenting your code, and in recent years, a number of systems have been developed that allow you to easily produce complex documentation that combines handwritten narrative sections with parts that are automatically extracted from the code. While the ability to automatically extract and generate documentation is valuable and important, we stress that it is critical that your projects have at least a modicum of *narrative* explaining the purpose of your tools, their scope, how to use them—with examples—and how the various concepts fit together. This kind of information cannot be gleaned from automatically extracted fragments that refer to individual function calls, and without it, your tools will be much less useful as a building block of robust scientific practice.

A number of tools exist for the generation of documentation, from the well-known LaTeX to systems focused on the generation of source-based documentation such as Doxygen[*] and newer ones designed for a combination of narrative and automatic documentation, like Sphinx.[†] There are other such systems, but Doxygen and Sphinx are widely used in the software world, actively developed and with rich toolkits that support complex documentation tasks.

Before discussing these tools, it is important to note that in recent years, new formats for *authoring* documentation have emerged, in particular reStructuredText[‡] (often abbreviated as reST) and Markdown.[§] These formats have slightly different philosophies, but they both aim at being more friendly to manual authoring and reading than LaTeX, while supporting more convenient integration with HTML output. They both share the basic philosophy of looking like plaintext with simple visual markup for commonly used tasks, for example, marking emphasis and boldface with asterisks (e.g., `*italics*` → *italics* and `**boldface**` → **boldface**). Markdown is aimed at the production of HTML and is a strict subset of HTML; it defines only a few special markup rules and leaves more complex tasks to be done by hand in pure HTML. In contrast, reST is a highly extensible format, where new commands (called "roles" and "directives") can be created and where the user can define entire new output pipelines by adding plugins written in Python to the processing stream. For example, the *SciPy Conference Proceedings*[¶] are written in reST and the PDF version is generated by a custom LaTeX translator written in Python.

So while Markdown is simple and easy for the production of simple HTML, it is not well suited to the generation of complex multipart documents with rich internal cross-referencing, bibliographic support, etc. Both

---

[*] http://doxygen.org.
[†] http://sphinx-doc.org.
[‡] http://docutils.sourceforge.net/rst.html.
[§] http://daringfireball.net/projects/markdown/syntax.
[¶] http://github.com/scipy/scipy_proceedings.

Markdown and reST support LaTeX for mathematical expressions, and with the right toolchain for rendering the output, they can generate a final PDF document that has been typeset by LaTeX. In talking about new documentation formats, it is important to mention the universal document converter, pandoc.* Pandoc is capable of translating between many document formats, including taking Markdown or reST input and producing HTML, LaTeX, and many other formats. It is an invaluable tool in managing a modern documentation workflow.

Returning our attention to systems that produce final output based on these formats, Doxygen has its own syntax that combines HTML with special commands for many tasks specific to computer source code, such as the specification of variable types, function arguments, and return values. It also supports Markdown, allowing users to use this more readable and concise syntax for the generation of common HTML markup. Sphinx, on the other hand, is designed around reStructuredText: it supports the basic format and provides a number of additional extensions aimed at the documentation of software projects. Sphinx was originally developed to produce the official documentation for the Python programming language but has become much more widely used. For instance, the SciPy community has developed an online wiki-like documentation editing system [33] on top of Sphinx that leverages the documentation standard discussed in Section 6.3.4, dramatically increasing the extent and quality of the NumPy and SciPy documentation.† Today, most Python projects use Sphinx as their documentation system, and because of the flexibility and extensibility of reST, it has also become widely used as a way of creating rich, complex documents with a strong computational base even beyond Python. There are even some statistics courses taught in R, which use Sphinx to create web-based notes with embedded R code and automatically generated output.‡

We note that all the formats we have discussed here, LaTeX, HTML, Markdown, and reST, share one critical feature: they can be handwritten in a plaintext editor, and they are stored in files amenable to version control with the tools described earlier. This stands in contrast to the binary formats of Microsoft Word and similar tools that lead to a terrible version control experience and which we avoid in computational workflows.

6.4 Collaboration

Open-source developers build on one another's work just as scientists build on each other's work. Since development communities are geographically

---

* http://johnmacfarlane.net/pandoc.
† http://docs.scipy.org/doc.
‡ http://www.stanford.edu/class/stats191.

spread and often dependent on contributions from volunteers, there has been careful attention paid to efficient and productive tools and processes for managing collaborations. As scientific practice becomes increasingly computational, it is imperative that we learn from the collaborative practices used in the open-source world.

6.4.1 Distributed Version Control

Earlier we discussed how VCS should be the foundation of a reproducible research workflow, even for a single investigator working in isolation. But the true power of these systems comes when considering the need to collaborate with others. Modern systems such as Git and Mercurial were designed from the ground up for large-scale distributed collaboration: Git was written by Linus Torvalds, the creator of the Linux kernel, to coordinate its development. The Linux kernel is arguably the largest and most complex open-source development project today: version 3.7 of the kernel included roughly 12,000 distinct sets of changes affecting over 1,100,000 lines of code by nearly 1,300 individual contributors.* Git's entire design aims to make collaboration on this scale smooth and efficient, and it succeeds admirably. Scientists can benefit from this power as well for any project that requires collaboration, whether it is the development of an open-source research code or the writing of a manuscript or grant proposal with multiple authors.

Git and other tools like it are called *distributed* version control systems (DVCSs) because they don't depend on a central server for their functioning, instead maintaining the entire history of a project inside every repository. This is in contrast to legacy systems such as CVS and SVN that made a distinction between the "working copy" that users would work on and which contained only the most recent version of files and a special repository hosted on a central server that had the entire project history. The centralized model does enable collaboration, but it also creates a number of problems that the distributed model addresses. In a DVCS, there is no single point of failure, as every repository carries all the project history and therefore serves as an automatic backup.

More importantly, a DVCS enables anyone who can *clone* a repository (the term used to indicate getting a full copy of an existing repository to start new work off of it) to develop their own history with new commits, even if they don't have write permission to the original source from where the repository was cloned. This means that once you have a clone of a repository (which could be someone else's or yours from a different computer), you can start working on that copy and building new history even if you are disconnected from the original system, such as when working on a plane or train without network access. If at a later stage you decide to merge your

---

* http://lwn.net/Articles/526748.

new history with the original repository, the merge capabilities in all DVCS make this straightforward.

This model of cloning an existing repository, building new history in isolation, and then merging it back into a common history is the basis for how these systems enable a fluid workflow for collaboration. When the time comes for a merge operation, DVCS can communicate the necessary changes even via e-mail attachments, but the simplest way to do so is to have a special repository* in a common location that all parties have access to and where the changes are *pushed*. Pushing, as the term suggests, means sending the set of changes from one repository into another; once the changes have been put into this central repository, all parties can *pull* them into their personal copies to synchronize their states and continue working again. So in practice, the simplest and most common collaboration workflow with a DVCS is one where each person has a copy they develop on, and they all connect in a star topology to a central node where a shared copy exists that is used for synchronization.

6.4.2 Code Review

In recent years, a number of web services have appeared that play the role of this central node; the most popular of them by far is GitHub,[†] but others such as BitBucket[‡] and Gitorious[§] play similar roles. GitHub has had a tremendous impact in the open-source community, reaching in a few years millions of active users and gaining rapidly popularity in scientific circles. We can attest to the power of this platform with our own experience: IPython moved its development to GitHub in early 2010 and immediately saw a rapid uptick in the pace of contributions. The workflow for collaboration enabled by GitHub was so much smoother than all previously available tools that many people were more willing to send contributions, while the core team was able to review and integrate these contributions at a much more rapid pace.

The core element of the collaborative process on GitHub is known as a *pull request*, and it is something akin to a public peer review of a set of changes to a manuscript. Let us illustrate how it works with a simple example: Alice wants to contribute to IPython, a project available on GitHub[¶] but to which she does not have write access. She can do so by getting her own personal copy of the IPython Git repository where she makes all the

---

* We note that this central repository does not change the distributed nature of the process: while it plays a special role for purposes of *synchronization*, the central repository is otherwise completely symmetrical to everyone's personal copy in the information it holds and can be replaced at any time in case it is lost or damaged from anyone's copy.
[†] http://github.com.
[‡] http://bitbucket.org.
[§] http://gitorious.org.
[¶] http://github.com/ipython.

changes she wants, and once she is ready to share them with the IPython team she can publish them on her GitHub user account.* At that point, she can click on a button to create a pull request for these changes: this contacts the IPython developers and creates a special page on the website that summarizes her changes as well as allowing everyone to begin a discussion about the changes. This discussion page allows the developers to ask Alice questions (even making comments on specific lines of her new code), and she can respond to these questions, update her code with new commits in order to address any required improvements, etc. Once the IPython developers are satisfied with this review and discussion (which may happen immediately or may require a lengthy back-and-forth process, depending on the changes), they can apply the changes to the official IPython repository with a click of a button. Once the changes are merged, they become part of the official project source and every individual commit that was merged is credited to Alice from the time she made it while she was working on her personal copy. Furthermore, even closed pull requests remain available on the website to inform future discussions, making the entire collaboration process an open one.†

The pull request process allows for a dynamic and open peer review process of all proposed changes to a project. The only special role that the official project authors have is the ability to approve the final merging of new changes, but otherwise everyone participates on an equal footing in terms of access to tools. This highly symmetrical structure proves to be extremely beneficial in encouraging a meritocratic process of contribution and review, where there are few points of special authority and where the discussions can remain focused around the contribution that initiated the pull request. Paraphrasing how some of the GitHub employees describe the process in public presentations: "a pull request is a conversation that starts with code."

From a scientific research perspective, we should consider these ideas in a broader context that goes beyond code: while peer review is one of the pillars of how the scientific community moves forward, in practice, modern scientific peer review is often an opaque, arbitrary, and limited process. The open, dynamic, and ongoing process of peer review enabled by the GitHub pull request system (or the equivalent ones that exist on other similar services) stands in sharp contrast to some of our institutional traditions, and our community could benefit significantly from adopting these ideas in our own review practices [11].

It is worth noting that by using a DVCS, authors can maintain private branches in the context of a publicly available project; this can be useful if new work needs to be developed in private prior to publication and subsequent public release. By tracking the public repository but keeping a private

---

* http://gitub.com/alice/ipython. if her GitHub user name is `alice`, for example.
† http://github.com/ipython/ipython/pull/1732. is an example of pull request and the entire, recorded review process.

branch, they can maintain exclusive access to their new work until it is published, while continuing to develop the openly accessible code with the rest of the scientific community. Once the code is ready to be made public, the new contributions can be seamlessly merged with the public version, and their entire provenance (including information such as time of invention and individual credit within the lab) becomes available for inspection. This simple observation shows how these tools can be used to balance the sometimes valid requests for privacy that may exist in a research environment with the desire for subsequent disclosure and publication, without losing any of the benefits of version control with regard to attribution and provenance tracking.

6.4.3 Infrastructure Redux

Once we have adopted tools that allow for distributed collaboration (e.g., Git and GitHub) and our computational machinery has tests and scripts that allow for automated installation and execution of the test suite, we achieve a number of important benefits that we can think of as *machine collaboration*. That is, once we have described in a standard way how our software must be installed or tested, then not only can our colleagues do that as they start collaborating with us, but so can machines. The Travis CI system, for example, can be configured to automatically run a project's test suite on every pull request created on GitHub. This means that when the humans come to review the proposed code, a report is already attached to the pull request that indicates whether the test suite passed or not (and provides details of any failures). This can save enormous amounts of time and make the collaboration much smoother, as reviewers don't need to wait before starting a new review for the tests to complete on their system and may even review when they are away from a development machine capable of actually running the tests. In the IPython project, we have seen the value of having this information always ready, as it reduces the small but persistent amount of "friction" we had before when each reviewer was responsible for running all tests first locally for each new pull request. While we still have tools for that and occasionally run tests beyond what Travis does (as Travis doesn't install every optional library we require), saving even 5 minutes for each review can make a huge difference for a project that sometimes has to process multiple pull requests in a day.

In a similar vein, the ReadTheDocs* project does for documentation what Travis does for CI. ReadTheDocs hosts documentation built with Sphinx but, more importantly, can be configured to automatically build it when new commits are made to the project at GitHub. In this way, users can always find a fully updated build of the project documentation without developers having to spend time on this.

---

* http://readthedocs.org.

Automated CI testing and documentation building are only two aspects of the benefits that can be gained from building on a foundation of distributed version control, well-automated processes, integrated test suites, and documentation generation. Once all these elements come together, a virtuous cycle can be sustained where the focus of the scientists or developers can be on producing new results (be they text, code, or computational outputs), and this machinery ensures that everything is validated and documented along the way.

While some of these points are more easily applied in the context of pure software development, the critical thing is how these ideas and tools work in concert to produce an environment of robust, reproducible results. Adopting this viewpoint, it is always possible to adapt to the specifics of any given project and apply only what is relevant. We conclude noting that the practices, tools, and ideas described in the previous two sections (Sections 6.3 and 6.4) may be put to use relatively quickly, but writing high-quality, trustworthy, scientific code is not easy. Mastery and expertise in developing reliable code that can be trusted to provide valid results takes sustained focus and deliberate practice.*

6.5 Communication

> Instead of imagining that our main task is to instruct a *computer* what to do, let us concentrate rather on explaining to *human beings* what we want a computer to do.
>
> *Literate programming (1984)*
> Donald Knuth

Whether engaging colleagues in data analysis, educating students about numerical algorithms, or publishing computational results, scientist need to ultimately convey their computational work to others—not just the artifacts of that work but the specific details of how those artifacts arose. We begin with a brief description of some existing tools for literate programming as a backdrop to present a more recent approach we refer to as literate computing. Again, our view is shaped by the desire to tackle the life cycle of computational research described in Section 6.2.1 in an integrated way. From this perspective, we argue that the literate computing approach is a better fit to the needs of reproducibility in computational research than traditional literate programming tools and will present the IPython Notebook as an example implementation.

---

* http://norvig.com/21-days.html. is a recommended reading.

6.5.1 Literate Programming

Donald Knuth proposed *literate programming* in the early 1980s, and a complete description of this approach to computer programming can be found in his later book of the same title [19]. Knuth's concern was the development of a better approach to documenting computer software; he devised a process whereby programmers would write literate source files that describe in full prose the ideas underlying a given program, interspersed with the code fragments implementing the actual computations. Knuth developed tools that can process these input files to produce two different representations: a *tangled* code file meant for compilation and execution by a computer and a *woven* file containing the formatted documentation. Knuth's original implementation, the WEB system [18], was focused on producing Pascal code and LaTeX documentation, but this basic idea has been extended to many other programming languages and documentation systems.

The R community has embraced the ideas behind literate programming, and a mature implementation of the concept exists for R in the Sweave system [20]. Sweave is one of the central elements of the Bioconductor system [8,10] for computational biology and bioinformatics. All Bioconductor packages must be accompanied by at least one *vignette*, a literate program that contains executable code illustrating the tasks the package is meant to perform. Vignettes can be read in PDF format, but functions exist to automatically extract all the R code for immediate execution. The journal *Biostatistics* encourages authors to use literate programming tools such as LaTeX and Sweave when submitting articles they wish to be designated reproducible [28].

A new entrant to the R community that is gaining rapid adoption is the knitr package.* Knitr can be seen as a highly evolved Sweave with a number of improvements, but still within the conceptual lineage of literate programming tools. The use of literate programming tools is gaining increasing traction in statistical education as well. For instance, at UC Berkeley, students taking computational classes in both the Statistics Department and the Division of Biostatistics are encouraged to use LaTeX with Sweave or LaTeX (or R Markdown) with knitr.

As the earlier examples suggest, literate programming has been most commonly adopted when the desired final document is intended primarily for human consumption. A few, if any, large software libraries are written this way. In fact, the most prevalent use of literate programming has been among scientists to communicate computational ideas and results to one another. These ideas have also influenced open-source software projects where tools have been developed to automatically generate project documentation based on source files and to create *live documentation* containing

---

* http://yihui.name/knitr.

the output from embedded code run during document generation (see Section 6.3.5).

6.5.2 Literate Computing

Tools described in the previous section for literate programming are mature and have been used to great effect to improve the quality of documentation in scientific programs and data analysis, especially in the R community. But they remain rooted in the original model proposed by Knuth, of authoring a literate file that is then postprocessed by various tools to produce either documentation or executable code.

In this section, we present an alternate approach to improving the connection between code and documentation that we refer to as *literate computing*. Our choice of terminology emphasizes the act of computing itself rather than the writing of code, as the systems we describe are all centered around *interactive environments* where the user can enter code for immediate execution, obtain results, and continue with more commands that produce new results based on the previous ones. A literate computing environment is one that allows users not only to execute commands but also to store in a literate document format the results of these commands along with figures and free-form text that can include formatted mathematical expressions. In practice, it can be seen as a blend of a command-line environment such as the Unix shell with a word processor, since the resulting documents can be read like text but contain blocks of code that were executed by the underlying computational system.

The earliest full-fledged implementation of these ideas is the graphical user interface of the Mathematica *Notebook* system, which dates back to early versions of Mathematica on the NeXT computer platform and took advantage of the superior graphical capabilities of NeXT. Today, a number of other systems (both open-source and proprietary) provide similar capabilities; on the open-source front, we notably mention the Maxima* symbolic computing package, the Sage[†] mathematical computing system, and the interactive computing project IPython, on which we will focus the rest of our discussion.

6.5.3 IPython Notebook

In 2011, a web-based notebook was developed in IPython that connects to the same interactive core as the original command-line shell but does so using a web browser as the user interface, automatically enabling either local or remote use as the system running the web browser can be different from that

---

* http://maxima.sourceforge.net.
[†] http://www.sagemath.org.

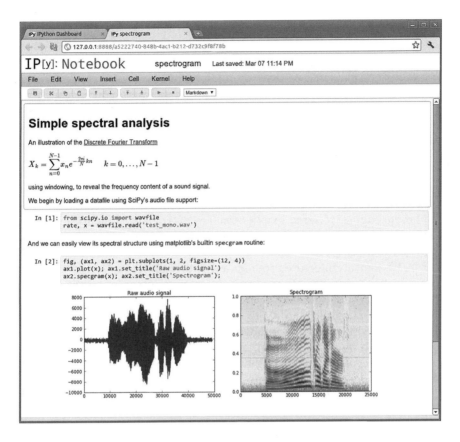

FIGURE 6.1
The web-based IPython Notebook combines explanatory text, mathematics, multimedia, code, and the results from executing the code.

executing the code, with all communication happening over the network. Figure 6.1 shows a typical notebook session with code, text, mathematics, and figures.

The driving idea behind the IPython Notebook is to enable researchers to move fluidly between all the phases of the research life cycle described in Section 6.2.1. If the environment where we conduct our exploratory research can also support all subsequent stages of this cycle and does so while smoothly integrating with the version control and process practices we've previously espoused, the likelihood that a final published result will be reproducible increases significantly. The Notebook system is designed around two central ideas: (a) an openly specified protocol to control an interactive computational engine and (b) an equally open format to record these interactions between the user and the computational engine, including the results produced by the computations.

Before diving into the specifics of these two ideas, we note that the previous design is independent of the Python language: while IPython started its life as a Python-specific project, the vision of the Notebook system is language-agnostic. First, while working in Python, users can mark the entire code blocks for execution via a separate language by using a special syntax on the block's first line: a user can, for example, start a block `%%R`, `%%octave`, `%%bash`, or `%%ruby` and IPython will execute the entire block with the respective system. The development community is also busy implementing similar support for new and experimental scientific languages such as Julia, enabling a user to control from a single IPython Notebook, a workflow that combines the most commonly used high-level languages in modern scientific computing. Second, an *entire notebook* can be executed in a different language if a remote engine (referred to as a *kernel*) exists that implements the interaction protocol. As of this writing, prototype kernels are being developed for Ruby, JavaScript, R, and Julia.

The IPython architecture provides a way to capture, version control, reexecute, and convert into other output formats, any computational session. Notebooks can be shared with colleagues in their native form for reexecution or converted into HTML, LaTeX, or PDF formats for reading and dissemination. They can be used in slideshow mode to give presentations that remain connected to a live computation and can be exported into plain scripts for traditional execution outside of the IPython framework.

The IPython protocol consists of messages in JSON (JavaScript Object Notation) format that encode all actions that an interactive user can request of a computational kernel, such as executing code, transferring data, or sending results, among many others. While this protocol is implemented in IPython, it can be independently implemented to provide new kernels also able to interact with the notebook interface and clients. The notebook file format is a simple JSON data structure that contains a series of one or more worksheets, each of which is a list of cells. A cell can contain either text or code, and code cells can also have the output corresponding to the execution. All substructures in the notebook format (the entire notebook, the worksheets, and the individual cells) have attached flexible metadata containers; this metadata can be used by postprocessing tools. The file format stores the communication protocol's payloads unmodified, so it can be thought of as a structured and filtered log (since the user chooses what to keep while working interactively) of the computation.

The IPython project has taken elements pioneered by the Mathematica and Sage Notebooks and created a generic protocol and file format to control and record literate computing sessions in any programming language. This was a deliberate choice in contrast to the literate programming approach: by providing a tool that operates close to the live workflow of research computing (in contrast to the batch-processing mode encouraged by classic literate programming tools), the resulting documents are immediately reproducible sessions that can be published in their own right or as companion materials to a traditional manuscript. Given how IPython also includes support for parallel computing, which we don't discuss here in the interest of conciseness, the system provides an end-to-end environment for the creation of reproducible research.

The real-world possibilities this offers were demonstrated during a collaboration in 2012 between the IPython team, a microbiology team led by Rob Knight from the University of Colorado and Greg Caporaso from the University of Northern Arizona, and Justin Riley from MIT who created the StarCluster* system for deployment and control of parallel resources on Amazon's EC2 cloud platform. As part of an NIH-funded workshop to explore the future of genomics data analysis in the cloud, this combined team collaborated on creating a fully parallelized analysis comparing the predictive behavior of different sizes and locations of gene sequence reads when reconstructing phylogenetic trees. The microbiologists had developed a serial prototype of this idea using their Qiime libraries [6], but a large-scale analysis with a full dataset would require roughly a month of CPU time on a single workstation. By locating the IPython Notebook server on Amazon cloud instances, the entire team was able to log into a single instance and by editing the code directly in the cloud, in a single day turn this prototype into a set of production Notebooks that would execute the analysis in parallel using multiple Amazon servers. Once the parallel code was tested, it became evident that there was not only an interesting example of using cloud technologies for rapid development of research ideas but also a biologically relevant finding; within a week, the team had completed a more extensive run using 24 hours of execution on 32 nodes and submitted a manuscript for publication [31]. This paper is now accompanied by all of the IPython Notebooks that enable any reader to fully reproduce our analysis, change parameters, and question our assumptions, without having to reimplement anything or be hampered by lack of access to the code and data. We have made available not only the final notebooks but also the Amazon Virtual Machine Images (data files that represent a virtual computer on Amazon's cloud platform), so that the entire analysis can literally be reexecuted under identical conditions by anyone with an Amazon account.

This example, anecdotal as it may be, indicates the validity of the vision we propose here: that by providing tools that encompass the entire cycle of research, from exploration to large-scale parallel production and publication, we can provide the scientific community with results that are immediately accessible to others and reproducible, seeding the continued evolution of the research process.

---

* http://star.mit.edu/cluster.

The IPython project has also developed tools to make it easy to share and disseminate content created as notebooks in a variety of forms. The Notebook Viewer* is an online service that renders *any* publicly available IPython Notebook as a web page. This enables users to share notebooks by simply putting them online and pointing colleagues to the rendered web page. The same technology that powers the Notebook Viewer service can also generate HTML files suitable for inclusion in other websites, in particular, blogs. Since a lot of rapid technical communication is happening today on the Internet via blogs, this is an important aspect of linking reproducible research to the rapid feedback cycle of web-based discussion. With a single command, a user can convert a Notebook file into HTML ready for posting to a blog, and this is already being used by scientists to write both short technical posts and also more complex materials: Jose Unpingco, a researcher with the US Department of Defense, is currently working on a book titled *Python for Signal Processing*, and this book is available during writing as a GitHub repository.[†] This repository contains a series of IPython Notebooks so that readers can directly execute the code in the book, and they are also being published as a series of blog posts as they become available,[‡] so readers can comment and discuss with the author throughout the process of book development, and they can do so based directly on the actual code that creates all the examples in the book.

The signal processing book is, to our knowledge, the first example of a full book being written as a collection of executable IPython Notebooks, but this follows a tradition created by Mathematica, whose documentation is itself a collection of executable notebooks. Furthermore, in recent years, Rob Beezer, from the University of Puget Sound, has developed a popular Introductory Linear Algebra book [2] that is based on the Sage system and also combines the mathematics and text with code that can be directly executed and modified by the readers. This ability to "close the loop" between what the authors had on their screens and what their readers can execute themselves is an important element of the movement toward reproducibility in research.

As a concrete implementation of the ideas of reproducible research using the tools we've described in this chapter, during the ongoing process of research itself, we can point to work being carried by a collaboration where one of us (FP) is a member, on novel ways to model the mathematical structure of the signal generated by MRI devices in the imaging of water diffusion in the brain. This work, as yet unpublished, is being developed as an open repository on GitHub[§] where all code for our research is posted during writing, all computational experiments are created as IPython Notebooks,

---

* http://nbviewer.org.
[†] http://github.com/unpingco/Python-for-Signal-Processing.
[‡] http://python-for-signal-processing.blogspot.com.
[§] http://github.com/fperez/spheredwi.

and submitted manuscripts are created directly from the code and notebooks (along with additional narrative written by hand).

The aforementioned tools are also playing a central role in the last stage of the computational research life cycle, education. We will increase our chance that the next generation of scientists adopts improved reproducibility practices if we educate them with the same tools that we use for everyday research, and a couple of modern efforts that aim to bring improved computational literacy to scientific research have adopted the IPython Notebook. Software Carpentry* is a project funded by the Alfred P. Sloan Foundation and led by Greg Wilson at the Mozilla Foundation whose motto is Richard Feynman's famous "What I cannot create, I do not understand." They produce, with rigorous follow-up and assessment, workshops aimed at working scientists (typically graduate students and postdoctoral researchers, but always open to broad audiences) and whose purpose is to instill in them a collection of skills and best practices for effectively using computing as a daily research tool. The Software Carpentry workshops cover topics ranging from the basics of the Unix shell to version control, Makefile automation of processes, and basics of scientific Python including data analysis and visualization. They have recently adopted the IPython Notebook as the base system for teaching the scientific Python parts of their curricula and provide the IPython team with direct feedback on its strengths and weaknesses as an educational tool. In a similar vein, Josh Bloom from the astronomy department at UC Berkeley has led, for a number of years, 3-day workshops on the use of Python as a tool for scientific computing.[†] These are open to the entire campus community and followed by an optional for-credit seminar where students learn more advanced skills for using Python as a research tool. F. Pérez and other members of the IPython team at UC Berkeley regularly lecture in the bootcamps and courses, where the notebook is the means for the delivery of course materials and interactive lecturing. While we have identified a number of weaknesses and areas for improvement, we have also found this environment to be markedly superior to all previous tools we had used in the past for teaching in similar contexts.

As these capabilities in IPython reach wider usage, with scientists now developing complete books and lecture series based on the system, we are considering a number of new challenges and questions introduced by these capabilities. The interactive computing model is a fluid and natural one, but we need to find ways to extend it into the development of longer-term production codes that are robust, documented, tested, and integrated into reusable libraries. This means bridging the gap between a *scripting* mentality and a *developer* one, and while we have already made progress on that front in IPython, many questions remain open for the future.

---

* http://software-carpentry.org.
[†] http://pythonbootcamp.info.

6.6 Conclusion

As research grows increasingly dependent on computing, it becomes critical for our computational resources to be developed with the same rigor, review, and access, as the results they support. In particular, we believe that reproducibility in computational research requires (1) sharing of scientific software, data, and knowledge necessary for reproducible research; (2) readable, tested, validated, and documented software as the basis for reliable scientific outcomes; (3) high standards of computational literacy in the education of mathematicians, scientists, and engineers; and (4) open-source software developed by collaborative, meritocratic communities of practice.

Achieving these goals will not be easy. It requires changing the educational process for new scientists, the incentive models for promotions and rewards, the publication system [25], and more. In this chapter, we focused on the need for an open-source ecosystem for scientific computing developed by communities of practice. We then introduced several tools and practices necessary—but not sufficient—for reliable code that can be the basis of reproducible research. We illustrated these ideas with examples of how they have been applied and advanced in the open-source scientific Python community. Finally, we presented the IPython project's powerful combination of interactivity, distributed and remote computing features, and literate computing functionality as a natural integration point for the computational research life cycle in order to make it more fluid, efficient, and reproducible.

We emphasize that the mechanical reproduction of computational results is not an end in itself. The ultimate goal is to bring the rigor, openness, culture of validation and collaboration, as well as other aspects of reproducible research to our everyday computational practices. This is not a goal we will happily attain one day and then move on to pursue another; it must become and remain an ongoing part of our scientific practice.

Acknowledgments

We thank all the members of the scientific Python community as well as the many scientists from various labs whose work and ideas have inspired what we have presented here. John D. Hunter, to whom this chapter is dedicated, created the matplotlib graphics library that has been a central pillar of the scientific Python ecosystem; his tragic early passing in 2012 was a personal blow to the authors as well as a loss to our community. Brian Granger and Min Ragan-Kelley have worked closely with FP on the development of IPython for a number of years and are responsible for many of the ideas that the project embodies. Matthew Brett has through long collaboration and patient discussion helped clarify and refine many of the ideas presented in this chapter. Titus Brown, Vincent Carey, Paul Ivanov, Jean-Baptiste Poline, and Stéfan van der Walt provided valuable feedback on drafts of this chapter.

References

1. D. A. Aruliah, C. T. Brown, N. P. Chue Hong, M. Davis, R. T. Guy, S. H. D. Haddock, K. Hu et al. Best practices for scientific computing. arXiv:1210.0530, September 2012.
2. R. A. Beezer. *A First Course in Linear Algebra*. Congruent Press, 2012 http://linear.ups.edu
3. D. Boswell and T. Foucher. *The Art of Readable Code*. O'Reilly Media, Sebastopol, CA, 2011.
4. C. T. Brown. *An Extended Introduction to the Nose Unit Testing Framework*. 2006.
5. C. T. Brown and R. Canino-Koning. Continuous integration. In A. Brown and G. Wilson, eds., *The Architecture of Open Source Applications*, pp. 77–89. Creative Commons, Mountain View, CA, June 1, 2011.
6. J. G. Caporaso, J. Kuczynski, J. Stombaugh, K. Bittinger, F. D. Bushman, E. K. Costello, N. Fierer et al. Qiime allows analysis of highthroughput community sequencing data. *Nature Methods*, 7(5):335–336, 2010.
7. M. Doar. *Practical Development Environments*. O'Reilly Media, Sebastopol, CA, 2005.
8. S. Dudoit, R. C. Gentleman, and J. Quackenbush. Open source software for the analysis of microarray data. *Biotechniques*, 34(13):S45–S51, 2003.
9. M. Fowler. *Refactoring: Improving the Design of Existing Code*. The Addison-Wesley Object Technology Series. Addison-Wesley, Reading, MA, 2000.
10. R. C. Gentleman, V. J. Carey, D. M. Bates, B. Bolstad, M. Dettling, S. Dudoit, B. Ellis et al. Bioconductor: Open software development for computational biology and bioinformatics. *Genome Biology*, 5(10): R80, 2004.
11. S. S. Ghosh, A. Klein, B. Avants, and K. J. Millman. Learning from open source software projects to improve scientific review. *Frontiers in Computational Neuroscience*, 6:18, 2012.
12. T. Hey, S. Tansley, and K. Tolle, eds. *The Fourth Paradigm: Data-Intensive Scientific Discovery*. Microsoft Research, Redmond, WA, 2009.
13. A. Hunt and D. Thomas. *The Pragmatic Programmer: From Journeyman to Master*. Addison-Wesley, Reading, MA, 2000.

14. J. D. Hunter. Matplotlib: A 2D graphics environnent. *Computing in Science & Engineering*, 9(3):90–95, May–June 2007.

15. J. D. Hunter and M. Droetboom. Matplotlib. In A. Brown and G. Wilson, eds., *The Architecture of Open Source Applications*, Vol. II, pp. 165–178, 2012.

16. D. Joyner and W. Stein. Open source mathematical software. *Notices of the American Mathematical Society*, 54(10):1279, 2007.

17. B. W. Kernighan and R. Pike. *The Practice of Programming*. Addison-Wesley Professional, Boston, MA, 1999.

18. D. E. Knuth. The WEB system of structured documentation. Stanford Computer Science Report CS980, Stanford University, Stanford, CA, September 1983.

19. D. E. Knuth. Literate programming. CSLI Lecture Notes Number 27. Stanford Center for the Study of Language and Information (CSLI), Stanford, CA, 1992.

20. F. Leisch. Sweave: Dynamic generation of statistical reports using literate data analysis. In W. Härdle and B. Rönz, eds., *Compstat 2002—Proceedings in Computational Statistics*, pp. 575–580. Physica Verlag, Heidelberg, Germany, 2002.

21. R. Mecklenburg. *Managing Projects with GNU Make*, 3rd ed. O'Reilly Media, Sebastopol, CA, November 2004.

22. S. McConnell. *Complete: A Practical Handbook of Software Construction*, 2nd ed. Microsoft Press, Redmond, WA, 2009.

23. K. J. Millman and M. Aivazis. Python for scientists and engineers. *Computing in Science & Engineering*, 13(2):9–12, March–April 2011.

24. K. J. Millman and M. Brett. Analysis of functional magnetic resonance imaging in python. *Computing in Science & Engineering*, 9(3):52–55, May–June 2007.

25. C. Neylon, J. Aerts, C. T. Brown, D. Lemire, K. J. Millman, P. Murray-Rust, F. Pérez et al. Changing computational research. The challenges ahead. *Source Code for Biology and Medicine*, 7(1):2, 2012.

26. T. E. Oliphant. Python for scientific computing. *Computing in Science & Engineering*, 9(3):10–20, May–June 2007.

27. A. Oram and G. Wilson. *Making Software: What Really Works, and Why We Believe It*. O'Reilly Media, Sebastopol, CA, 2010.

28. R. D. Peng. Reproducible research and biostatistics. *Biostatistics*, 10(3): 405–408, 2009.

29. F. Pérez and B. E. Granger. IPython: A system for interactive scientific computing. *Computing in Science & Engineering*, 9(3):21–29, May–June 2007.

30. F. Pérez, B. E. Granger, and J. D. Hunter. Python: An ecosystem for scientific computing. *Computing in Science & Engineering*, 13(2):13–21, March–April 2011.

31. B. Ragan-Kelley, W. A. Walters, D. McDonald, J. Riley, B. E. Granger, A. Gonzalez, R. Knight et al. Collaborative cloud-enabled tools allow

rapid, reproducible biological insights. *International Society for Microbial Ecology Journal*, 7(3):461–464, 2012. http://qiime.org/home_static/nih-cloud-apr2012

32. M. J. Turk. How to scale a code in the human dimension. arXiv:1301.7064, 2013.

33. S. J. van der Walt. The SciPy Documentation Project (technical overview). In G. Varoquaux, T. Vaught, and K. J. Millman, eds., *Proceedings of the 7th Python in Science Conference*, pp. 27–28, Pasadena, CA, 2008.

34. S. J. van der Walt, S. C. Colbert, and G. Varoquaux. The NumPy array: A structure for efficient numerical computation. *Computing in Science & Engineering*, 13(2):22–30, March–April 2011.

7

Reproducible Bioinformatics Research for Biologists

Likit Preeyanon*, Alexis Black Pyrkosz*, and C. Titus Brown

CONTENTS

---

* These authors contributed equally

7.1 Introduction

7.1.1 Computational Analysis in the Pregenomic Era

At the dawn of computational biology in the 1960s, datasets were small. Protein sequences were first distributed in the printed Dayhoff atlases [29] and later on CD-ROM, with bioinformaticians eyeballing entire datasets and shuffling data by hand. By the 1990s, bioinformaticians were using spreadsheet programs and scientific software packages to analyze increasingly large datasets that included several phage and bacterial genomes. In 2003, the pregenomic era ended with the online publication of the human genome [7,14,26] and the National Institutes of Health invested heavily in sequencing related organisms to aid in annotation. By the mid-2000s, Sanger sequencing was replaced by faster and cheaper next-generation sequencing technologies, resulting in an explosion of data, with bioinformaticians racing to develop automated and scalable computational tools to analyze and mine it [3].

7.1.2 Computational Analysis in the Era of Next-Generation Sequencing

As sequencing becomes ever more affordable, the grand genomic and transcriptomic datasets that were the dream of many pregenomic era biologists have become commonplace. A single experiment in a small research lab can inform on thousands of genes or entire genomes, and small genomes can be sequenced and assembled in a few hours. Initiatives such as ENCODE [4,19], 1000 Genomes Project [2], Cancer Genome Project [10], Human Microbiome Project [15], Eukaryotic Pathogen and Disease Vector Sequencing Project [20], Clinical Sequencing Exploratory Research [23], Centers for Mendelian Genomics [22], Environmental Genome Project [21], and HapMap Project [24] make vast amounts of data readily available for download and analysis. As the field comes closer to achieving the $1000 genome [16], waves of individual genomes will inundate the public databases, providing a rich information source for researchers to analyze with a wide array of tools. Further, proteomics, metabolomics, medical imaging, environmental conditions, and many other kinds of data are becoming readily available for integration. As scientists continue to push the edge of data analysis and integration, the integration of these different data types is increasingly required. The field has advanced far from the eye/hand methods of the pregenomic era and outstripped the spreadsheets and single software packages of the early postgenomic era. Modern computational analyses are a major part of biological studies [32] and require analyzing gigabytes or terabytes of data in complex computational pipelines, which typically involve running several distinct programs with custom scripts porting data between them. These pipelines start from quality control of raw data (or by downloading primary data from public databases) and pass the data through many steps of calculation, validation, and statistics. They end with summarization and graphical visualization to aid end-users in comprehending the complex results. In short, modern biological studies require datasets that are so large, that scientists must use advanced computational tools to perform useful analyses.

Genomics has expanded the drivers of science from hypothesis (devise a question and design/conduct experiments in response) to include discovery (sifting through large datasets in search of patterns). With this greater emphasis on statistical analyses of large datasets and data-driven modeling, even wet-lab biologists are increasingly finding themselves at the computer instead of the bench [31]. However, many biologists lack a strong background in mathematics or computer science [6,27], and struggle to transition from a graphical computer desktop environment to the command-line interface required for many analyses. Further, while they usually have been trained in good wet-lab practice, they often have minimal experience with computational practice and lack the knowledge necessary to efficiently perform high-quality reproducible computational research [13,31]. Effectively, many biologists lack the computational skills they need to perform modern biological studies.

7.1.3 Concerns in Bioinformatic Research

Biologists' lack of computational experience is a significant hindrance as biology expands to include data- and model-driven methodologies. While the obvious solution is to limit the computational aspect of biology to trained bioinformaticians and computer scientists, this is impractical for two reasons: not enough skilled bioinformaticians are available (only a minuscule percentage of US universities have bioinformatics undergraduate programs), and many computer scientists are uninterested/uneducated in science. Moreover, the substantial background in biology required to make appropriate use of data blocks computer scientists from quickly moving into bioinformatics. As a result, wet-lab biologists have begun to venture into computation in increasing numbers [8,9]. They are usually long on data and short on time, so they focus on learning the computational tools needed to analyze their specific data, concentrating on rapidly processing data with the tools as opposed to understanding the tools' underlying assumptions. Even more troubling, there is a cultural and social gap because many labs and programs do not consider bioinformatics essential for their biologists [13]. A researcher seeking to analyze large datasets when few or none of his/her coworkers or superiors have computational expertise may have no clue where to begin, and be given very little time to find or develop appropriate tools. This situation fosters a dangerously ad hoc approach to bioinformatics.

The effects of this lack of expertise can be dire:

1. Many researchers download computational tools from the Internet or collaborators and use them on large datasets without first running a known test set (the computational equivalent of a control). Many programs contain technical or scientific errors that will be readily apparent when running test sets, but will be missed otherwise [9]. Errors will be carried into downstream analyses, costing hundreds of hours of compute and bench time, and potentially requiring retraction of papers when the errors are caught [31].

2. Many tools only run on the command line, are difficult to install, lack documentation, etc., and therefore software may be selected based on ease of use rather than accuracy and scientific relevance.

3. With the trial-and-error approach used to create custom pipelines, biologists can lose track of which tools they ran, the order in which they ran them, and the parameter sets used for each. Many biologists

have not carried the standard scientific practice of painstakingly recording wet-lab procedures in laboratory notebooks over into their computational research.

4. Many biologists use software with the default parameters. The defaults are frequently selected by the original programmers to optimize processing of the original test data or were based on a set of assumptions that was correct for the original study, but may not be appropriate for the different biologist's data or research question. While some parameters are relatively insensitive such that the defaults are sufficient, others will produce widely different results if varied slightly. A single parameter can be the difference between one group's results being correct and another's being wrong.

5. Those biologists who program their own tools must decide how to release and support their code. Some labs post their software on a website but rarely update it. Frequently, as soon as a programmer leaves the lab, their code becomes unsupported and joins the online graveyard of dead and obsolete code. Some labs refuse to release their code [18], leading reviewers and collaborators to wonder if inaccuracies are being hidden.

The good news is that all of these problems can be addressed using tools and practices that are already available.

7.1.4 Reproducible Research Is Attainable in Bioinformatics Using Modern Tools

Many tools for reproducible computational research exist and are already being used in computer science, physics, and engineering. These tools are routinely used to quality control the data analysis process, facilitate useful collaborations, and maintain laboriously developed programs and pipelines in the long term. While these tools may be new to many biologists, they have been in production for many years and are well-tested with tutorials and online documentation. Investing time to learn the tools and establishing good habits of using them yields a larger benefit: errors are consistently detected and corrected early instead of being discovered only after time-consuming downstream analyses and attempted wet-lab verification (or after a paper has been submitted or published [17]). The more computational methods employed by a laboratory, the more essential the tools are to efficiency, correctness, and reproducibility [30]. Further, while use of these tools is currently optional in biology [12], researchers can expect that within a decade, most journals and granting agencies will require the appropriate use of tools and methods in computational biology research; the National Science Foundation already requires detailed discussion of data management.

7.1.5 Guidelines for Getting Started

Our goal is to help those biologists who have zero or little background in computation to get started with good practices and tools for computational science. The following sections are structured to provide introductory knowledge for those biologists venturing into bioinformatics and the command-line interface for the first time; intermediate knowledge for those biologists ready to start programming; advanced techniques for seasoned programmers for improving programs and automating pipelines; and a related tools section that names tools and concepts that readers can seek out once they have established a basic foundation for reproducible research.

Note: The following sections give overviews and simple examples of the tools, but readers are encouraged to use the resources listed at the end of the chapter to find specific information and step-by-step tutorials. Essentially, this chapter tells readers about existing tools and why they are important to use, but is not itself a course in computational research. If readers are interested in more hands-on experience with some of these topics, the Software Carpentry project (http://software-carpentry.org) offers free online videos as well as 2-day workshops in these and related areas.

7.2 Beginner

Here, we discuss simple practices that beginners can use to establish a strong foundation for making their computational research reproducible, emphasizing those that are practical for scientists who primarily run other people's pipelines and are making the switch from a graphical user interface (GUI) to the command line. We describe basic practices such as working on the command line and selecting a text editor.

7.2.1 Computing Environment: The New Benchtop

Just as much wet-lab biology work is done on a lab bench with routinely available tools such as micropipets, shakers, and spectroscopes, computational work is usually performed on a computer with routinely available data parsing and analysis tools. In this section, we will discuss computing environments (also sometimes called operating systems or platforms) and the general tools that stand ready on the computational benchtop.

7.2.1.1 UNIX/Linux Operating System

Most biologists are aware of two primary computing platforms available: Windows and UNIX systems (which include Linux and Mac OS X). In the United States, most people learn basic computer skills on Windows machines. However, developers of bioinformatics software primarily use

UNIX systems because of the large existing ecosystem of tools, most of which are open source—meaning that anyone can freely use, modify, and redistribute resources under open-source licenses. The open source or free-software community has long attracted programmers and other technically oriented people who creatively solve problems. The Free Software Foundation [25] has numerous open-source collections of development tools, libraries, licenses, and applications for the GNU/Linux system. Consequently, biologists can get an operating system and all the bioinformatics tools they need for free for all of their computers and clusters. Further, UNIX systems have traditionally been easier to use remotely than Windows machines, whether the machine is a desktop computer in another room, an institution's high-performance compute cluster, or the cloud.

Making the switch from GUI-based software to command-line software is useful because the power to remix and combine tools at the command line outstrips that of most GUI-based software; as an analogy, GUI-based software is similar to a lab that relies exclusively on commercial kits, whereas the command line is like a lab that is also equipped with chemicals and instruments that can be used to supplement the kits or develop novel techniques. Also, many GUI-based software packages require users to manually click through an analysis, while the command line can be used to write complete instructions for an analysis and run many datasets simultaneously. Automation ensures that each dataset is run according to the same instructions (avoiding human error) and unshackles the biologist from the computer. Some GUI and web-based programs are available for those biologists who want to build pipelines without using the command line (such as Taverna, Pipeline Pilot, and Galaxy). Further, most cutting-edge bioinformatics tools lack a GUI, partly because building GUIs is a time-consuming task that is difficult to fund. As a result, biologists need basic command line navigation skills to use the latest bioinformatics software. Biologists who use Windows but wish to take advantage of this cache of software and tools can download and install a free program such as Cygwin (or MSYS+Mingw32 or Microsoft Interix) to emulate a UNIX system on their Windows machine, or can use PowerShell.

Note: Modern Macintosh machines are a good compromise for modern research labs because they have a UNIX-based system with a friendly GUI for casual use.

7.2.1.2 UNIX Tools: The New Benchtop Tools

Just as wet-lab biologists use simple tools like pipettes, centrifuges, incubators, and gel boxes individually and then combine them to perform a specific procedure, so bioinformaticians use simple UNIX tools that each perform a specific task and string them together into a pipeline. Many UNIX tools come preinstalled on UNIX systems or are freely available online. These tools are invaluable for the beginning bioinformatician, particularly if that researcher has no programming experience, because they will perform tasks with speed, customizability, and reliability that custom scripts cannot easily match. Here, we introduce some of the most basic, useful UNIX tools for bioinformatics.

7.2.1.2.1 Shell

The shell is a language as well as interpreter that reads and interprets commands from a user. Any biologist who has opened a terminal or command-line interface and typed a command has used the shell. There are several types of shells available for UNIX, but bash is predominant. Bioinformaticians often use shell commands to run tools and automate tasks such as running pipelines, backing up data, and submitting jobs on a computer cluster. While shell languages can be used to develop full-fledged programs, this can be time consuming because the shell is designed around operating system tasks rather than data analysis. Bioinformaticians usually use shell commands to perform routine tasks such as sorting large datasets, searching for specific data in a group of files, or sifting through a large log file and printing only the data relevant to a given project; they write programs to perform more complex tasks. Detailed next are some of the most general and useful tools on UNIX systems.

7.2.1.2.2 grep, sed, cut, and awk

grep, *sed*, *cut*, and *awk* are tools for parsing text files. *grep* is used to quickly search through a text file for a given word or sequence motif, similar to using a *find all* command in a word processor. *sed* is useful for replacing words or phrases, similar to using the *find and replace* command in a word processor. *cut* is used for selecting a column of data in a text file. *awk*, among many other uses, can search through files containing many columns of data and only print those lines or columns that are needed for a given application. Each tool is useful when the user needs to perform a single task quickly, create a simple pipeline to accomplish a combination of simple tasks, or process files that are too large to open in a spreadsheet program. *grep*, *sed*, and *awk* understand regular expression syntax (covered in Section 7.5), which offers more robust pattern searching options.

7.2.1.2.3 apropos and man

apropos is a program that displays a list of programs related to a keyword. It is useful for biologists who need to find a tool to perform a specific function without knowing the tool's name. For example, if a biologist wanted to archive files, he or she might use the *apropos* command to search for an appropriate tool:

```
$ apropos archive
```

Note: $ indicates the command prompt or where the user would begin typing. The user would not actually type the $.

The output will vary depending on system, but should look like:

```
jar(1) - Java archive tool
libtool(1) - create libraries ranlib - add or update the table
of contents of archive libraries
tar(1) - manipulate tape archives
unzip(1) - list, test and extract compressed files in a
ZIP archive...
```

To learn more about each program, the biologist can look at the standard manual for each program using a *man* command (*man* is short for manual).

```
$ man tar
```

In this case, a *man* command will display a standard manual page, which typically includes the name of the program, synopsis, detailed description, and options, as well as some examples. Here is an example of the first few lines of the standard manual for the tar program:

```
NAME
    tar -- manipulate tape archives

SYNOPSIS
    tar [bundled-flags <args>] [<file> | <pattern> ...] tar -c
[options] [files | directories] tar -r | -u -f archive-file
[options] [files | directories] tar -t | -x [options]
[patterns]

DESCRIPTION
    tar creates and manipulates streaming archive files. This
implementation can extract from tar, pax, cpio, zip, jar, ar,
and ISO 9660 cdrom images and can create tar, pax, cpio, ar,
and shar archives...
```

Note: apropos and man commands serve as a reference, not a tutorial on how to use a particular command. Biologists may need to search in Google, Wikipedia, and other resources (such as `software-carpentry.org`) *to find tutorials, examples, and other information.*

7.2.1.2.4 History and Script

The shell automatically keeps a record of all commands used in a session. Typing *history* will print the list of commands. This tool is useful when a biologist is developing a computational procedure. Once the biologist has determined which commands and parameters are necessary to perform a required task, he or she can use the *history* tool to view and save the commands for future use (see next section).

If a biologist needs to save an interactive session at the command line, he or she can use the *script* tool. A record will be generated for all data output to the terminal window.

7.2.1.3 Saving Commands

One of the advantages of the command line is that biologists can save the exact commands and parameters used to perform a computational procedure, as opposed to a GUI-based procedure where it is difficult to record which buttons and options were used and the order in which they were clicked. At the most basic level, biologists can write the commands in their bound lab notebook. Another option is save the commands in a text file as a rudimentary electronic notebook so the biologist can search for a procedure later.

Example of shell commands to be written/typed in a bound or electronic notebook:

```
bowtie-build DataSet001.fa DataSet001.Index
bowtie -m 1 DataSet001.Index DataSet001Reads.fq
 DataSet001.map
```

In the example, the biologist is using software that aligns short reads to a reference sequence database. To use the software, two commands are required: (1) call the program bowtie-build to read the reference file *DataSet001.fa* and return the output files with the prefix *DataSet001.Index*, and (2) call the program bowtie with the parameter -m 1 using the files from the previous step and the read file *DataSet001Reads.fq* as input and name the resulting alignment file *DataSet001.map*. By recording these commands exactly as typed in a notebook, the biologist will know the procedure used to generate the alignment file. If the biologist has more datasets, then these can be run using the same commands, changing only the file names. Further, when the biologist is writing up the results several months later, he or she can include the procedure so the results are credible and reproducible.

The most useful option is to create a short shell script. While this may seem daunting to a beginner, it is no more difficult than programming lab equipment (e.g., creating a PCR program on a thermocycler), and just as the PCR program is recorded in full in a notebook and used for all subsequent experiments, so can the shell script be painstakingly written and then simply used and referred to in later analyses.

Example bash script *alignMyRnaSeqData.sh*

```
#! /bin/bash
dataName='DataSet001'
params='-m 1'
bowtie-build $dataName'.fa' $dataName'.Index'
bowtie $params $dataName'.Index' $dataName'Reads.fq'
 $dataName'.map'
```

In the example bash script, the biologist has converted the previous example procedure into a series of commands with the name of the dataset and parameter list turned into variables. When the biologist needs to run it, he or she will type *bash alignMyRnaSeqData.sh* at the command prompt and the instructions will be executed automatically. The beginner can then manually edit the dataset name or parameter list with a text editor (see next section) each time it is run. Users who need to run the shell script on tens or thousands of files will benefit from learning a few extra commands so they can make the script loop through a list of file names. Using the script ensures that each time a dataset is run, the procedure remains the same, the output files are named systematically, and the biologist saves time by not having to retype commands or troubleshoot errors from typos.

Bash scripts are particularly useful when testing scientific software with different parameter sets. Biologists may not know that unlike laboratory commercial kits that have been rigorously tested on a variety of samples by experienced technicians, a significant number of scientific programs are written by graduate students and other academic researchers who are trying to solve a specific problem and optimize the parameters to that particular system. These default parameter sets are usually untested with other types of data unless the software has an active community of users who have found many of the problems. Therefore, biologists using new software should start by running a dataset with a known result (a control) on the default parameters and then vary the parameters one by one to determine their sensitivity. (When possible, the parameters should be looked up in the documentation and through Internet searches to determine whether they are set to their optimal values for the current sample type.) Bash scripts are useful for testing parameter sets because biologists can set default parameters, run the script, change one parameter, and run the script again, confident that the only change to the procedure is the one they deliberately made. In the previous example, a biologist would vary the parameters listed in the params variable for each run. The results are more comparable and reproducible, and automating the procedure speeds up the parameter optimization process.

7.2.1.4 Text Editors and IDEs: The Bioinformaticians' Word Processors

Text editors are similar to word processors in that they are used to open, create/modify, and save text files and source code, but different because they do not save formatting characters or binary information in the file. Therefore, they produce the clean, simple files that are needed for running programs and analyzing data. There are many freely available editors with features such as GUIs, programming language-specific syntax highlighting, and advanced text parsing commands. We recommend that biologists learn cross-platform editors (i.e., can be used on Windows, and UNIX systems), particularly those biologists who use Windows machines locally and UNIX systems remotely.

Two editors are particularly popular among computational scientists: Emacs and vi (or Vim). Both editors are cross-platform and have productivity-increasing features such as macros that can automate tasks, regular expressions to tailor/speed search and replace, etc. Emacs and vi can also be modified/customized using their internal scripting languages. This feature makes these two editors highly expandable and flexible. Numerous free plugins for both editors are developed and maintained by a large user-based community, which also provides free support for new users.

Integrated development environments (IDEs) provide tool sets, including an editor, usually with advanced features, debugger, package manager, and numerous plugins. The most popular IDEs such as Eclipse and Netbeans support many languages, including C++, Python, and Ruby, as well as HTML, PHP, and JavaScript for web development. IDEs also provide a nice GUI, which is built on top of command-line tools. Moreover, online tutorials are freely available for users of all levels. For biologists who intend to invest heavily in programming their own tools, IDEs may be a convenient step up from text editors.

7.3 Intermediate

This section addresses programming, the keystone of bioinformatics research. Biologists will be introduced to programming languages (and how to select one), good programming practices for developing less error-prone and more efficient code, program documentation for self and general use, version control systems as electronic notebooks and distribution methods, and controls for testing homegrown code. These sections are each intended as an overview of the tools and practices for writing scripts that are shorter than a page or two in length. References and tutorials for learning the languages and tools are included at the end of the chapter.

7.3.1 Programming

Programming is one of the most valuable skills for bioinformaticians. For example, just as an experienced wet-lab biologist might quickly pour a gel to purify a new sample, a bioinformatician will write a small script to filter a raw data file. Some scripts only contain a few lines of code, which are written for immediate use and then discarded. Therefore, most custom scripts are not documented, tested, or maintained. Major problems arise when hastily written scripts are blindly reused for other projects or different datasets without proper quality control. In this section, we discuss programming tools and approaches that are important for good computational lab practice.

7.3.2 Programming Languages

Programming languages are sets of human-readable instructions that are translated into machine code to instruct a computer to perform a task. They are generally categorized as *interpreted* and *compiled*. Interpreted (scripting) languages (such as Python, Perl, R, and Ruby) use an interpreter program to run programs in one step, when the user is executing the program. Compiled languages (such as C, C++, and Java) use a two-step approach; the first step, compilation, runs during code development, and produces an executable file that contains only machine code. This file is then executed by the user. Both types of languages are widely used in bioinformatics. While compiled languages are generally more difficult to learn and time consuming to use than their counterparts, the programs usually run faster because the code is translated in advance. This makes compiled languages more suitable for processing large datasets or performing complex or repetitive calculations. Interpreted languages are easier to learn and use, especially for beginners; for processing small datasets or relatively simple tasks, the difference in execution speed is negligible, while the advantage in development time can be significant.

Biologists with no programming experience should consider using a widely used interpreted language such as Python or Perl. Both languages have a large user-based community, reference materials, and high-quality third-party libraries for a wide range of application domains. Another widely used, more domain-specific language for data analysis is R. R is a language of choice for many statisticians and bioinformaticians because of built-in data structures that are suitable for data analysis and a large number of libraries for complex statistical analyses and graphics. Biologists who work on high-throughput data analysis, such as microarrays and next-generation sequencing, may need to use libraries written in R or to run the analysis in R environments. Therefore, a knowledge of both R and Python or Perl languages is strongly encouraged.

While academia in particular has a culture of developing tools from scratch even when alternatives are available, this practice is increasingly challenging because pipelines are now too complex for a single novice programmer to develop quickly and accurately [11]. Third-party libraries are developed by programmers other than the developers of the language itself, and extend the language with custom written functions. For example, biologists who want to develop a bioinformatics web application can use Django, a Python third-party library that already contains most of the code required to build a Web application. The use of libraries is encouraged because they reduce errors in a program (i.e., a new programmer need not develop code that is susceptible to bugs when polished, well-tested code is already available). It also reduces program size, which increases maintainability (the programmer's ability to fix bugs and update code as upgrades to software and scientific methods become available). Many libraries are actively maintained, developed, and used by a community of programmers and scientists; therefore, they are well-tested and fairly reliable. Many libraries also support users via tutorials, online web forums, and mailing lists.

7.3.3 Good Programming Practices

Just as biologists follow standard lab practices in the wet lab, they should follow standard computational practices when writing code. For example, biologists frequently write short scripts to perform simple tasks. While writing a script, the meaning of each command and program logic is easy to understand. However, after a month or two, it can be nontrivial to determine the code function, inputs, etc., that were once so clear. Furthermore, as scripts change over time, multiple versions of each script may be scattered over multiple computers. This problem is akin to a biologist performing a quick procedure at the bench, and not taking the time to clean up or write down the procedure in detail.

In this section, we discuss basic programming practices that help programmers organize their code and make their scripts more reusable for other projects. We also introduce version control software, which facilitates code distribution among collaborators.

7.3.3.1 Code Documentation

All code should be documented in plain English. The main purpose is to inform users about code function, expected input and output, and usage details, which are collectively called a code description. This is similar to laboratory equipment being packaged with standard operating procedures and troubleshooting guides that are available to biologists using the equipment. The programmer should remember that just as the wet-lab biologist will be more interested in using an instrument to do an experiment than opening the control panel and tracing circuits, so will users be focused on using a program to process data rather than reading source code, and should write the documentation accordingly.

Professional programmers conventionally write a code description at the beginning of each short program. Code descriptions should be short, providing maximum information in a minimum of words. For simple scripts, this may be two lines: one for the description and one to describe the usage. Some people might include the name of the author, date created, and date modified in the code description; however, when using a version control system (introduced later in this section), this practice is redundant.

When beginning programmers graduate to bundling scripts, programs, and libraries together, they should add a README file, which is a text file containing documentation for the entire code repository. The file usually includes the name of the author, contributors, installation, usage, and licenses for all programs in the repository.

Code readability is important. Experienced programmers customarily use two practices to improve readability: descriptive variable/function names and concise but clear comments. For example, a new programmer might write the following code and comment (comment shown in italics):

```
x += 1 # add 1 to x
```

Readers with minimal knowledge of Python will recognize that 1 is added to a variable *x*. The comment does not provide any information about the purpose of the statement or describe *x*. Instead, the code should be commented like this:

```
x += 1 # increase the count of DNA sequences read in
```

This way, a reader will quickly understand that *x* is the number of DNA sequences read and the number is being incremented by one. A more useful way to write this statement is

```
sequence_count += 1 # increase the number of DNA\
 sequences read by one
```

After choosing a more descriptive variable name, the comment is now largely redundant and could even be omitted. This is a simple example of how code readability can be improved by choosing appropriate variable names and commenting code sparingly.

7.3.3.2 *Managing Code/Text with a Version Control System*

A common change tracking nightmare is when a programmer creates a script and sends a copy to a collaborator, who we will call Adam. After a period of time, Adam finds a bug in the first version of the script and informs the author. The author fixes the bug and sends Version 2 to Adam, not knowing that Adam previously sent Version 1 to other collaborators, Beth and Celia, who are not privy to Version 2 and therefore are likely to have unknowingly generated erroneous results with it. Further, Adam may inadvertently confuse Version 1 with Version 2 because the script name is the same for both and the only difference is in the code, which Adam cannot or does not read. Later, Beth (with Version 1) might try to compare results on a similar dataset with Adam's results from Version 2, and spend considerable time tracking down the reason for the differences. This would result in (at best) a rediscovery of the bug and (at worst) attempts to publish the erroneous comparison in a journal. The author, who has meanwhile upgraded to Version 5, is unaware of anything except that Adam was sent Version 2 a long time ago. This convoluted comedy of errors is one reason why a technique known as version control is essential in collaborations.

A version control system (VCS) is a program that tracks changes made to a file or set of files in a specified directory and records them on a central server. Users can add, remove, or edit files and the version control program will compare each file with the previous version and record the differences. For a single programmer, a VCS can be a rigorous, efficient electronic notebook of changes to a program: as the programmer creates/updates scripts, with a simple command, each change is meticulously recorded, dated, and archived. For a team of programmers, a VCS is a group notebook, distribution tool, and collaboration aid. Each programmer can access the latest version of the code and make changes. If multiple changes are made by different programmers to the same file(s), the changes are noted and automatically merged if possible. Conflicts (in which two programmers modify the same line or region of code) are flagged for manual resolution. For nonprogrammers, a VCS is a useful means of obtaining up-to-date software tools, as the VCS usually contains the most recently updated version of the software. Users can download code from the VCS and know exactly which version they are downloading.

Note: VCSs are also commonly used when collaborators are writing a paper using LaTeX or a similar document markup language. Each collaborator has immediate access to the most recent version of the paper, can make changes directly to the document, and commit them, as opposed to trading various versions via email or having one author decipher scribbled-on printouts.

This is particularly useful when a research lab has one programmer and multiple users because the programmer can create/edit scripts in the VCS and then users can check that VCS to determine if they need to download new/updated tools. When publishing computational analyses, study authors can note the version numbers of the scripts used so biologists wishing to reproduce the study can be given the correct version. Also, when the programmer leaves the lab, the latest versions of the tools are in the VCS so a newly hired replacement can immediately access and start maintaining the code.

Note: While setting up a VCS will require an initial investment of time, subsequent use is usually limited to a few simple commands. Some VCSs also include a GUI to help beginners and nonprogrammers use the system.

Some major version control systems (SVN, Git, Mercurial, etc.) are associated with websites that host repositories for free. For example, Github.com hosts more than 200,000 free code repositories for open-source projects. These can be accessed freely from any part of the world. Using Github simplifies distribution because individual research labs do not need to constantly update lab websites with the latest version of a program; they can point lab members, collaborators, and blog readers to the online repository that the lab programmers are already updating. It is becoming common to publish a link to a GitHub repository to fulfill bioinformatic journals' requirement of open access software, which is advantageous because the site is separate from university or business websites that may change over time.

Note: Biologists who prefer to keep their code for internal use only can set up Git or Mercurial to work locally only, pay Github for private repositories, or use Bitbucket. Alternatively, they can set up their own secure server (or have a network administrator set it up) and use Git or Subversion (SVN).

7.3.3.2.1 Real-World Example

In our lab, every programmer has a Github.com account for their projects. Some projects on Github are also linked to the lab repository https://github.com/ged-lab, which serves as the main repository for all source code and other materials written by lab members. We include a link to the main repository in each publication so anyone can download our source code and materials for using or reproducing results without first contacting us. In addition, anyone who finds a bug or wants to contribute to the project can do so by simply cloning the project, editing or adding code, and submitting a request to merge a change to the project. This opens up opportunities for improving the quality of scientific software as well as collaboration. This method has been proven quite successful in the open-source community.

7.3.3.3 Basic Code Testing

Roughly 1–10 programming errors occur per thousand lines of code [5,12]. In this section, we discuss two techniques to help programmers find obvious bugs upfront: assert statements and doctests. We introduce more advanced tools such as unit tests and automatic testing systems in Section 7.5.

7.3.3.3.1 Assert Statements

The purpose of an assert statement is to compare a calculated value with an expected value and return true or false based on a programmer-defined condition. Assert statements can be used to test if code works as expected. They are particularly handy when testing *edge cases* such as when a user uses unexpected parameters or data files in unrecognized formats. For example, if we write a function *count_gc* in Python that returns the number of G and C nucleotides in a sequence, we could use assert statements to test the function:

```
assert count_gc("ATGTC") == 2
assert count_gc("ATTTTA") == 0
assert count_gc("") == None
```

The first line tests whether the *count_gc* function correctly counts the number of Gs and Cs in the normal case or a mix of all four nucleotides. The second line tests if the function can correctly handle a calculation where there are no Gs or Cs present, which is also expected to be a common case. It is good programming practice to always test cases where zero is the expected result to ensure that it is correctly calculated and reported. The third line tests if the function recognizes that it has been passed an empty sequence and correctly reports an error; *None* is a defined value in Python that indicates "no result." In Python, a programmer can also specify an error message if an assert statement fails:

```
assert count_gc("") == None, "Empty sequence, should
  return None"
```

In this case, if the function does not return *None*, the assert statement fails and prints "Empty sequence, must return None" on the computer screen. This error message alerts the programmer that the *count_gc* function is not handling the case correctly.

Theoretically, assert statements should check all possible input values; however, this is not usually practical. In the aforementioned example, it would be impossible to generate every possible sequence that a user may input to the function. Therefore, a programmer will usually design a representative set of input data to systematically test the code. The previous example demonstrated this by testing both common and uncommon cases. The more assert statements added, the more likely an existing error will be found.

Note: Assert statements are also useful when the programmer is modifying the code later. Well-tested programs can be more easily modified and extended because the tests ensure that changes that break the existing code are more likely to be immediately discovered.

7.3.3.3.2 Doctest

Doctest is a useful feature in Python and several other languages that helps the programmar document and test his or her code simultaneously. This is particularly useful when writing documentation for developers planning to use or extend functions because the doctests can ensure that the documentation examples are correct. Basically, doctest compares output from the Python interpreter with user-defined output. The test fails if they do not match. The doctest for the previous function would look like this:

```
>>>count_gc("ATGTC")
2
>>>count_gc("ATTTTA")
0
>>>count_gc("")
None
```

7.3.3.4 Code Testing in Real Life

Effective testing catches errors. However, the human programmer can almost never consider and write tests for all possible ways of breaking code when he or she is developing the first version of a program; bugs are both inevitable and common. Therefore, writing tests should be incremental; each newly discovered bug should prompt the addition of a new test.

Tests are useful on many levels, but some programmers still do not write tests [11]. One reason is that test writing is not formally taught in most undergraduate computer science courses and therefore many programmers, let alone biologists, lack the required knowledge or experience. Another reason is because it is time consuming and not considered a critical path activity [9,13]. Rigorous tests may contain more lines of code than the actual code. However, there is some evidence that programmers who write tests spend less time debugging and produce higher quality code (see Chapter 12 of [1]). Moreover, time spent repeating an analysis because of a bug is usually far costlier than time spent writing tests.

7.3.4 A Solid Foundation

For many biologists, the guidelines introduced in Sections 1.2 and 1.3 are sufficient to build a strong foundation for reproducible computational research. Beginners should be ready to investigate the tutorials and resources listed at the end of the chapter to build knowledge and experience with the command-line interface: the new benchtop with open access, high-quality UNIX tools for data processing (without programming), simple bash scripts to generate reproducible procedures and optimize parameters, and text editors/IDEs to create and manipulate files across platforms. Intermediate users will be able to investigate various programming languages to find the one with the most high-quality libraries and support for their research area, document their code so it can be easily read and understood by other researchers, use a version control system as an electronic notebook and up-to-date distribution system for their evolving code, and write systematic tests to catch bugs early in the development and analysis process.

For most biologists, effectively using already developed software pipelines and writing small scripts to port data between them or processing large results files with UNIX tools is all they will need to complement their bench work. By incorporating these tools into their computational research and observing the computational lab safety practices, biologists can work effectively on the new benchtop, produce timely, accurate results that are simple to repeat with bash scripts or short, well-tested programs, and can distribute their programs per journal requirements using online version control so the research community can spend less time reinventing and fixing code and more time advancing science.

7.4 Advanced

Once a biologist has built a strong foundation for reproducible computational research, he or she may wish to progress to more complicated analyses, which accordingly require more complex calculations. The resulting programs can contain hundreds or thousands of lines of code, more instructions than a single human can keep in his or her head. An eager biologist who has been developing scripts with less than 50 lines may jump in and create a single file of several hundred lines. However, once a program has advanced beyond the simple script, new programming practices need to be followed to produce usable and maintainable code. This is analogous to chemists running a small reaction in the lab vs. chemical engineers scaling up a reaction to run in a chemical plant; process and resource management become significantly more important. As in the previous sections, these practices facilitate reproducibility, productivity, and frequently help maintain the programmer's sanity. In this section, we will discuss modularity (the practice of writing code in small blocks), refactoring/optimizing code performance for use with modern huge datasets, and using the IPython notebook as an interactive notebook/computing environment for integrating different programs and platforms and performing a complex analysis from start to finish.

Biologists/bioinformaticians grapple with a major problem before they start designing a new program: absence of detailed program specifications [13]. In the bioinformatics lab, programmers frequently have only a piece of the problem laid before them and minimal input from lab members. Bioinformaticians need to quickly develop a program that reads in data, performs a calculation, and writes results so lab members will react to the results, try to validate them, mention specifications/expectations that were not stated initially, or rethink the problem. The programmer is then expected to refine the program and show new results to solicit more feedback iteratively.

Productivity is the key to using this evolutionary approach to problem solving. Writing robust, reusable, and maintainable code is traded for writing code quickly because the programmer assumes that most code will be modified or discarded during the development process. Therefore, biologists/bioinformaticians should write code that is (1) functional, (2) readable, and (3) testable. Functionality is most important because if code does not work, then it is considered worthless [9]. Readability is necessary so the code can be understood by all programmers and users on the project. Testability is required because larger programs have more ways to break and therefore even more tests are needed.

7.4.1 Modularity

The first good practice for writing high-quality large programs is to divide code into small modules. A module is generally a small block of code that performs one specific task such as reading FASTA files, ensuring that a DNA sequence contains only the characters A, C, G, and T, or calculating the average of a set of numbers. The short scripts produced by intermediate biologists/bioinformaticians can easily be converted into modules; experienced software developers frequently write scripts that are simultaneously both. The purpose of creating modules is to take advantage of all the tools and practices discussed in Section 7.3. It is also easier for a programmer to write logical and organized code when creating several small modules and linking them together than when writing one long linear program. These modules can then be bundled together to form a programmer's custom library. A large program ideally should consist of a main program file that accepts user input and then passes it through a series of modules or library functions.

There are several additional advantages to this practice:

1. Simplify and speed up programming. For example, a program may need to read several FASTA files to function. In a linear program, the code to load a file would need to be copied and pasted several times, which decreases the readability of the code. If the program is modular, the programmer need only create one module in the library and then reference it as many times as needed in the main program.

2. Library modules are easier to maintain. In a long, linear program that reads multiple files, if a bug is found when reading the first FASTA file, then the error is likely to be in all instances of the code. Novice programmers frequently only fix those instances where an obvious error is shown, leaving silent errors in other parts of the program. In the modular example, the bug is fixed once in the module. Similarly, the module need only be tested once whereas good testing of a linear program would require many more tests.

3. Modules are easier to reuse for other projects. Once a programmer has written and tested a module and added it to his or her custom library, it can be used for all future projects. If the program is linear, the programmer must copy and paste lines of code from an old program to a new one, and then diligently test the code, or the programmer will need to reinvent the wheel by writing fresh code to load FASTA files for each new program.

4. The programmer can easily combine custom modules and third-party libraries to quickly and efficiently create large programs.

5. A team of programmers can easily collaborate on a large program when each of them is writing/testing different modules.

7.4.2 Code Refactoring

Refactoring is the process of changing a program so the code is different but the results are same. This is similar to changing a wet-lab procedure so it uses fewer consumables, less dangerous chemicals, and less time while still generating a result of similar quality. Once a programmer has a well-tested program that produces the desired results, he or she can refactor the code so it is more readable and simpler, with emphasis on the readability. Before beginning, test codes should be written to ensure that the code still runs properly after refactoring and the code itself should be self-documenting (see Section 7.3).

When refactoring, there are several preferred practices:

1. Remove programming language-specific idioms or overly complex statements to increase readability. Many programming languages have similar structures and syntaxes (i.e., a Perl statement can be read by a C++ programmer as long as it does not use Perl-specific syntax). Also, overly complex statements tend to contain errors.

2. Divide large functions or modules into smaller ones. As discussed in the modularity section, smaller modules are easier to code, test, and reuse.

3. Remove dead/obsolete code. Because of the evolutionary process discussed previously, nonfunctional code from earlier versions might be lurking in the code. Removing obsolete lines will both increase readability and ensure that compute resources are not wasted on useless processing.

7.4.3 Code Optimization

As mentioned previously, interpreted languages are often used to develop bioinformatics software because they reduce development time with the acceptable trade-off that they are slower than compiled languages. In some cases, however, performance is critical and the program may be considered useless if it cannot achieve a particular speed. Optimization involves identifying the bottlenecks (the slowest sections) and modifying them to use different algorithms or compiled languages. This is similar to a wet-lab biochemist taking a multistep synthetic pathway, determining the rate-limiting steps, and either substituting those steps with new reactions or using catalysts. Generally, only the bottlenecks should be optimized; opening one or a few bottlenecks is usually sufficient to achieve the desired performance without spending time optimizing the entire program. The bottlenecks can be found using a tool called a profiler, which reports the time and the number of times a particular function or a method is called. Examples of a profiler are GNU gprof for C/C++ and profile, cProfile, and pstats module for Python.

Note: Some common bottlenecks in bioinformatics programs involve loading data multiple times, unnecessary reading/writing data to disk, and inefficient searching using nested loops. These can often be solved by using more efficient algorithms or data structures. Novice programmers are prone to writing algorithms that mimic how a human would perform a task instead of harnessing the abilities of the computer and the specific programming language. Consultation with other bioinformaticians or with an online forum is a good path to improving code efficiency or solving specific problems.

Optimization depends on the languages being used. For a bioinformatics program written in Perl or Python, a bottleneck function can be rewritten in a compiled language like C++ or Fortran, and then wrapped so the interpreted language can use it. This method can increase performance by several magnitudes, although poorly written code in a compiled language may not function as well as well-written code in a scripting language. Experienced bioinformaticians utilize data structures and libraries that are built into the scripting and compiled languages, which have been optimized by professional software developers. For example, in Python, many built-in data structures and functions are actually implemented in C and wrapped so they can be called using Python code. A biologist/bioinformatician need only find an appropriate method that has already been optimized and implemented in a compiled language.

7.4.4 Research Documentation

As stated previously, a complete bioinformatic analysis usually consists of running several third-party software packages, scripts that port data between them, and visualization tools to represent the final results. To reproduce the results, a biologist must repeat every step in the analysis in the correct order with the appropriate parameters. Until recently, providing a complete set of instructions was not trivial. Because the biologist/bioinformatician is frequently using an evolving procedure, the bookkeeping required for recording detailed procedures can become complicated. Fortunately, scientists and companies have developed tools to simplify the process, allowing researchers to conduct computational analyses while simultaneously building the final set of instructions. In this section, we introduce the IPython Notebook, which has become popular due to its support of Python, shell commands, and R (a statistics and graphing language).

7.4.4.1 IPython Notebook

IPython Notebook is a combined electronic notebook and programming/ computing environment. Users can create a notebook for a project and link all scripts, programs, and shell commands, and parameters used to the notebook, including the order in which they are to be run. Users can then run the analysis from start to finish in the notebook and view/save output at each step as well as add textual notes and comments. This high level of organization can boost biologists/bioinformaticians' productivity while giving them the tools to run and rerun their analyses reproducibly. A single click will run an entire pipeline, expediting parameter optimization, replicate runs, and reruns after fixing bugs. Automating the process in the notebook minimizes mistakes from typos and other human errors. Moreover, IPython Notebook can be run on a remote server, making it suitable for computer clusters or cloud computing systems. The notebook can be distributed to collaborators who can rerun all commands and see identical results on their computer without the programmer writing any additional documentation.

The notebook is not efficient for running processes that require days to finish. Therefore, we write shell scripts to perform the laborious number-crunching and use the notebook for everything else.

The IPython Notebook is built on top of IPython, an advanced Python shell for interactive Python programming, but can support many more programming languages. For example, with the rmagic plugin and Python RPy2 library, users can use R libraries such as those from Bioconductor (see *Resources*) and store the results in Python data structures, which can be further analyzed with Python. This feature reduces the number of steps in data transformation, which is a common problem in bioinformatics research. Another useful feature of IPython Notebook is an in-line plot, which allows users to use Matplotlib library or R to make and visualize a plot, and then save it in IPython Notebook with comments and code to create an executable document. Finally, the notebook is stored as a plain text file that can be version controlled and distributed as discussed earlier. IPython has many more plugins and shortcuts that support scientific computing analyses. Following is an example of boosting productivity by using a shortcut:

```
>>> expression_values =!cut -f 2 expression.dat # read in\
 a value from the second column of a text file
```

Expression values are contained in expression.dat, and *cut* is used to select the second data column (*-f 2*). This data is then assigned to the Python list variable *expression_values* for later use. Running this single line in the notebook can take the place of running the *cut* command at the command line, saving the results to a file, opening that file in Python, reading the data, and then assigning it to the list variable. Throughout this chapter we have emphasized the idea that biologists/bioinformaticians can benefit greatly by utilizing existing well-tested tools as opposed to reinventing the wheel. IPython has been used by many scientists for several years; therefore, it is not surprising that there are many commands, shortcuts, and plugins that perform common tasks elegantly, accurately, and expeditiously. IPython also allows users to create their own plugins to extend its functionality.

Note: Similar "notebook"-like tools exist for R, as well.

7.4.4.1.1 Real-World Example

The senior author (C. Titus Brown, MSU) used the IPython notebook to complete parts of a bioinformatics analysis for a recent manuscript, available at https://github.com/ged-lab/2012-paper-diginorm.git. The author also provides a tutorial on running the pipeline and reproducing the results using IPython notebook at http://ged.msu.edu/angus/diginorm-2012/pipeline-notes.html. The analysis was tested on Amazon cloud service with *ami-61885608*, which has all the required programs preinstalled. Anyone can follow the pipeline and use the notebook to reproduce the analysis from start to finish with identical results with minimal effort.

At the NIH Cloud Computing for the Microbiome workshop in 2012, a team of researchers from different backgrounds used IPython notebook and StarCluster as collaboration tools to produce publishable results in record time [28].

7.5 Related Topics

For biologists/bioinformaticians who have progressed through sections (Sections 7.2 through 7.4), we briefly describe more advanced computing topics that can facilitate accurate, efficient computational research.

7.5.1 Using Online Resources

A wide range of topics relevant to programming, bioinformatics, and data analysis are discussed online in blogs, web forums, and Twitter. Perhaps the single most useful approach to problem solving available is to do an online search for your problem; if it is a problem or bug that has been encountered in a popular piece of software, a solution will almost certainly have been posted.

Two particularly useful web forums for bioinformatics are BioStars and Seqanswers. We strongly suggest that novice bioinformaticians search these forums for discussions of tools. Both of these forums also support asking questions, and the online bioinformatics community is generally very friendly and willing to help; we encourage you to first search to see if someone has already asked your question, and if not, to then post the question on one or both of these forums.

7.5.2 Advanced Tools

7.5.2.1 Regular Expressions

Regular expressions are tools for searching text for a particular pattern of letters/numbers/symbols. For example, a bioinformatician can search for short DNA sequence motifs in a data file with an unstructured or unusual format. Regular expressions have their own syntax for defining a specific pattern, which to the casual eye can look like an unintuitive shell language. For example, `logy\b` defines a pattern for a word that ends with *logy*, matching biology, physiology, technology, etc. In programming, regular expressions are used to concisely and quickly search for patterns in data.

7.5.2.2 Debuggers

Code frequently contains errors or bugs that cause unexpected results. While syntax errors will be caught by the interpreter or compiler, logic errors often go undetected. A debugger allows programmers to interact with their program by running and pausing execution, stepping in and out of functions or loops, and changing values in variables during execution to locate a bug. Most programming languages have at least one debugger. GDB is a standard GNU/Linux debugger that can be used with most compiled programming languages. Moreover, major IDEs such as Eclipse and Netbeans have built-in debuggers with GUIs.

7.5.2.3 Unit Tests and Automated Testing

Unit testing consists of writing code that tests individual units of a program, such as functions or modules. Each subunit is tested in isolation; therefore, tests on a given subunit will not be affected by bugs from other subunits. This procedure helps locate errors in a large program. Unit testing libraries are available for most major languages and help users create test suites. Some libraries also provide a test runner to run tests automatically. An important advantage of automatic testing is that users can test the program to ensure that the installation process is bug-free. Unit testing also promotes a test-driven development process, which helps guarantee that every function works as expected and tests are written for every function in a program.

7.5.2.3.1 Real-World Example

In our lab, large programming projects have separate folders for test code. However, each project uses different libraries for testing; for example, Gimme (https://github.com/likit/gimme) uses a Python unittest module whereas khmer (http://github.com/ged-lab/khmer.git) uses user-defined functions, which are recognized and run by the nosetest module. Instructions for automatically running tests are included in the corresponding README files.

7.5.3 Advanced Programming Topics

7.5.3.1 Object-Oriented Programming Paradigm

Object-oriented programming (OOP) is a standard programming paradigm that creates "objects," which usually consist of data and specific methods for operating on that data. This is useful when a large program needs to have standardized methods for data-processing but also uses several different types of data. In addition, objects can be reused or extended without rewriting them from scratch. This practice has been extensively used throughout the software industry and in scientific programming because it promotes code maintenance and expansion.

7.5.3.2 Algorithms and Data Structures

Most major languages used in scientific computing provide libraries supporting well-optimized algorithms and data structures. However, using a preimplemented algorithm without understanding the underlying concepts is unsafe. Most algorithms have strengths and weaknesses that should be evaluated based on the specific application. Basic knowledge of algorithms will help biologists make a correct decision.

7.5.3.3 Compiled Languages

While time consuming, learning a compiled language such as C/C++ gives insight into how a computer functions because it requires machine-based knowledge. Scripting languages are designed to abstract away many low-level details to improve programmer productivity. As a result, it also abstracts away important concepts of computing such as memory management. A basic understanding of how to program in a compiled language will help biologists write better code using scripting languages because of a greater understanding of the underlying mechanisms. In addition, many bioinformatics software packages still use code in compiled languages to do rapid processing.

7.6 Conclusion

The general problem in the bioinformatics field is not an absence of tools and good practices, but rather that many researchers lack knowledge and training with them [31]. Embedded in a scientific culture that is relatively inexperienced with good computational practices, many biologists make their first foray into bioinformatics with only intuition and the Internet to guide them. The tools and practices discussed here are intended to help those biologists build a solid foundation in reproducible computational research. Because it would be impossible to condense several years of scientific computing and data analysis training into a single book chapter, we have focused on describing those tools and practices that are particularly useful for novice bioinformaticians, emphasizing their contribution to productivity and reproducibility, with the intention of giving biologists the introduction they need to then seek specific information and step-by-step tutorials elsewhere.

Investing in learning computational tools and practices will yield incalculable return over the course of a biologist's career. With the enormous potential for discovery available in this era of Big Data, datasets can be expected to continue expanding. A biologist who starts learning and applying general tools and good computing practices now will eventually save *years* of time. Moreover, the biologist who invests time in writing tests, using version control, and automating/distributing analyses in the IPython Notebook will both be compliant with journal and granting agency policies for distributing code and avoid the aggravation of other scientists repudiating his or her unreproducible results.

While the list of tools and practices to learn might seem overwhelming to biologists with no prior computational experience, we encourage them to take a systematic approach to the education process. Just as a wet-lab researcher is trained on one instrument at a time, practices using it for his or her current project, and then moves on to more complicated methods, so too can he or she learn computational techniques. Installing software for running a command-line interface and looking through a text file of data is a good start. This simple task can build confidence with using the command line, and soon the biologist will be ready to learn simple *grep* commands to make looking through that text file easier and faster. Each skill will build upon the last, and the biologist will soon be applying these skills to his or her research, finding experiments where a new option or tool will accurately process data in seconds that would otherwise have required hours of mind-numbing clicking. Once the biologist gains experience with the beginner tools and practices, he or she will be ready to tackle the intermediate (Section 7.3), and the tools and practices described will seem like a natural progression. As the biologist becomes a more practiced bioinformatician with a well-stocked toolbox, so too will his or her research advance.

A single biologist using good computational tools and practices can produce a lifetime of biological breakthroughs and innovation. A team of skilled biologists can work in parallel to push the limits of biological knowledge in a particular area by several lifetimes. As more biologists practice reproducible computational research, the sheer breadth and depth of their work will collectively move the entire field of biology into a new era of scientific discovery.

Acknowledgments

The authors thank Greg Wilson, Steve Haddock, Hans Cheng, Randy Olson, Eric McDonald, Jason Pell, and Cari Hearn for reviewing early drafts of this manuscript.

This chapter is maintained online at http://reproducibility.idyll.org/, where it is being continuously modified and updated by the authors and members of the scientific community. Readers are encouraged to visit the associated forum and leave questions/comments.

Available Resources

Books

UNIX/Linux Tools

- Haddock & Dunn *Practical Computing for Biologists*.
- Newham, Cameron *Learning the bash Shell* [O'Reilly].
- Robbins, Arnold and Dougherty, Dale *sed & awk* [O'Reilly].
- Cameron, Debra et al. *Learning GNU Emacs* [O'Reilly].
- Robbins, Arnold et al. *Learning the vi and Vim Editors* [O'Reilly].
- Neil, Drew *Practical Vim: Edit Text at the Speed of Thought* [Pragmatic Bookshelf].
- Chacon, Scott *Pro Git* [Apress].

Python

- Campbell, Gries, Montojo and Wilson *An Introduction to Computer Science Using Python* [Pragmatic Bookshelf].
- Lutz, Mark *Learning Python* [O'Reilly].
- Model, L Mitchell *Bioinformatics Programming Using Python* [O'Reilly].
- Vaingast, Shai *Beginning Python Visualization* [Apress].
- Arbuckle, Daniel *Python Testing: Beginner's Guide* [Packtpub].

Others

- Joe Pitt-Francis, Jonathan Whiteley *Guide to Scientific Computing in C++* [Springer].
- James Tisdall *Beginning Perl for Bioinformatics* [O'Reilly].
- James Tisdall *Mastering Perl for Bioinformatics* [O'Reilly].
- Ellie Quigley *Perl by Example* [Prentice Hall].
- Peter Cooper *Beginning Ruby* [Apress].
- Joseph Adler *R in a Nutshell* [O'Reilly].
- Peter Dalgaard *Introductory Statistics with R* [Springer].
- Paul Teetor *R Cookbook* [O'Reilly].

Online Resources

UNIX/Linux Tools

- GNU Operating System
  http://www.gnu.org
- Cygwin (Linux emulator for Windows)
  http://www.cygwin.com
- MSYS+MinGW
  http://www.mingw.org/wiki/MSYS
- Vi and Vim
  http://www.vim.org/index.php
- Emacs
  http://www.gnu.org/software/emacs/
- Github: Git online repository
  http://github.com
- Git tutorial
  http://git-scm.com
- Mercurial
  http://mercurial.selenic.com/
- SVN
  http://subversion.apache.org

Python

- Python: Python official website
  http://python.org
- Python style guide
  http://www.python.org/dev/peps/pep-0008/
- The Zen of Python: A guideline for Python coding
  http://www.python.org/dev/peps/pep-0020/
- Learn programming by visualizing code execution
  http://www.pythontutor.com/
- Python doctests
  http://docs.python.org/library/doctest.html
- Python unittest
  http://docs.python.org/library/unittest.html
- IPython: Advanced Python shell
  http://ipython.org
- Scipy: Scientific tools for Python
  http://www.scipy.org/
- Matplotlib: Python plotting library
  http://matplotlib.org/
- Python: Speed and performance tips
  http://wiki.python.org/moin/PythonSpeed/PerformanceTips

- Performance analysis of Python programs
  http://www.doughellmann.com/PyMOTW/profile/

R

- R Official website
  http://www.r-project.org/
- Rseek: Search engine for R related materials
  http://rseek.org
- Bioconductors: R packages for bioinformatics
  http://bioconductor.org

Web Forums

- BioStars: Bioinformatics answers
  http://www.biostars.org/
- Seqanswers: Bioinformatics answers
  http://seqanswers.com
- Stack overflow: General programming
  http://stackoverflow.com/

Others

- Software carpentry: Online training
  http://software-carpentry.org
- Rosalind: Learning Bioinformatics
  http://rosalind.info/problems/as-table/
- Reproducible research
  http://reproducibleresearch.net
- Analyzing next-generation sequencing data
  http://bioinformatics.msu.edu/ngs-summer-course-2012

References

1. A. Oram and G. V. Wilson, editors. *Making Software: What Really Works, and Why We Believe It*. O'Reilly, Farnham U.K., 2011.
2. D. Altshuler, R. M. Durbin, G. R. Abecasis, D. R. Bentley, A. Chakravarti, A. G. Clark, F. S. Collins et al. A map of human genome variation from population-scale sequencing. *Nature*, 467(7319):1061–1073, Oct 2010.
3. W. J. Ansorge. Next-generation DNA sequencing techniques. *N Biotechnol*, 25(4):195–203, Apr 2009.

4. No authors listed. The ENCODE (ENCyclopedia Of DNA Elements) project. *Science*, 306(5696):636–640, Oct 2004.
5. B. Boehm, H. D. Rombach, and M. V. Zelkowitz, editors. *Foundations of Empirical Software Engineering: The Legacy of Victor R. Basili*. Springer, Berlin, Germany, 2005.
6. W. Bialek and D. Botstein. Introductory science and mathematics education for 21st-century biologists. *Science*, 303(5659):788–790, Feb 2004.
7. F. S. Collins, M. Morgan, and A. Patrinos. The Human Genome Project: Lessons from large-scale biology. *Science*, 300(5617):286–290, Apr 2003.
8. D. Heaton, J. C. Carver, R. Barlett, K. Oakes, and L. Hochstein, The relationship between development problems and use of software engineering practices in computational science and engineering: A survey. Website, 2012. http://www.software.ac.uk/sites/default/files/softwarepractice2012_submission_10.pdf
9. D. Kelly and R. Sanders. Assessing the Quality of Scientific Software. Website, 2008. http://secse08.cs.ua.edu/Papers/Kelly.pdf
10. S. A. Forbes, N. Bindal, S. Bamford, C. Cole, C. Y. Kok, D. Beare, M. Jia et al. COSMIC: Mining complete cancer genomes in the Catalogue of Somatic Mutations in Cancer. *Nucleic Acids Res*, 39(Database issue):D945–D950, Jan 2011.
11. G. Wilson. Where's the real bottleneck in scientific computing? *Am Sci*, 94(1):5, 2006.
12. D. C. Ince, L. Hatton, and J. Graham-Cumming. The case for open computer programs. *Nature*, 482(7386):485–488, Feb 2012.
13. J. Segal. Some problems of professional end user developers. *IEEE Symposium on Visual Languages and Human-Centric Computing*, Coeur d'Alène, ID, 2007.
14. E. S. Lander, L. M. Linton, B. Birren, C. Nusbaum, M. C. Zody, J. Baldwin, K. Devon et al. Initial sequencing and analysis of the human genome. *Nature*, 409(6822):860–921, Feb 2001.
15. C. M. Lewis, A. Obregon-Tito, R. Y. Tito, M. W. Foster, and P. G. Spicer. The Human Microbiome Project: Lessons from human genomics. *Trends Microbiol*, 20(1):1–4, Jan 2012.
16. E. R. Mardis. Anticipating the 1,000 dollar genome. *Genome Biol*, 7(7):112, 2006.
17. G. Miller. Scientific publishing. A scientist's nightmare: Software problem leads to five retractions. *Science*, 314(5807):1856–1857, Dec 2006.
18. A. Morin, J. Urban, P. D. Adams, I. Foster, A. Sali, D. Baker, and P. Sliz. Research priorities. Shining light into black boxes. *Science*, 336(6078): 159–160, April 2012.
19. R. M. Myers, J. Stamatoyannopoulos, M. Snyder, I. Dunham, R. C. Hardison, B. E. Bernstein, T. R. Gingeras et al. A user's guide to the encyclopedia of DNA elements (ENCODE). *PLoS Biol*, 9(4):e1001046, Apr 2011.

20. NIAID and NHGRI. Eukaryotic Pathogen and Disease Vector Sequencing Project. Website, 2012. http://www.niaid.nih.gov/labsandresources/resources/dmid/gsc/pathogen/Pages/default.aspx

21. NIEHS. Environmental Genome Project. Website, 2012. http://egp.gs.washington.edu/

22. NIH. Centers for Mendelian Genomics. Website, 2012. http://www.genome.gov/27546192

23. NIH. Clinical Sequencing Exploratory Research. Website, 2012. http://www.genome.gov/27546194

24. NIH. International HapMap Project. Website, 2012. http://hapmap.ncbi.nlm.nih.gov/

25. No authors listed. Free Software Foundation. Website, 2012. http://www.fsf.org/

26. US Department of Energy. Human Genome Project Information. Website, 2012. http://www.ornl.gov/sci/techresources/Human_Genome/home.shtml

27. P. Pevzner and R. Shamir. Computing has changed biology–biology education must catch up. *Science*, 325(5940):541–542, Jul 2009.

28. B. Ragan-Kelley, W. A. Walters, D. McDonald, J. Riley, B. E. Granger, A. Gonzalez, R. Knight, F. Perez, and J. G. Caporaso. Collaborative cloud-enabled tools allow rapid, reproducible biological insights. *ISME J*, 7(3): 461–464, Oct 2012.

29. B. J. Strasser and M. O. Dayho. Collecting, comparing, and computing sequences: The making of Margaret O. Dayho's Atlas of Protein Sequence and Structure, 1954–1965. *J Hist Biol*, 43(4):623–660, 2010.

30. Greg Wilson, D. A. Aruliah, C. Titus Brown, Neil P. Chue Hong, Matt Davis, Richard T. Guy, Steven H. D. Haddock, Katy Huff, Ian Mitchell, Mark Plumbley, Ben Waugh, Ethan P. White, and Paul Wilson. Best practices for scientific computing. *arXiv*, abs/1210.0530, 2012.

31. Z. Merali ...why scientific programming does not compute. *Nature*, 467:775–777, Oct 2010.

32. I. B. Zhulin. It is computation time for bacteriology! *J Bacteriol*, 191(1): 20–22, Jan 2009.

8

Reproducible Research for Large-Scale Data Analysis

Holger Hoefling and Anthony Rossini

CONTENTS

8.1 Introduction

8.1.1 Disclaimer

The opinions expressed in this chapter are solely those of the authors and not necessarily those of Novartis. Novartis does not guarantee the accuracy or reliability of the information provided herein.

8.1.2 Punchline

We report on the experiences of applying and adapting the literate statistical analysis methodology [6,7] to the work activities surrounding the statistically oriented components of a large-scale corporate research and development project. This complex, multiyear, multiclient/stakeholder, multidata analyst (with team members and clients mixed across multiple continents) presented challenges to the assigned data analysis workgroup in terms of accomplishing the data analysis using the standard available programming tools. The difficulties in planning this project prodded the data analysis workgroup (statisticians and bioinformaticians) to explore alternative group working practices both at the beginning and with respect to changes during the study. These explorations were constrained by the realization that there were to be more than one data analyst and more than one client specifying the form of results over the project lifecycle. There was also a requirement to ensure documentation and reproducibility of prior and ongoing results across the group and to ensure comparability and, when possible, coherency of results across subprojects and deliverables, some of which originated from completely different motivations.

Our conclusion from this experiment was close to what we had hoped to achieve—computationally driven research and development, of a large enough scale to require a disparate and heterogeneously formed and spatially located data analysis workgroup, can be made reasonably transparent and reproducible through application of appropriate work practices. In our situation, this would be assured of taking place through the application of LSA methodologies. However, the final form that the steady-state work process and environment took was very different than what we had originally intended and hoped for. The major goal of this chapter is to communicate the overall experience, the challenges, and the lessons that we learned, in the hope that we might influence, both internally and externally, thinking surrounding the development of appropriate toolchains and work practices for large-scale computationally reproducible and transparent statistical activities.

8.1.3 Audience

Writing such an article always poses a certain challenge with respect to the intended audience, which is certainly also true in our case. As we are reporting on an especially large-scale project in the context of literate programming (LP), a certain familiarity with these tools is helpful in understanding the particular challenges involved. Our report will likely be the most useful for people attempting or contemplating, making their research more reproducible, or who have some experience in generating small-scale reports.

Therefore, in order to learn the most from this article, a limited familiarity with an LP tool is assumed, but we will not assume any additional knowledge about other tools we will be using. Our initial selection of a toolchain consisted of R for the statistical programming, org-mode for the source documentation, HTML and LaTeX for communication–artifact generation, Git for recording and sharing, and Emacs (for ESS and org-mode) as the programming environment. These tools were dictated by local conditions; others can easily be substituted and justified based on technical, cultural, practical, and religious reasons.

8.1.4 Project

We need to start at the beginning in order to understand the contrast to how we think and would proceed in the present time. The work-practice philosophy that we originally intended to adhere to, originating when the project had just a single data analyst and two stakeholders, was that there should be an evolving LP-based scientific technical report. From this technical report, all of the communications and deliverables would be generated, from the initial to the final. Very idealistic indeed!

Having tangible research results be understandable and completely reproducible is the focus of this chapter, and for us, this means the technical report. Derived from this are a range of "advertisements" and artifacts, including journal articles, presentations, and smaller communications that originate from the material generated for the technical report and that should also be generated in a reproducible and transparent manner. Reflecting back on that, it seems like a grand and unachievable goal!

We knew at onset some of the final data analysis deliverables, and there were guidelines and suggestions regarding the shell format that would be suitable. These artifacts would include core interim results on the observational study and a final report on the study. We also knew that there could be the potential for other communication deliverables, depending on requests and intermediate findings. These could include typical internal corporate presentations to other teams in the division, presentations to groups within the corporation but outside the division, communications to senior division management teams on findings and status, as well as internal and external scientific communications on findings (or lack there-of) in journals and conferences.

The core activities of this project consisted of statistical and bioinformatic analyses of an observational trial of a few hundred subjects, with the goal to support research and development of a prognostic molecular diagnostic. In addition to standard clinical trial demographic and health-status information, there were approximately 100 GB of raw data across a range of "-omic" technologies, including mRNA, SNP, and a few other similar genetic and genomic molecular diagnostic-appropriate readouts, with multiple types (and times) for samples per subject.

The data collection for this project covered several years with one early interim analysis, study monitoring reports, and a final analysis. For the final analysis, the time frame available was roughly 3 months, after which the results had to be presented. In our team, three people were "resourced" to make the analysis possible with the given amount of time. They were split over two locations on different continents, with e-mail and telephone being the most common forms of communication. In addition, external consultants were hired that worked semi-independently to provide a certain level of outside validation of the results, apply their own novel methodologies, and contribute their expertise.

The data analysis methods employed for this project were too computationally and memory intensive for a standard single/dual-core desktop machine. However, they were still small enough so that a few well-equipped powerful servers could handle the tasks.

The people coming into the project had a diverse background in terms of their work practices (computer programming development environments). The project manager was an experienced user of org-mode in Emacs and version control systems. However, the other team members had previously used vi or similar basic text editors and had only limited experience with version control. As an additional hurdle, the external consultant resisted and did not use our recommended tools, making exchange of code and results more difficult and time consuming than anticipated.

Team members typically supported a few completely different projects concurrently, so one assumption that we wanted to assess was whether we could devise an approach that minimized the ramp-down and start-up time required in switching between projects. We also hoped to better document the different contacts, requests, and information originated from the greater project team. Finally, there needed to be a common place for project planning, task responsibility, and time lines. Unfortunately, this last component never really got off the ground.

8.1.5 Outline

In the following sections, we will discuss our setup, mainly with respect to how we documented our code and ensured reproducibility. In the form as we describe here, some of the discussion only applies for projects such as ours—for example, large-scale exploratory research. Especially for smaller projects or reports, other conclusions could be drawn and some of the needs, assumptions, observations, and issues we describe in the succeeding text will not hold. We will still focus mainly on our project as a complicated case for reproducible research and thus in our mind also more instructive for practices that can be universally applied in the settings of large-scale data analysis challenges.

As in our project we used LP, we will briefly introduce it and its "extension" toward reproducible research, LSA. Our experiences using this approach in our project will be outlined and discussed. We introduce some other tools that we used to solve our challenges and finally end with a discussion of reproducible research tools and other remaining issues.

The reader should note that we do not claim that our solution is the only possible one. Other users in similar circumstances may find other approaches more practical and useful. Nonetheless, we think that many people will find our experiences and insights valuable for their own projects in order to support selection of an efficient and productive setup without losing time for experimentation.

8.2 Literate Programming and Reproducible Research

At the beginning of 2011, we started out with a set of tools and work instructions for performing reproducible research in the context of statistical and bioinformatic analysis, and our initial setup was strongly motivated by the theory of LSA. Over the course of the last 2 years, we have tuned, refined, and experimented with this setup on a number of projects that were relatively diverse, from relatively small dataset with a set of well-defined analyses to exploratory analysis on large-scale data as mentioned earlier.

It was our intention at the beginning to use LSA as the central tool for reproducible research for every aspect of the analysis, that is, the documentation and writing of code, the execution of code, as well as the generation of the final reports. Over time, it became clear that especially for the more exploratory and large-scale projects, LSA has a number of drawbacks and other tools commonly used in software development are, in our opinion, better suited for this area. However, before going into more detail here, let us review some history.

8.2.1 Literate Programming and Its Relationship to Reproducible Research

LP originated with the WEB system in the early 1980s for the development of TEX [1] and, since then, has been extended to other application domains. Some subsequent implementations target support for specific languages, such as CWEB, while others are language independent, such as Spiderweb, noweb, and nuweb. The original intent of the documents written in this literary style was to document the implemented computer program algorithm in a mathematically reasonable way, as well as automate indices and similar structural guides to the program structure. To this end, natural language describing the logic of the analysis steps is being mixed with code chunks performing the actual operations. In order to execute the code and produce the final results, the code chunks are put together in a certain defined way to produce the source code ("tangled") and the documentation is extracted as well ("weaved"). This way, one ends up with one file containing the documentation and a set of files with the source code.

In LP, the natural language is the central point, and the documentation and ordering of the code chunks is written in a way that follows the human thought process rather than the usual flow of a program. This way, the code is supposed to be more understandable for human readers. Another important aspect is that, other than in the case of most programming languages, the code is located in a single file, organized by sections relating to the logical components of the project. This setup can be very helpful for following the underlying logic of the analysis; however, as such it is not sufficient to ensure reproducibility of a result. In the strict sense, LP only documents the code in a certain way and structures it according to a more "literal" thought process. For this setting, several frameworks exist that implement LP in a language agnostic manner. Here, we only want to mention *noweb* [5] as one of the several examples. noweb has a very simple syntax, can contain code of any programming language, and export the weaved documentation into several output formats, among them LATEX and HTML. Its usage is very simple as it essentially only has the two commands *tangle* and *weave* (with several customization parameters).

After the introduction of LP, it was extended to LSA [6,7], which is the process of ensuring that data analysis is clear and transparent in terms of describing the exact computer programs used for generation of results. We assume that such clarity and transparency are sufficient for reproducibility. As such, its main focus is not the computational and mathematical implementation as with LP, but the whole statistical analysis. This ranges from the origins of the data, the employed statistical methods with caveats and assumptions, the code and programs used, to the results of the analysis and their interpretation. LSA is a reproducible research method in that it ensures that the computations are clearly defined and reproducible; however, it goes beyond mere reproducibility in that it should show the approach, assumptions, and results in a clearly expressed and detailed manner. While ESS (Emacs mode for programming in R) provided editing support for such practices starting in 1997, the first software tool to incorporate these principles in a direct and automated fashion in the statistical domain was *Sweave* [2].

On a purely practical level, in addition to allowing the mixing of code and natural language as in LP, LSA also intends for tables and figures that are the results of the code to be imported back into the document. It concentrates more on the data analysis as a whole instead of just the programming of computer code and the corresponding computational and mathematical aspects. This provides the basis for a system that allows to ensure that the analysis is completely reproducible as well as easy to understand. In LSA, the whole document is usually written with the final report in mind as it is essentially the weaved documentation together with the inserted results. As such, similar to LP, it is usually contained in a single file or, for larger projects, split up by sections. This way of writing is designed to ensure that the flow of code follows the narrative of the final report and it furthermore intended that the code chunks should be kept short so that their behavior is explained by the surrounding text. Of course, in some situations, the report does not conform to this structure. In order to accommodate this, it is also possible (depending on the implementation) not to export the documentation of certain sections and/or even reorder sections from the original LSA document. All in all, LSA has a lot of flexibility but many challenges that we will discuss later. First, we want to talk about the literary style of our LSA document and the software tools we use.

8.2.2 Approaches for Literate Coding

As with literary styles, there are many different styles in which literate coding can be done. The particular one used for a project depends on the requirements of the project itself as well as the background of the researchers. For example, in a consulting project, it would be important to make notes about the initial conversations, the analysis plan, preprocessing of the data, the actual analysis, and the conclusions. In other situations such as knowledge transfer, a different approach is needed. There, a higher emphasis on previous state of the art on the problem in question and the newly added methods would be needed. In the analysis of a clinical trial for submission purposes, special attention would have to be paid to the analysis plan, any transformation of the data, and very detailed descriptions of the analysis methods.

In our case, the plan was to do an exploratory/signature development analysis of high-dimensional genomic data. Therefore, the main parts of our documentation concentrated on the preprocessing of the data, univariate and multivariate analysis of the genomic markers, and the production of tables and figures for the report. As the code used in the analysis was quite extensive, a major goal for the literary style that we use is to make orientation in the code base easier.

8.2.3 Tools for Literate Statistical Analysis

The requirements on the tools for LSA are somewhat higher than those used for LP due to necessary processing of the code and the importing of the results. *Sweave* was the first that implemented this concept, based on the LP system *CWEB*. It allows R (or S) code to be mixed with LaTeX documents, executes the code, imports results (output and plots), and then converts the resulting LaTeX file with the output into ps, dvi, or pdf file (and other options such as exporting to HTML are also possible). Afterward, it has been extended to other languages (such as SAS), but the standard version is very closely linked with R and LATEX. There are several other tools of a similar type available, building and improving on the earlier ideas. An example for this is the R package *knitr*, which is also explained in a chapter in this book.

Instead of these, we decided on using org-mode, an Emacs mode that was originally written for organizing notes, tracking to-do items, agendas, etc. Later, an option to include source code and execute it was added, making it into a tool for LSA. With respect to programming languages, it is not restricted to any one in particular; code blocks can be written in a wide variety, among them R, SAS, Python, and Perl (more can be seen on the org-mode manual). This is very convenient when more than one programming language has to be used (which was not the case for us). It also provides a good programming environment, as Emacs has specialized modes for editing many languages and org-mode allows the user to treat every code block essentially as a separate file for editing purposes. Other than *Sweave* and *knitr*, which can be written, edited, and used in many editors, org-mode is strictly restricted to Emacs, which may be quite a high hurdle in some cases—although one that is worthwhile to climb over in our opinion. Of course, at this point, we do not want to hide the strong affinity of one of the authors of this article to ESS, the statistics mode in Emacs, which may also have played a role in the choice of tools.

Overall, all of these LSA tools would have done a roughly equivalent job for our project, given that we were using R. However, for someone who is not averse to Emacs and wants flexibility, *org-mode* is a good choice. *Knitr* on the other hand is well integrated into RStudio and *Sweave* has a lot of support due to its long history.

Apart from org-mode in Emacs, we used *R* as our programming language of choice due to its great package support in biostatistics and bioinformatics as well as because all members of the team were already fairly comfortable with it. In the following, we will take our choice of programming language as a given. Subsequently, the article will concentrate on discussing LP and LSA in the context of R and our solutions to some issues; however, we think that it will similarly apply to many other languages that are used for data analysis as well.

8.3 Experiences and Recommendations for Practical Literate Coding

As we already mentioned earlier, we were determined to adhere to the strategy outlined by LSA, that is, to write programs that follow our flow of thought, not the requirements of the programming language. So we tried to write a lot of documentation that was interspersed with code blocks that implemented the steps we described verbally before. The intention also was to use org-mode to execute all code blocks directly in the org-document and write the final report at the same time, which was to be generated out of the same document.

8.3.1 Multiple Reports and Presentations

In a project such as this, usually there are several different "advertisements" that are required during the development phase as well as at the end. These usually take the form of interim and final reports as well as presentations. In addition to the final report, a technical report may be produced that contains more details on the results and techniques. In the spirit of LSA, all of these documents and presentations should be generated using the same source document.

In case of a single output document, this is relatively simple and it is here that the big advantage of automatic report generation is best illustrated. In addition to every update to the results being immediately included in the final output, the origin of every table or figure is also easily traceable as it is only necessary to look at the code at that position of said table or figure. All in all, it is very simple and allows for easy reproducibility and understanding.

However, meeting these requirements is already much more difficult when both, a report and a presentation, have to be produced. In terms of their formatting, literary style and amount of information included, they are so different that they have to be written separately. However, how should the code then be kept close to figures and tables in several "advertisements" at the same time? Either, the code would have to be duplicated as well (which would be extremely hard to maintain) or at least some of the advertisement would be produced without the corresponding code. In the latter case, the immediate question arises as to why the code should not be kept separate completely—however, still written using LP.

A related issue was the organization of the code base. The logical organization of code does not necessarily correspond to the outline of the related report. Some parts of the codebase do not correspond directly with any report section at all (e.g., the preprocessing). In these cases, it is not clear if it is an advantage to order the code so that it corresponds to the outline of the report as it can be directly related to the text or if it is a disadvantage as it breaks a very logical approach for a programmer to think about the problem.

8.3.2 Tables/Figures and Listings

The most important direct contributions of a statistical analysis to an article or report usually come in the form of tables and figures that summarize the results. LSA allows for these to be directly included in the final document without the need to explicitly save them to disk and import them.

This approach is very much in the spirit of LSA as it forces the code that generates the figure or the table to be exactly where it will be placed in the report. From a practical point of view, however, this also entails some problems. As they are not necessarily written to disk, the tables and figures may only be available in the final report. So every time the analyst wants to look at the details of the results in R, the particular code chunk has to be evaluated interactively—and possibly, very many others. Furthermore, if the same output is to be used at other places (e.g., a PowerPoint presentation), it would have to be copied out of the final report, which could be potentially difficult or degrade the quality (e.g., on a plot). For these reasons, we decided to write out all tables and figures to disk. This, of course, does not prevent a user from including a table or figure into a report from its location on disk—but the direct connection between the code and the imported results has been severed. This is especially the case in the context of projects with a very large number of predefined tables, figures, and listings (TFL) or in data-driven, exploratory analysis that produce many TFLs, the exact nature of which is unclear at the beginning of the project. In these circumstances, it is usually the case that many more figures and tables are being created then should/could be included in the final report and only a representative plot or a subset a table is finally used in the end. Therefore, writing out all of them to disk is essential to be able to evaluate all results later.

8.3.3 Execution of Code

One of the major obstacles to using LSA tools for large-scale projects is the execution of the code. Sweave, knitr, and org-mode are set up to tangle out the code and then execute all code blocks in a document every time the document is being exported into a pdf. For projects where the code has a short run-time, this is fine. However, for computationally intensive projects, it is a much bigger problem. Executing all code may take many days, so that the execution of all parts as intended in these tools becomes infeasible. This problem has of course already been recognized with some solutions developed and implemented, including the caching of results of code blocks and reexecution of code blocks only if changes have occurred. These are good approaches, but they certainly fall short in some areas, for example, when it comes to complex dependency structures between code chunks, displays, and changes in the input data.

Taking a step back and evaluating the problem, it becomes clear that it is a rather old and well solved already. What is needed is essentially a makefile that controls the execution of the various code blocks, knows about the dependencies between the code chunks and the underlying data, and only reexecutes those code parts where truly something has changed—either in terms of input data, the code itself, or the underlying software (R or the packages). Of course, a system similar to a "make" program could also be included as well in org-mode or Sweave. However, this would just require the duplication of well-maintained and tested make programs. Furthermore, using a make program that is external to the LP framework also has the added advantage of allowing an outside user to execute the code without having to know how to use the literate analysis framework, overall lowering the barrier for entry (at least if they know how to use make).

8.3.4 Cooperation with Other Team Members

Finally, there are certain issues with the key output of the whole work—the final report. Its review by people outside of the data analysis workgroup can be a challenge if tools that are outside of the usual Microsoft^{TM} office products are used. In cross functional teams (e.g., on clinical trials), project managers or clinicians regularly review the reports before they are officially approved. Most of these report, however, are usually written in Microsoft Word or similar word-processing software. During the iterative review process, changes are commonly written directly into the final report, allowing the reviewer direct control of their intended revisions and comments. This becomes problematic with the automatic generation of reports into pdf files or HTML pages, as changing the report actually does not change the source code. The source code, however, is in these cases written in a form most nontechnical people are not comfortable with. For example, in Sweave, the report is commonly written in LaTeX—which as a markup language few people are familiar with. But even for org-mode, which is much simpler and could be edited with any text editor with minimal knowledge, giving access to the org-document can be an issue as other team members are not used to the format or tools (Emacs in particular) and can be frustrated by "content over presentation" challenges. In the end, we settled on providing everyone with the pdf file and asked for comments to be inserted into the pdf. While this is an overall acceptable solution, it does certainly require more time from the analysis team as they have to make all changes themselves and essentially copy and paste the input from other team members back into the org-mode document. This of course makes it a lot harder to know what the comments and changes were and it is very easy to lose track of who the comments originated from. It is furthermore also not really satisfactory to the commenters, as they lose direct control over what they want to say.

From a practical perspective, the workflow is a lot less convenient than with standard word processors. Inserting comments into a pdf is not generally supported by pdf readers, much less directly editing the text in the pdf itself. Second, Microsoft Word has a sophisticated "track change" mode that makes it very easy to see who has made which alterations in a document. While some of this functionality can also be found in other tools, we are not aware of anything as comfortable, easy to use, and well known by all members in such a diverse team as found our actual research project.

Another challenge we faced is that in large commercial organizations, certain standards are mandated by corporate policy and must be followed for efficiency reasons. It may be necessary that reporting documents have a certain file type (e.g., docx)—or even be produced by a specific program (e.g., Microsoft Word) using specialized, restricted formatting options and/or specific templates. Similar requirements may hold for presentations. Even if there is no strict requirement, it is fairly common that other people reuse a few slides from old presentations, for example, to give an overview on the current stage of all programs in a department. In such a case, it quickly becomes inconvenient when different tools have been used, making the copying cumbersome, say when reusing formatted text from a pdf presentation in PowerPoint.

8.3.5 Our Implementation

In the end, we decided to largely split the code from the final report for these reasons and have one org-document with the code of the project and a separate document with the final report. This approach has the advantage that in the code org-file, it is easier to focus on the code itself during programming and concentrate on its sensible ordering and a reasonable level of comments. Of course, this approach somewhat deviates from the philosophy of LSA. However, the disadvantages are mainly that it is somewhat harder to link a table or figure in the final report back to the code that produced it. This can be addressed by, for example, inserting the name of the script that produced the object in the footer of the plot or the caption of the table.

Another area where we deviated from the principle of LP is the length of the code blocks. In particular, and this also may just be the authors preference, we found it difficult to follow a program that heavily intersperses verbal text and code blocks that are not self-contained, making it overall harder to read than a single code block with a "sufficient" amount of comments in it. The main reason for this is that for short, not self-contained blocks, the dependencies were not immediately clear, that is, can it be executed on its own or do several other code blocks have to be executed first—and if yes, which? In an LP framework, a code block is intended to be tangled out with all other code blocks before execution. From looking at the org-document, it is hard to see, however, which other of the potentially hundreds of code blocks are also strictly needed to be executed first (e.g., to provide needed data or some preprocessing step) and which blocks are not needed. This becomes even more complicated when several hierarchical levels are used to structure the analysis (e.g., Level 1: preprocessing, with level 2 subheading for mRNA, SNP, methylation, and miRNA). In the end, we decided to go with an approach that is somewhat closer to the traditional programming practice but incorporates some aspects of LP:

- The whole analysis was written in one single org-mode document using different level headings to structure the analysis into logical parts.
- Functions that are reused at different parts of the code are to be written into separate code chunks and documented there. They should be tangled out into separate code files so that script files can load these functions as needed.
- Code chunks should be self-contained and if possible not be split. Each code chunk should either be a function that is used by some other code chunk (and loaded there as a library) or be self-contained in the sense that, at the beginning, all necessary data are loaded and processed and the results written out at the end.
- The code chunks should, if possible, not be longer than 100 lines.
- Where possible, functions should be written in abstract form and loaded as a library.
- If a code chunk is split because it is long and easier to document and understand in several parts, these parts should not span several sections of the code.
- Before each code chunk, a short description of the task to be done should be provided.
- Each self-contained code chunk should be tangled out into a separate file.

This way, we want to get the best of both worlds, the traditional programming practice as well as the advantages of LP. The code and script files when tangled out are each relatively short and perform a specific task that can be understood, if necessary, without the org-mode document and can also be executed independently as well. On the other hand, org-mode provides us with an easy way to structure all these files, giving an outline of the project that can easily be followed as well as additional documentation and explanations that would not easily fit into any single code file.

8.4 Additional Tools in Use in Our Project and Other Considerations

Aside from the issues with the LSA framework discussed earlier, there are several other connected aspects that we would like to discuss, namely, how to ensure reproducible execution of (long running) code, the usage of version control systems, the availability of the software used in the future, as well as alternatives to the LP style we used for documentation.

8.4.1 Code Execution

Considering the discussion earlier and taking into account the computation time of some of our analyses (about 24 h for parts), it was very important to have a more robust way of only executing the parts that have really changed. There, we used a make program. Originally intended for the compilation of software projects, it is just as useful for executing scripts in R in batch mode. In our case, the input to a script was usually a source or derived dataset; the outputs were tables, figures, or derived datasets. Different script files would depend on each other through their derived data. These dependencies can be entered into the makefile—the instructions for executing all files—and this way, it can be ensured that only those part of the scripts are executed for which the script itself or some input into the script has changed.

This approach did not work as smoothly as intended at the beginning. First, when tangling out code blocks into files, the timestamps on the corresponding files always got reset to the newest date. As most make programs use the timestamp of the files to decide if something has changed, the standard make programs would decide to reexecute all code after each tangling. This would, of course, defy a large part of the purpose of using a make program in the first place. After some investigation, we found that a make program exists that does not rely on timestamps and instead on hash codes of the files. This way, only the part of the code gets executed for which the dependencies have really changed and not just the timestamp. The make program we used is called makepp [3] and is free, mature, available on a large number of platforms, has a rich feature set, and performed well in our project.

Second, this approach using makefiles works best if the analysis can be broken up into self-contained parts with limited run-time. Therefore, as much as was feasible, we saved intermediate results of our analysis and broke up our analysis into smaller script files that only depended on intermediate output of previous scripts. This way, reexecution could be limited to parts of the analysis that changed, and it also became easier to understand the underlying code. And if at some points it really was not possible to reduce the run-time of a script, we used dedicated packages like *R.cache*, which provide reliable caching (if used correctly) in R itself rather than in the LSA toolset.

8.4.2 Report Generation

Of course, with every report being contained in a separate document, the question of the reproducibility of all the included figures and tables remains. In order to ensure that all these results are updated when the code is executed and some data has changed, the documents have to be written in a way so that automatically the new output is being used. This is easily achieved, for example, using LaTeX, as figures as well as tables can be imported from external files—although with tables it specifically has to be in a LaTeX-conforming format. Another option is to write the report documents again in org-mode (without the source code), as similarly to LaTeX, it can also import results from files. Each of these methods can export the final report into several formats, including HTML, pdf, and open-source word-processing formats. When including the compilation of these files in the makefile, even all reports get reproduced upon execution.

8.4.3 Version Control System

When several people collaborate on a project, managing changes to the code or final document becomes quickly difficult and error prone without a version control system (or a rudimentary, specialized version of it such as track changes in Word). For version control systems, essentially two different main types are available—client–server and distributed systems. One of the most well-known open-source client–server systems is Subversion, but there are many others, both commercial and open source. Examples of distributed systems are mercurial, bazaar, and git, among others.

In our project, we decided to use git for reasons that are mostly related to it being a distributed system and any of the others mentioned earlier would have been an adequate replacement. For a client–server model, the server has to be set up, made constantly available, and has to be maintained even after the end of the project for archiving purposes. In large organizations, additional hurdles exist if an internal server has to be accessible from the outside for external collaborators—in fact, standard operating procedures (SOPs) and security concerns make this almost impossible in some organizations. Distributed systems do not face many of these challenges as they have lower requirements. A central server is not necessary as each user has a complete copy of the repository on their own drive. In order to exchange code, only a shared file system is needed, which in many organizations is already the case and is centrally provided and maintained. This also makes external collaboration easier as solutions for exchanging data with someone outside a company usually exist as well. Aside from these advantages, the user does not have to be online to commit a change to the repository, and as a "commit" is only a local operation on the users file system, it is usually very fast. The same is true for creating branches, which can be very useful for organizing development in different parts of the project when used correctly.

Another problem in a large-scale project is the organization of the source data. Due to its size, checking it into repositories in the usual manner may not be possible or even desirable. However, data can also change over time and if code is written for a particular version of the data, this should be tracked as well. Furthermore, including the data in the repository would make it easier to copy it to other people and ensure that it is in the correct place. Depending on its size, making one copy available for every user may be impossible and then it should not be included in the repository. Our case of about 100 GB was the borderline in that respect. We decided to include it in git using the git-annex extension, making it available in a similar way as the code to all users. However, this brought its own set of problems. An alternative to this approach would have been to keep all source data in a single directory (with possible subdirectories) and making it available to everyone else on the same machine using symlinks, only copying it if it is needed on other machines. Of course, this way, different versions of the data are not associated with the appropriate code. To our knowledge, there is no system that would serve every situation well, and every project has to find a way to manage these problems taking the particular circumstances into account.

8.4.4 How to Ensure Future Availability of Software

The issue of software availability in the future is an important and difficult problem for projects with life spans across multiple software generations, which is fairly common in the health-care sector and its regulatory restrictions. If one wants to ensure that the results can be reproduced even years later, many challenges can occur, especially for large-scale projects with many dependencies. Today, open-source analysis software (e.g., R) have new versions being published every 6 months or so, which are not always backward compatible. Additionally, very often additional libraries, packages, or modules are needed to complete an analysis, which are regularly updated as well. In our case, we are using R ([4], version 2.14.0) and over 40 external packages from CRAN and Bioconductor. Especially for Bioconductor packages, it is very important that the version of R and the version of a package are compatible as otherwise unexpected errors can occur. However, these issues generally occur in any computational environment, and it is important to ensure that the right software version together with correct versions of supporting packages is saved for future use.

Another source of difficulty can come from the choice of parallelization framework, which can be deeply embedded in the operating system and in general be difficult to store for the future. For example, if Sun Grid Engine is used for parallelization—should its version be stored with the project? What about dependencies? In case of Linux—as source or a package? Are compilers necessary for compiling the source with all libraries? Should the whole Linux distribution on which it was run be stored as well? As we can see, this can quickly become very complicated. If one wants to address most/all of these concerns, using a separate virtual machine for the project is probably the easiest and most comprehensive option.

In our case, we contended ourselves with storing the source versions of the packages we used and documented the version of R, but did not go so far to develop a complete virtual machine.

8.4.5 Alternatives to Literate Programming

In our discussion earlier, the issues are mainly related to literate statistical analyses rather than LP in a strict sense, and throughout the project, we adhered to this concept. Indeed, it was very useful to have all code in one file and the organizational features of org-mode complemented this framework very well.

However, we still want to mention other techniques that can be used, specifically documentation generators (DGs). They are available for many programming languages and provide automatic tools to extract comments from the source code (usually written in a special format or at certain places) and collate them into a documentation document for the software project in various formats such as HTML or pdf. These can come in the form of built-in systems such as javadoc for java but efforts to provide a general framework across a large number of different programming languages exist as well (Doxygen is an example of such a project).

These differences also have implications for the tools that are used for LP versus DGs. Using documentation extraction tools only imposes minimal requirements on the tools or workflow used by a software developer. In general, any development environment is compatible with DGs as they only extract comments that are already supported in the programming language used. In order to use a DG, the user only has to change the style of commenting to conform to the DG, and everything else is done by a postprocessor that is used on the source code. Such systems are also available for statistical data analysis projects. For the statistical programming language R, the package Roxygen ports the Doxygen tool to R and provides other useful features for understanding a codebase apart from the documentation generation. Of course, the documentation is not in the format of a normal analysis report nor does it contain any results of the analysis. Therefore, the final report has to be written separately. Furthermore, DGs do not provide for a direct way to order the source code of separate files into a logical ordering. Thus, while such an approach is certainly appealing to a software developer due to the low overhead and little changes to the normal workflow, it goes only partway toward a full LP project and does not provide the internal logical structure.

8.5 Discussion

LSA is well established for small-scale projects (e.g., articles and homework) and works very well in these areas. However, its usability in large-scale statistical projects has not been thoroughly discussed to our knowledge, and with this article, we want to contribute our experience in such a setting.

At the beginning of our project, we had every intention to follow the philosophy of LSA very closely. However, during the course of the analysis, we found that very strictly adhering to it comes with its own set of problems and pitfalls as described earlier, and we found it to be practically impossible in a project such as ours. Overall, we found that LP is a very good way of structuring a large source codebase and document it at the same time. However, its extensions for reproducible research as implemented in many LSA tools were a lot more problematic, and after some experimenting, we relaxed or abandoned many of them.

Reproducibility of a project can also be achieved with LP but without the other features in LSA tools discussed earlier. Instead, we used a way of ensuring reproducibility that does not rely on a single tool but instead on a set of tools that is build up in a modular fashion:

- Org-mode for code development and documentation (can be replaced by Sweave, knitr, many IDEs + document generators)
- Any make-program to ensure reproducible execution of code (compatible with the tool earlier)
- Org-mode for dynamic report generation (again, can be exchanged against LaTeX or knitr)
- Any version control to follow the history of the code

Overall, we have gained valuable insight into how to do reproducible research for large exploratory projects and will continue to use these tools in the future. One takeaway message however also is that, despite reproducible research being discussed for over a decade in the statistical community, there are still a number ways the experience could be improved from a practical level.

8.5.1 Improvements for the Future

8.5.1.1 Microsoft Office

Although it is not so much the case in the informatics/engineering/statistical community where LP and other reproducible research tools originated from, Microsoft Word and its document formats *doc* and *docx* are the de facto standards for writing reports in many fields in academia and even more so outside of academia. The situation is similar for presentations, where PowerPoint is in a similar role. However, support for these formats has been somewhat lackluster. Exporting to Word is often only possible through the "Open Document Format for Office Applications" that is mostly natively supported by open-source word processors. Exporting a presentation written in LaTeX or org-mode to PowerPoint is even more difficult. Working in a big company where these formats are the de facto standard is a big impediment for reproducible research.

From our perspective, useful solution would be tools that can dynamically insert figures and tables into Word as well as PowerPoint documents. Some support like this is already possible (e.g., in R through the R2wd package), but they usually rely on Word/PowerPoint being available on the same computer (and therefore, these solutions do not work in a Linux environment). If they would be more widely available, cross platform, and easier to use, this would be a big step forward in our mind.

8.5.1.2 Table Support

Another issue that appeared is the support for tables in different document formats. First, let's look at the situation for graphics. There, support for PNG as well as JPEG graphics is relatively universal, both for creating them as output in data analysis languages such as SAS^{TM}, R, or MATLAB$^®$, as well as including them into reports in Word, LaTeX, HTML, etc. Using JPEG for photo-like images (smooth contrasts) and PNG for figures and plots (sharp contrasts) covers most needs for graphical output.

The situation is very different for tables. There, each document format has essentially its own way of describing tables but the support between them is rather poor. Directly including, for example, an HTML table into a LaTeX document or vice versa is not possible. Other possibilities, such as reStructuredText are also usually not supported. For Microsoft Word and PowerPoint, the situation is pretty much the same. To be sure, some tools exist that can convert between these formats (although we did not investigate how reliable they are), but this still requires to have several different versions of the same table on disk in order to be flexible w.r.t. the final report format.

In R, this can be a problem as there is no universal table format that can be used and easily exported to most others. Instead, it is up to each package author to support different output formats, and depending on your application, you may be lucky that, for example, org-mode or reStructuredText is supported—or not.

Of course, we also do not have a simple solution for these problems, but feel that improving this situation would certainly benefit reproducible research activities overall.

8.5.1.3 Headers and Footers

In the preceding sections, we have talked a lot about different ways of making research reproducible. One component of this is to allow a reader who sees some output (table or figure) to easily figure out which code produced it and thereby go back and check how the results come about. LSA does this by keeping the output and code closely associated, but it is also possible in LaTeX and org-mode by backtracking the name of the imported object to its source code.

Of course, if a figure is copied from one presentation to the next by someone, this link is broken and retracing the steps becomes a lot more difficult.

A very simple method of addressing this is to put a header and footer on each output that specifies the source code file that produced it and other information such as research program and date. Software such as SAS has very good support for these concepts, allowing for header, footers, and titles on figures as well as tables.

This is not the case in R. As mentioned earlier, a standard table format with these features is not available, and most packages that use tables at most give the option for a caption, but not a title, nor header or footer. The situation is similar for plots. Of course, it is not hard to add a header or a footer on a plot, but the lack of native support discourages its general use and impedes widespread availability. In our opinion, native support for these simple concepts in R would be the lowest hanging fruit to achieve if not reproducibility, then at least better traceability for results in the statistics community.

8.5.2 Summary

All in all, we have successfully used LP with data analysis tools to make large-scale, corporate, exploratory data analysis reproducible. Although the experience was not smooth, there is no good reason not to make research reproducible, even for large projects. In fact, we have had the experience where a colleague's use of literate techniques allowed for weeks of work to be reproduced in hours, with a large-scale project being pulled from an unsuccessful result to an extremely powerful and commercially viable result, and that has further driven our motivation toward making these work practices efficient.

We hope that our discussion will help some readers who contemplate using reproducible research more in their own analyses.

References

1. D. E. Knuth. *Literate Programming*. Number 27 in CSLI Lecture Notes. Center for the Study of Language and Information, Stanford, CA, 1992.
2. F. Leisch. Sweave, Part i: Mixing R and LATEX: A short introduction to the Sweave file formate and corresponding R functions. *R News*, 2:28–31, 2002.
3. D. Pfeiffer. Makepp: A compatible but reliable and improved replacement for make. http://makepp.sourceforge.net. Version 2.0. (Accessed on 2012.)
4. R Development Core Team. *R: A Language and Environment for Statistical Computing*. R Foundation for Statistical Computing, Vienna, Austria, 2011.

5. N. Ramsey. Noweb—A simple, extensible tool for literate programming. http://www.cs.tufts.edu/nr/noweb. (Accessed on 2012.)

6. A. Rossini. Literate statistical practice. In *Proceedings of the 2nd International Workship on Distributed Statistical Computing (DSC 2001)*, Technische Universität Wien, Vienna, Austria, 2001. http://www.ci.tuwien.ac.at/Conference/DSC.html

7. A. Rossini and F. Leisch. Literate statistical practice. *UW Biostatistics Working Paper Series* 194, University of Washington, Seattle, WA March 2003.

9

Practicing Open Science

Luis Ibanez, William J. Schroeder, and Marcus D. Hanwell

CONTENTS

9.1 Introduction

Science is a system for gathering knowledge and developing explanations and predictions about the universe in which we live. A central tenet of this system is that something is known only when multiple, independent observers agree on a common experience. That is, experiences (which are more commonly called experiments) are reproducible. As such, the scientific method is by definition open; it is only when independent parties precisely replicate an experience that experimental results are considered valid.

Despite this basic and obvious tenet of openness, the pressure to close science is growing. Due to the exceptional innovative power of science, commercial interests, and personal career pursuits, many scientists and research institutions are heading down the path of secrecy and strong protective measures. In addition, publishers that garner financial benefit from controlling the dissemination of scientific knowledge have been reluctant to openly share information and are under increasing financial pressure to protect what they consider their intellectual property. Consequently, one of the tragedies of the current era is that the term "open" must be prepended to science despite the fact that openness is a fundamental requirement of the scientific method. Open science is a redundant, descriptive phrase, yet it has become necessary to remind ourselves that we must maintain openness if we are to effectively practice science.

Countering the pressure to close science is of course the emergence of the Internet. This ongoing web revolution has given rise to near-zero-cost methods of disseminating information, meaning that the ability to share the results of scientific research has been greatly enhanced. It is more than a simple matter of playing nice and sharing with others, the increasing complexity of modern science demands sharing and collaboration, since large teams with multidisciplinary expertise are required to address current challenges and gather advanced knowledge (see Section 9.1.2). Thus, it may be that scientific progress will stall without greater openness, and scientists will have no choice but to share, and share more effectively.

The conflict between sharing and secrecy has been present since the earliest days of scientific practice. Some scientists have routinely hidden or encoded their data and often released it only as necessary to support their work or once personal career achievement was assured. However, it is clear that the practice of science is changing rapidly, with key players such as publishers and societies, as well as scientists themselves, under significant pressure to change their ways. Thus, the conflict is taking on deeper meaning and is nothing less than a revolutionary reevaluation of how we practice science.

9.1.1 Evolution of Scientific Community

As the scientific method was developed, the demands of reproducibility, and hence the need to share information, quickly gave rise to scientific societies and publications. For example, the Royal Society was established in 1660 with the enviable motto (translated from Latin) "Take nobody's word for it." Effectively, the role that this society and the many following it took was one of *community building*. Early on, meetings were held in which experiments were jointly performed (the earliest form of peer review), and eventually results were codified, published, and shared. Given the technology of the time, this process rapidly evolved into a journal-based system in which communications between scientists were collected and distributed (for a fee) to subscribing recipients (see Figure 9.1).

Fast forwarding to the current day, scientific publishing has grown into a multibillion dollar industry. It has served science well over the centuries by gathering information and sustaining scientific communities, including sponsoring conferences and facilitating the peer-review process. However, the Internet has unleashed new ways to grow and support communities; as a result, the publishers are feeling the inevitable financial pressures and demands for process change. The current, ponderous model of peer-review and paper-centric publication has been viewed by many, as far too slow, and limited in the amount of information that is conveyed. The main deficiency is that journal articles do not always provide the information necessary to reproduce a result.

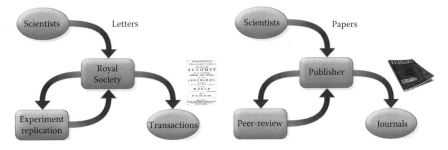

FIGURE 9.1
The evolution of the practice of the scientific method, from a society-oriented process of reviewing and verifying reproducibility (left) to today's peer-review-based publishing process (right). Without openness, the peer-review process cannot verify reproducibility.

The notion of what is a scientific community is also changing rapidly. Conventional publishers might choose to maintain their old ways (and profit margins) if they had a choice, but with the low cost of exchanging information and serving communities, it is clear that change will continue unabated. In the end, the publishers will only survive by returning to their roots: serving the scientific community. This may mean taking a supportive role by adopting new methods for curating, organizing, and coordinating scientific knowledge, as well as continuing the support of communities through various interaction forums, in particular, conferences, data hosting, and repositories, to name a few. In the mean time, many scientists and institutions are taking matters into their own hands and using reproducible methods such as those described in this chapter to further the reach of science.

9.1.2 Sharing Is Essential

Scientists are taught that the scientific method rests on three pillars of practice: experimentation, theory, and computation. Recently a fourth pillar has been proposed, data-intensive computing [1], although many consider it part of the computational pillar. Until recently, the standard publishing model that uses written articles to describe experimental apparatus, articulate theory, and codify computation was reasonably sufficient to support reproducibility and therefore scientific progress. Unfortunately, this model is no longer adequate: the complexity of experiment, theory, and computation is such that a brief paper cannot even begin to capture the detail necessary to describe an experiment to the point where it could be reproduced by a third party.

Consider a paper in computer science, an area in which the authors are familiar. A typical eight-page paper, or even an extended paper, can never describe the nuance behind complex algorithms. For example, an advanced algorithm may require many dozens of control parameters, not to mention internal data structures that can greatly affect performance and accuracy. In our experience, we have found that actually reproducing such an algorithm may require years of effort, and in doing so, we invariably fall back on the help of the original author who sheds light on "implementation details" that are frequently omitted in journal articles. The time demands to verify reproducibility are so large, that if we as authors were to reimplement algorithms for research purposes, it is unlikely that we would find the time to develop our own line of research. As a result, many experiments are never reproduced (especially in peer-reviewed documents), with the additional burden on researchers who spend inordinate time reimplementing what has been done before, due to lack of access to the original software implementation of published works. Thus, without the necessary sharing, the scientific endeavor is choking on complexity and resting on unstable foundations.

There are further, selfish motivations to practice openness: there is correlative evidence supporting the notion that sharing furthers a scientific career. Recently, [2] posted an article that suggests that sharing materials results in greater citation of the published material. Further, some argue persuasively that collaborative teams who, by definition, share information are the future of science [3].

9.1.3 Reproducibility and Open Source

The goal of this chapter is quite practical: to share some of the methods that we use in our practice of science to ensure reproducibility and encourage community building. To that end, we have learned much from our participation in various open-source projects, of which we are all developers and contributors. Indeed, the principles and practices of scientific reproducibility have been imprinted within the DNA of the open-source movement since its inception. This is no coincidence; it is the consequence of the fact that the open-source movement originated in an academic environment and more specifically it was kindled in research laboratories. In open-source communities, reproducibility is ensured through the practices of code review, unit testing, continuous integration, public documentation, open mailing lists, and forums. In this chapter, we show how these methods, and extensions to them, can be used in the practice of science.

In general, open-source communities are far ahead of most scientific communities when it comes to the practice of reproducibility. In large part, this is due to the rapid evolution of the open-source movement in response to the growth of the Internet and the web. In the meantime, most of the scientific community has remained constrained by the limitations of a process entrenched in practices that date back to the introduction of the printing press.

One of the goals of this chapter is to describe how the practices of open-source communities are being brought back to mainstream scientific research. This is based on many years of work developing open-source software for scientific applications. During this time, we have regularly interacted with scientific research teams, providing software engineering support for them, bringing their algorithmic implementations up to the standards expected in modern software engineering, particularly with regard to testing, and facilitated the reuse of their software and data, through widespread sharing of resources.

These practical experiences are presented here with the aim of encouraging the scientific community to adopt them in their daily work. Such adoption typically involves working at several levels simultaneously. We have found that it is important to work in parallel at the following levels:

- Cultural
- Rewards/recognition

- Career building
- Funding
- Technical training

A common hurdle to adopting open-source principles is the confusion that individuals and organizations face when classifying their challenges. For example, it is common to find that a technical difficulty can be misinterpreted as a cultural challenge or that a disincentive in the funding space is mischaracterized as a technical problem. As we go through the topics in this chapter, we will attempt to properly identify the challenges and opportunities in the adoption of reproducibility verification and how they relate to the specific levels listed earlier.

9.1.4 In Pursuit of Open Science

The open science movement is, at its core, an attempt to correct behaviors in the scientific community and return to an environment where reproducibility is again at the center of scientific research activities. Practicing open science requires four fundamental ingredients:

- Open data
- Open source
- Open access
- Open standards

Each one of these ingredients is necessary to realize the core aims of sharing results and stimulating scientific progress. Open data provides the opportunity for verification, analysis, and subsequent publication of scientific results in new forms. Open source embodies scientific methods, so that new computational processes can be independently examined, reviewed, and reused. Open access facilitates the review and validation of research processes and results. Finally, open standards, while not absolutely essential to open science, simplify the process of exchanging data, methods, and publications thereby accelerating the research process.

We see a broader role for open science and its impact on society. The web is opening up new lines of communication, providing access to scientific results that can be viewed by virtually anyone with an Internet connection. This includes the general public who often have strong interest in scientific research, for example, when learning about current medical treatments. However, current practices such as pay walls and overly aggressive data rights limitations are impeding realization of its full potential. The current situation requiring authors to sign away copyrights to publishers, despite the fact that results are often produced with the support of public funding, simultaneously represents an impediment to progress as well as a significant public subsidy to narrow business interests. Instead, permissive data rights using open licenses, such as CC0 or CC-BY, are necessary to return the spirit of community to science and ensure its role as an effective driver of innovation and major contributor to societal progress.

9.1.5 Organization

This chapter is organized into two major parts. With the introduction behind us, the first part consists of four sections discussing in general terms issues relating to open data, open-source, open access, and open standards. Next, the second part (in Section 9.6) elucidates specific tools and practices that the authors use in their daily practice of science. Naturally, this section has a strong software orientation as the authors are computational scientists (of sorts) by training. However, as science is becoming ever more computationally driven, we hope this material will be of interest to scientists of all persuasions. Finally, we conclude with a brief section on the future challenges of *practicing open science*.

9.2 Open Data

To a significant extent, the scientific method concerns itself with gathering, analyzing, and deriving data, partially to perform the essential work of acquiring knowledge but also to buttress explanations and support predictions. Data play different roles in each of the three scientific pillars of experiment, theory, and computation and naturally support each during the process of scientific investigation. For example, gathering data through experiments or direct measurement is necessary to subsequent data analysis, typically to develop theories of causality and correlation. On the other hand, theories are used to inform computation, which generates predictive output data, which is typically compared to experimental results to *falsify* [4] theories and refine computational models. *Thus, data serve as the focal point in the scientific workflow, and unfettered access to it is required for the scientific process to proceed efficiently.* Without such open access to data, the power of science to produce knowledge, and thereby drive innovation and economic progress, is severely impeded.

Despite the obvious necessity of unfettered access to data in order to support the scientific process, there are several powerful forces that create barriers. Many scientists place career goals above sharing, as a valuable data set may generate an important body of work and hence citations. Some scientists also insist on withholding data (at least for a short period) while they verify its correctness (and hence preserve their reputation). While these reasons certainly have merit in the real world, it is easy for them to become unbalanced behaviors that significantly impede scientific progress or, in the worst case, violate the core principle of reproducibility.

Finally, commercial interests represent another set of growing pressures to withhold data or impede its reuse: data can be withheld because it is deemed valuable, or copyright may be used to limit its distribution. What is unfortunate about these barriers is that, not only do they interfere with potentially specialized lines of research, but they all prevent large-scale meta-analysis across potentially thousands of data sets. For example, consider the case of automated access followed by statistical analysis over dozens of disparate data sets from a variety of sources. Having to formally request access to data one instance at a time is not feasible; automated meta-analysis depends on ready access and, to a lesser extent, open APIs (see Section 9.5). Furthermore, restrictive licensing can prevent derived data sets from being published (or severely limit the breadth of the source data).

9.2.1 Plan to Share Data

Until recently, the central role of data was implicit to the scientific process. Data was modest in size and could be exchanged among the community through tabulation of published results. Then, in the very recent past, data grew significantly in size. This has led to significantly more complex data sharing, requiring computers and associated storage media, such as tapes, floppy disks, and hard drives, to exchange information. At about this time, scientific publications began referring to external data sets; and now, with the advent of the Internet, these data could be placed on a public site and distributed across the scientific community. While this process continues today, the sharing process implicitly depends on the scale of the data. Modest-sized data can be exchanged and, if necessary, reacquired (if scientists decide not to share it) at reasonable cost. However, this process is changing rapidly as data size increases—it is becoming increasingly hard to exchange large data, and the reacquisition of data is often prohibitively expensive.

Consider the current state of affairs. The cost to acquire data or generate (simulate/calculate) data on a supercomputer can be extremely expensive. For example, the original cost to sequence the human genome was nearly one billion dollars. The size of data is growing rapidly too, with terabyte data sets becoming common and with petabyte data sets emerging (and exascale is on the horizon). At this scale, data is too large to easily exchange (the copy operation can take weeks or months even with high-bandwidth links) and too expensive to reacquire. Thus, the costs and practical data transfer considerations are driving science in a direction that absolutely demands better sharing of data. In the past, data was exchanged by researchers who felt the ethical obligation to share information for the purposes of advancing science.

Unfortunately, data was occasionally withheld, or publication delayed, for the purposes of validation, or worse yet, due to competitive career motivations. On a small scale, this had modest impact on the advancement of science; however, at the current scale, sharing data has become vital to scientific progress. In our opinion, it is imperative that scientists include data sharing plans as part of future funding proposals; indeed, many US federal organizations have put in place requirements for such plans. This is the case for the National Institutes of Health [5], for example.

9.2.2 Data-Centric Computing

The practical concerns related to the cost, size, and scale of "big data," combined with the philosophical motivations to publish scientific materials across the larger community, have led to new models of data distribution and curation. Data-centric computing [1] is one such response. In this approach, data is central to the workflow (Figure 9.2); once acquired, generated, or computed, the data are left in place in a "central" repository (in practice, the data repository can be distributed across the web depending on where it is acquired or computed). Access to the data is enabled through the web. Client–server architectures are employed wherein the server resides directly alongside the data, and clients are used to access, analyze, visualize, organize, and otherwise curate data. In addition, it is expected that extended

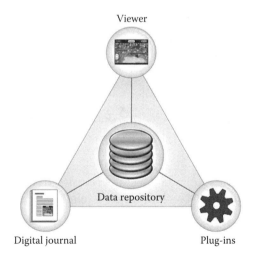

FIGURE 9.2
Data-driven scientific process. Data, once acquired or generated, are rarely moved. Rather visualization, analysis, and journaling process orbit the data repository. This requires special architectures (e.g., client–server) and to locate the computational resources close to the data.

research teams are working together on the data, meaning that simultaneous data access and collaboration must be supported.

The point here is that these data repositories represent significant scientific resources, and the work flow inevitably revolves around them. The scale of the data is so vast that multiple collaborative teams are required to ferret out useful information. With the expense and complexity of data, these data must be recognized as resources that are readily shared and are accessed through means of open standards and programming models.

9.3 Open Source

Much of today's science depends on computation, which to ensure reproducibility must be completely defined. Due to the complexity of computational methods, it is no longer possible for brief descriptions or pseudocode in a publication to properly characterize methods. The demands of reproducibility and hence open science require full disclosure, which means providing source code, execution parameters, and the computational environment—everything required to exactly reproduce an experiment.

A large portion of the practice of open science has been informed by the open-source movement. In this section, we describe the various methods used by open-source communities to ensure reproducibility. In particular, we focus on access to software through version control systems (VCSs) automated testing to ensure reproducibility and systems for code review to validate the correctness of the software.

9.3.1 Version Control and Provenance

Carefully tracking changes to scientific data, methods, and publications is essential to the scientific process, especially as part of the verification of reproducibility. This curation of scientific resources is also fundamental for educating future scientists, who will have the opportunity to inspect in great detail how particular experiments were performed in the past. In the open-source world, the ability to track changes is referred to as *version* or *revision control*.

Version control has been at the heart of the open-source software movement since its earliest days. Initially, version control was crudely implemented as a collection of tools to create patches on top of the original files, e-mail them, and apply the patches on the receiving end. Source code was freely available, approaches could be discussed, and changes proposed via these patches, clearly indicating changes made to the original. As this practice became widespread, more sophisticated tools were developed, but at their core, they were designed around moving patches more efficiently over the available transport options. Today, we are fortunate to have a wide array of sophisticated version control systems available, with many powerful and open-source systems under active development.

One of the most popular version control systems is Git, a project initiated by Linus Torvalds to manage the flow of patches produced by Linux kernel developers. It is perhaps one of the most sophisticated, offering a vast array of options that can be daunting for novices. It is part of a new generation of version control systems, called *distributed version control systems* (DVCSs). The major improvement over previous systems is that users can "clone" a repository to receive a full copy of the source code and all of the changes that were ever applied to it. This is in stark contrast to many alternative, centralized version control systems such as revision control system (RCS), version control system (VCS), and Subversion in which developers receive only the tip of the current branch and only the central repository server contains the complete history of the project.

There are many advantages to DVCS, one of the most important being the ability to easily create a local directory that can then be initialized and placed under version control, and then readily shared with others. New files can be locally committed and their contents stored. Later, when changes are made, these can be recorded in the form of patches applied to the original, giving the author total freedom to look at all previous versions of a given file. If the project grows, then sharing the project along with its history is simple, whereas centralized version control systems require up front planning and coordination with the central repository maintainer. Another important advantage to DVCS is that the source and all history can easily be mirrored at multiple locations, with private branches that can later be published in public repositories for all to see. This allows for work to take place in private when necessary, which can later be shared with full history.

Another often overlooked, but powerful capability of version control systems is the ability to track code provenance. Not only are the files and all modifications stored, the date and time of each modification is stored along with the author and a message detailing the reason(s) for the change. This tracking is implemented in the form of *commits*, which mark events in which a particular set of changes were applied to the content of the repository. Particular points in history can be *tagged* to indicate major events such as software releases and signed using encrypted keys to assure that a particular tag was signed by a given person using a cryptographically secure signature. Due to the nature of systems such as Git that uses special hashes to establish the identity of a commit, it is possible to detect alterations to previous commits that the signed commit depends upon, thus offering high levels of data integrity. It is not necessary to compare all files against a known good copy, just the hash of the commits you have to a signed copy that you trust. This offers a desirable degree of data provenance, using openly verified algorithms for establishing data integrity, that is difficult to obtain with other approaches.

9.3.2 Automated Testing

Moment of Zen:
What scientists call: *Experiments*
Open-source developers call: *Tests*

In this section, we equate the scientific concept of *performing an experiment* with the open-source practice of *running a test*. In today's world of scientific computational research, these two actions are one and the same.

The scientific principle of *verification of reproducibility* is implemented in open-source communities by relying on automated processes. The reason is simply that software systems have a natural tendency to develop into large and complex systems. In such an environment, the informal notion that *We attempt to replicate today an experiment that we did yesterday*, cannot be left to the fallibility of good intentions; it must be formalized.

The bottom line is that we are forced by practical necessity to script automated processes that can be run repeatedly. This is because the accumulation of *the things that we did yesterday*, and the ones done the *day before*, and the *day before that one*, rapidly become a combination of thousands of experiments. Attempting to repeat them by manual execution guided with notes or plain memory, simply does not scale and discourages practitioners from actually running all the experiments.

Automation not only makes reproducibility practical, it also makes it reliable. By capturing the process describing how to repeat an experiment, automation forces the practitioner to script every single detail that is relevant to its execution. This means leaving nothing to informal processes, local idiosyncrasies, or good intentions.

The practical way to encourage developers and researchers to automate their tests is to ask them to run them on a **daily basis**. With a set of even tens of experiments that must be run every day, a methodology to automatically run these experiments emerges quickly. This is an example of how a **cultural** requirement leads to **technical adoption**. Unfortunately, the converse is also true. That is, in a laboratory environment that does not automate its experiments, the staff quickly grow accustomed to not running the experiments on a regular basis, reinforced by the excuse that it will simply take too long to, do it. While there are other more urgent tasks to tend to, automated tests will be neglected. This chicken and egg problem can be solved by working first at the cultural level and developing a sense of reputation and pride in the craftsmanship of being *the one who runs their experiments daily*.

Developers' reputations are built in open-source communities through practices of transparency, peer review, and accountability in a meritocratic process.

Transparency is achieved by publishing, on public websites, daily test results. It quickly becomes obvious who does and who does not run tests on a regular basis.

Peer review is performed by the larger developer community, who routinely scan test results as part of their daily software development work. The more formal practice of *code review* drills down into the changes that another developer may have made to the system (a prerequisite of code review is that all tests are run). During this exercise, it is easy to expose whether the original developers actually ran the tests before and after making changes. When a reviewer finds that a developer failed to run the tests, it is culturally expected that a public admonishment is in order. This is typically done in a cordial way, and sometimes with a humorous tone. The intention is not to provoke a confrontation, but to enforce a social norm. Not running tests in an open-source community is simply a *bad etiquette*. It is frowned upon, the same way as if you were to sneeze on a colleague's sandwich.

Accountability is a follow to the transparency and peer-review practices in open-source communities. It comes down to the implicit rule that *if you broke it, you fix it*, as a way of redeeming your reputation, with the caveat that if someone else fixes it for you first, then your reputation is damaged and the reputation of the person who fixed it is enhanced.

The combination of social, cultural, and technical practices builds an environment in which to be a good member of the community, tests are diligently maintained and run frequently. As a consequence, the testing process is automated in such a way that they can be run with minimal effort.

The notion that a *gradual* system of sanctions must be implemented in order to enforce compliance with community-established rules is one of the elements that Elionor Ostrom (Nobel Laureate in Economics 2009) identified as a result of studying self-governing communities who manage common resources in fields as diverse as fisheries, underground water basins, forests, and irrigation systems. Her contention is that a gradual system of sanctions is essential for the successful self-governance of the Commons, in the absence of government intervention or the use of property systems [6]. These are indeed the conditions under which both open-source communities and scientific communities operate on a regular basis.

9.3.3 Unit Testing

Unit testing is the translation of the principle of Occam's razor to the daily practice of software development. In particular, it is the quest for the minimal explanation for a given behavior. The goal of unit testing is to empower developers to rapidly pinpoint the root cause of problems in the software.

In particular, it is important to not rely on complex tests that involve the execution of thousands of sections of the software project. Otherwise, when a complex test fails, it is extremely difficult to figure out the root causes of the failure.

Unit testing takes the approach of verifying the correctness of the most basic components of the system, and in the process, to build confidence in the behavior of each component to the point where it is possible to rapidly locate which one of the many pieces of a software package is causing a problem.

Unit testing is not just a software practice—it is *a state of mind*. The practice is motivated by the same principles at the core of the quest for reproducibility verification in the domain of scientific research: *Acceptance of the fact that errors are ubiquitous*. Therefore, the only way to keep errors at bay is to continuously set traps for them at every corner of every experiment. The presumption is that *errors are indeed present*, and therefore, it is important to put in place tests that check for the presence of errors at every point in the process. It is the *continuous failure to find errors*, combined with the thoroughness of the testing efforts, that builds confidence in the correctness of the overall process.

The daily practice of unit testing also leads to the principle of decomposition, by which complex problems are partitioned into smaller units, and then those units are implemented and tested independently. This practice leads to better-designed software, which is clearly organized and easy to maintain. Requiring unit testing as a cultural practice forces developers to stay away from building large and complex pieces of monolithic software and to instead modularize their designs and build more general, robust, and reusable components.

Practitioners who employ unit testing write the test at the same time they write the code, in a rapid iterative process. They start with an empty piece of code, and then write a test for the first minimal feature. The tests will at first fail, given that the feature is not yet implemented. The developer will then implement the feature and bring it to the level where it passes the test. Note that it is important to ensure that the test fails prior to implementing the feature, thereby validating the test itself.

The successful practice of unit testing is closely tied to the application of agile methodologies in software development (see Section 9.6.2). The practice of unit testing requires that one writes features in small incremental cycles, designing, implementing, testing the code, and then iterating back to revisit the design.

This way of working is conceptually no different from what any experimental researcher should do on a regular basis, for example, checking that the chemical reactions that they are about to use are pure enough, verifying that the thermometer to be used in an experiment is actually in a working state and correctly calibrated, and, overall, ensuring that the experiment is performed in a controlled environment with as few uncertainties and systematic errors as possible.

9.3.4 Code Review

Code review is a fundamental practice of quality control in which developers review the changes made by their peers, in the quest to spot potential errors and unnecessary features and ensure consistency with the overall design and style of a project. There are different methodologies for implementing code reviews, but many elements remain common.

One form of code review, which is used in many software projects, unfolds by having developers perform reviews in an ad hoc fashion with heavy reliance on the version control system. Review often takes place after changes are merged with the main code base, producing a stream of revision control commits. Developers subscribe to a mailing list that sends an e-mail with the contents of each commit made to the repository. Developers take time to read through these commits, checking those relevant to them and either fixing any problems they notice or e-mailing reviews to the relevant development list (or via private e-mail channels). This form of code review is quite common in projects using centralized version control systems for those developers who have commit rights.

Another form of review that has been employed for decades is the practice of e-mailing patches to a development mailing list for review. Developers then respond with high-level reviews and/or line-by-line comments and then iterate and modify the patch until it is deemed ready for commit. This practice not only serves to foster higher code quality, but it educates new developers in the expected code style, pitfalls, and common practice of the software project. Several variations of this basic procedure include attaching patches to bug reports, performing review in the bug tracking system, or using dedicated code review platforms where patches can be uploaded.

With DVCSes, an alternative model has emerged. Distributed version control enables a developer to develop in a new, private development branch and to apply a sequence of changes to that branch in the form of commits. Developers can create as many of these branches as they wish. Given that every branch contains an independent history of the project, this mechanism enables developers to undertake modifications to the project without interfering with the work of other developers, yet with the ability to share their work with any of their peers. The developer can also push branches of their choosing to multiple remote locations.

Software tools for code review, such as Gerrit, support remote repository locations where branches are pushed with proposed changes. These changes can then be displayed in a web application for the entire community to see, with an associated set of access control lists specifying permissions for developers of the project. Developers are then able to work freely on their code as they normally would, and when the code is ready to be merged into the main code base, or reviewed by a wider audience, it is pushed to Gerrit. Once pushed, Gerrit reviewers can be assigned to a topic, and the system will notify them. They can then make general comments about a commit or comment on particular lines with questions or comments. These comments are seen by all users of the system, along with the author of the topic. The author can then respond to the review, possibly uploading edited versions of their commits, until the code is approved.

Once the code is approved, it is given a score indicating approval; this is also recorded using a mechanism recently added to Git called notes and uploaded along with the changes when they are merged. This creates a permanent record of who reviewed the changes along with links back to the review. If bugs are later found, it is possible to go back to the original review if more detail is desired beyond what was recorded in the version control system. The code review process can also be significantly enhanced using various automated build, test, and analysis techniques (such as those described later based in Section 9.6.2). Pretesting before committing to the main branch enables developers to assess proposed changes before inclusion into the system proper.

The use of code review can seem like an unnecessary drain on resources, but it is usually much cheaper to review and catch mistakes before they are merged than to track them down afterwards. If good tests are written and careful code review is performed, it is much easier to bring new developers into a project and empower them to make significant changes with less concern for inadvertently breaking the system. Often, new developers fail to adhere to established practices, which if caught in an initial code review can be corrected very early on. If such problems are missed, weeks or even months of development effort may pass before the errors are detected and fixed, with the added cost that a good deal of the development that happened in the meantime will also have to be corrected to conform with expected standards.

9.4 Open Access

The public dissemination of scientific knowledge is essential to promoting social and technical progress. Making the results of scientific research readily available to other research groups (and the public at large) stimulates fact-based discussion, facilitates verification of reproducibility, and empowers others to build upon previous results. Public dissemination also fulfills an educational role, by enabling interested parties to become familiar with research without requiring direct participation in scientific process. The assumption that scientific literature is only intended for the scientific community is one that does not acknowledge the responsibility that scientific research has to society at large, particularly in the cases where research has been made possible using public funding. For example, patient advocacy groups are requesting greater access to the results of medical research; they argue persuasively that when research is paid for with public funds, the results need to be available to the public.

9.4.1 Open-Access Journals

When many people think of open science and initiatives to promote it, open-access journals are typically first on the list. Open-access journals are defined as scholarly journals available online "without financial, legal, or technical barriers other than those inseparable from gaining access to the internet itself" [7]. For many, open access simply means publishing the results of scientific research in journal form, paying for publication either by charging the author(s) a fee to publish or asking the authors to absorb the cost by self-archiving (or publishing) journal articles on their own website. Open-access journals are becoming quite popular, and there has been a flurry of new journals in the last few years [8].

While an important first step, this simple view of open access as an open journal does a disservice to the cause of open science. Publishing a journal article, no matter how easy the access nor small the cost, does not guarantee reproducibility. Without associated data (open data) and methods (open source), experiments described in an article typically cannot be easily reproduced. Thus, many open-access journals also require submission of data and source code (see Section 9.6.3 later in this chapter for more details).

There are other interesting features that open-access journals provide including version control and review, as described in the following subsections. Another important aspect that is often neglected is the choice of licensing, where some open-access journals prevent commercial use, or derivative work, thereby blocking important reuse of published articles.

9.4.2 Versioning Documents

Similar to the arguments made in the previous section on software testing (Section 9.3.2), errors are pervasive and to be expected throughout complex endeavors such as scientific research. Accepting this reality, and controlling it, requires a continuous process aimed at identifying and correcting errors. Consequently, the venues used for sharing scientific information and disseminating results must provide mechanisms for capturing community feedback, tracking changes, and providing access to current documents as well as the previous versions.

9.4.3 Open Reviews

The open-minded experimentation around open-access journals offers an opportunity to reconsider many of the long-standing practices of scientific publishing, some of which have become long time traditions and deserve to be revisited given the emergence of the web.

One of the key aspects in which open-access journals can improve communications in the scientific community is the modification of the typical publishing workflow. In traditional journals, publication is delayed until after articles have been vetted by reviewers. This painstaking process can take years from the time of first submission to the time of publication. This delay greatly diminishes the value of the final publication, particularly in topics that are related to the rapidly evolving domain of computational research. The traditional review process also privatizes the conversation between the authors and reviewers, and by doing so deprives the community of valuable discussions and from the benefit of observing scientific discourse.

An alternative to the traditional closed-door, anonymous peer-review process is the practice of open reviews. This is a practice inspired by the self-regulation and self-certification processes that many online communities have adopted to curate their materials and to perform quality control on their content [9].

Open reviews blur the distinction between readers and reviewers, since they both have access to exactly the same amount of material. A reviewer is simply a reader who feels compelled and motivated to provide feedback to the authors. This is in contrast to traditional reviews that are performed by a select group of individuals who are considered to be experts in a domain. The notion of a *peer* in an open-source community is anyone who participates distinctions are made based on contributions, and authority is defined by meritocratic recognition.

Open reviews more directly honor the concept of *peer review* by empowering all our peers, not only a narrow group of selected experts, to share their views on the content of published materials. By not relying on the authority of experts, open reviews are better aligned with the tenet of the scientific method: *"to withstand the domination of authority and to verify all statements by an appeal to facts determined by experiment."* [10].

9.5 Open Standards

Open science is usually described as requiring three basic components: open data, open access, and open source. Providing these elements is enough to reproduce an experiment assuming that all information is provided. However, many practitioners of open science also advocate for a another element: open standards. If reproducibility is the goal, why is this additional element important?

There are several answers to this question. Pragmatically, using standards, or helping to create them, is an indication that a researcher is earnest about sharing and hence practicing open science. It may be true that open data, access, and source enable reproducibility; they do not necessarily make it easy. Using standards generally results in more efficient science as information can be readily accessed and analyzed, making life easier for other researchers. Open standards also enable large-scale analysis in which multiple contributions are combined to form new insights or build new tools. Consider the following examples: standards simplify access to multiple data sets from different research groups that can be combined to support analysis of larger information pools. Open document repositories can be analyzed (using methods from text analysis) to identify emerging concepts and determine relationships between lines of research; such information is important to science as well as investors and technology managers. And finally, well-designed and implemented code can be reused and combined to build powerful and useful software.

There is another way in which open standards support the scientific mission. That is, to ensure reproducibility, it is important to run experiments under controlled conditions. Therefore, open standards can also be thought of as data, software systems, and/or publications that are certified at a known state. This enables researchers to build on well-defined foundations; thus, a particular open standard specifies one of the components composing the environment of an experimental process, for example, standard data sets, software libraries, and even computing platforms. Such control is necessary when comparing algorithms or otherwise evaluating the performance of a computational system. For that matter, even supporting laboratory software used to acquire data may produce different results under the same conditions if not carefully controlled. Therefore, using open standards can remove experimental uncertainty.

It is not possible to say exactly what standards to use. Different research fields, ontologies, data composition, and software systems require different standards to support research and foster sharing. Moreover, as knowledge expands, standards must evolve as well. Therefore, the use of standards is a delicate balance between the demands of innovation and the requirements of sharing. However, it is important to distinguish between standards that are open and those that are not.

Open standards promote sharing and support the scientific mission. Many nonopen standards (which may claim to be open) permit reasonable and nondiscriminatory patent licensing fees that erect barriers to sharing and hence reproducibility. Generally, open standards are developed by collaborative teams, use permissive licensing free of licensing entanglements, are thoroughly documented with reference implementations, and are meant to be widely used [11]. An interesting twist to some forms of open standards licensing is that predatory embrace-and-extend tactics may be prohibited to prevent organizations with influential control over a market or technology area to game implementations and impose restrictions on how others use the standards [12].

9.6 Open Science Platform

In the previous sections, we described many of the motivations and basic concepts that drive the practice of open science. In this section, we provide concrete details of several key components that constitute our daily practice and workflow.

9.6.1 Midas Platform

As discussed earlier in this chapter, data-centric computing is critical to the practice of open science (Section 9.2.2). For many of our applications, we use the Midas platform [13,14], which is an integrated, open-source toolkit that enables the rapid creation of customized, integrated applications with web-enabled data storage and management, advanced visualization, and processing (see Figure 9.3). The Midas platform is implemented as a modular PHP framework with a variety of backend databases (in particular PostgreSQL, MySQL, and nonrelational) that scales well to large data.

The Midas platform system can be installed and deployed without any customization; it has been designed with this capability in mind. Given that data-centric computing depends on diverse workflows and it is generally custom integrated depending on the needs of a project, there is no single solution that fits all possible applications. Therefore, the Midas platform supports additional extension mechanisms such as plug-ins and layouts to facilitate customization.

Some example customization efforts have led to the implementation of several different types of document database (see Figure 9.3) including the Optical Society of America's Interactive Science Publishing system [15] and the *Insight Journal* (described later in Section 9.6.3). The publication database is a specialization of the platform to support academic publications, for example, at the Surgical Planning Lab at Harvard Brigham and Women's Hospital, the NA-MIC project [16] uses the publication database to host content (papers, data, and images) from all contributors to the project [17]. The publication database is a digital repository for scientific papers and a computational infrastructure intended to facilitate the outreach activities of scientists. It provides a streamlined way to upload, present, and share the research and publishing activity from an institution. This is an example of a resource that can be used to implement *institutional repositories* and provide the mechanisms for practicing open access.

9.6.2 CMake-Based Software Process

As described earlier, the effective practice of open source depends on a rigorous software process. Our process relies heavily on the CMake, CPack,

(a)

(b)

(c)

FIGURE 9.3

(a) Midas is an open-source platform supporting data-centric computing. It has been used in a variety of data-intensive applications, ranging from (b) publication databases (which include data and images) to (c) advanced volume rendering.

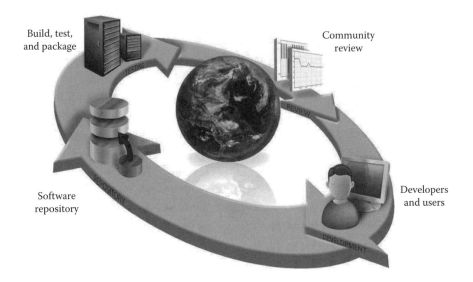

Build, test,
and package

Community
review

Software
repository

Developers
and users

FIGURE 9.4

A closed-loop software process depends on the CMake family of tools. CMake is used to build
software across multiple platforms. CTest tests software on a particular client platform; CDash
receives such test results and displays them on a web-based dashboard visible to the community. Finally, CPack is used to package and distribute code and executables for multiple
platforms.

CTest, and CDash family of tools [18], which we have organically developed
and refined over many years of developing large-scale open-source projects.
In addition, we prefer the Git DVCS, although we continue to use SVN (and
other VCS such as CVS when necessary). Basically, the software process
we use is highly automated, closed-loop, and convergent (Figure 9.4)—this
is vital to ensuring the stability of the software. As shown in the figure,
the software repository is constantly monitored for additions, and when
changes occur, the software is tested and the results displayed on a software-
quality dashboard (Figure 9.5). Developers and users monitor the dashboard
and correct any errors as necessary, pushing code changes to the repository and completing the cycle. The process runs continuously and hence
ensures reproducibility and informs users of the system of problems in a
timely manner.

In the following section, we describe the software process in more detail.
Along the way, we refer to several systems including the Visualization
Toolkit (VTK) [19] and the Insight Segmentation and Registration Toolkit
(ITK) [20]. These are examples of large-scale software systems that rely on
formal code review, with active communities of thousands of members, and
decades of use.

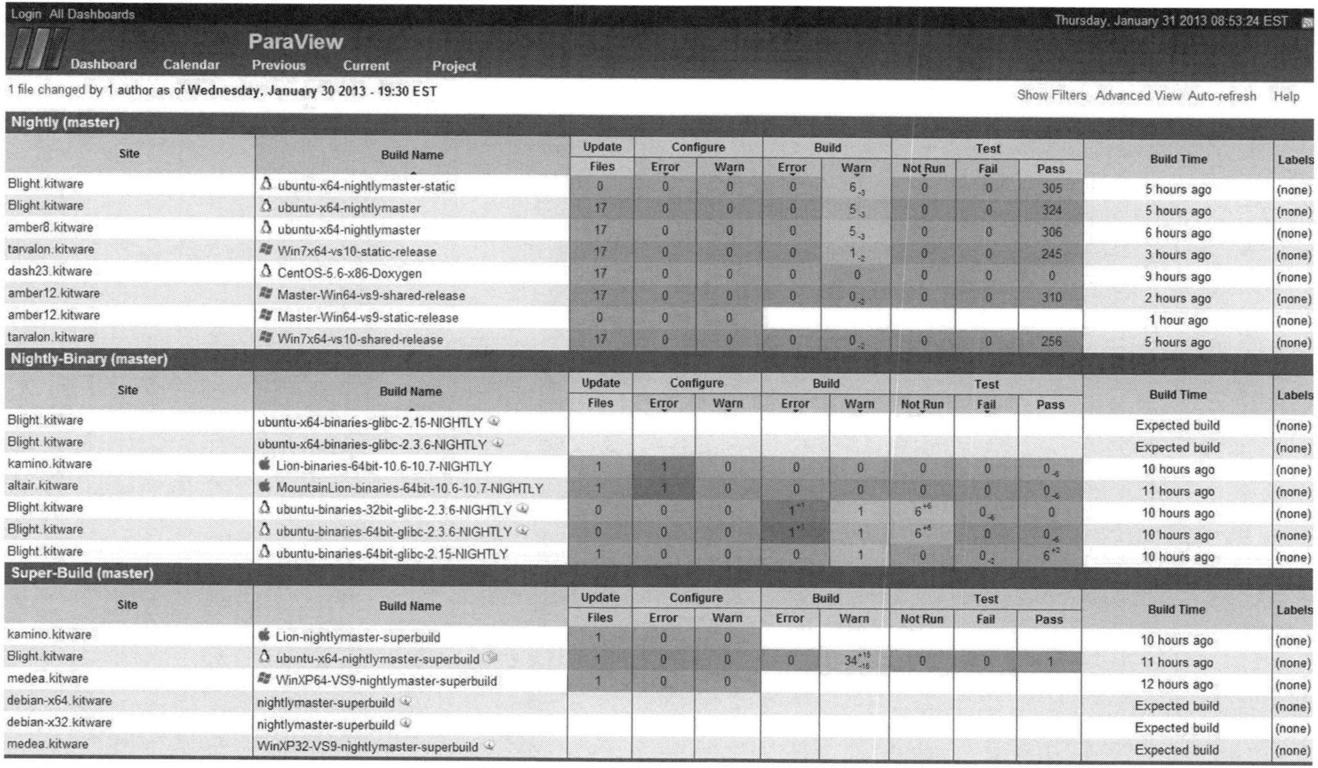

Nightly (master)

Site	Build Name	Update Files	Configure Error	Configure Warn	Build Error	Build Warn	Test Not Run	Test Fail	Test Pass	Build Time	Labels
Blight.kitware	ubuntu-x64-nightlymaster-static	0	0	0	0	6_{-3}	0	0	305	5 hours ago	(none)
Blight.kitware	ubuntu-x64-nightlymaster	17	0	0	0	5_{-3}	0	0	324	5 hours ago	(none)
amber8.kitware	ubuntu-x64-nightlymaster	17	0	0	0	5_{-3}	0	0	306	6 hours ago	(none)
tarvalon.kitware	Win7x64-vs10-static-release	17	0	0	0	1_{-2}	0	0	245	3 hours ago	(none)
dash23.kitware	CentOS-5.6-x86-Doxygen	17	0	0	0	0	0	0	0	9 hours ago	(none)
amber12.kitware	Master-Win64-vs9-shared-release	17	0	0	0	0_{-2}	0	0	310	2 hours ago	(none)
amber12.kitware	Master-Win64-vs9-static-release	0	0	0						1 hour ago	(none)
tarvalon.kitware	Win7x64-vs10-shared-release	17	0	0	0	0_{-2}	0	0	256	5 hours ago	(none)

Nightly-Binary (master)

Site	Build Name	Update Files	Configure Error	Configure Warn	Build Error	Build Warn	Test Not Run	Test Fail	Test Pass	Build Time	Labels
Blight.kitware	ubuntu-x64-binaries-glibc-2.15-NIGHTLY									Expected build	(none)
Blight.kitware	ubuntu-x64-binaries-glibc-2.3.6-NIGHTLY									Expected build	(none)
kamino.kitware	Lion-binaries-64bit-10.6-10.7-NIGHTLY	1	1	0	0	0	0	0	0_{-6}	10 hours ago	(none)
altair.kitware	MountainLion-binaries-64bit-10.6-10.7-NIGHTLY	1	1	0	0	0	0	0	0_{-6}	11 hours ago	(none)
Blight.kitware	ubuntu-binaries-32bit-glibc-2.3.6-NIGHTLY	0	0	0	1^{+1}	1	6^{+6}	0_{-6}	0	10 hours ago	(none)
Blight.kitware	ubuntu-binaries-64bit-glibc-2.3.6-NIGHTLY	0	0	0	1^{+1}	1	6^{+6}	0	0_{-6}	10 hours ago	(none)
Blight.kitware	ubuntu-binaries-64bit-glibc-2.15-NIGHTLY	1	0	0	0	1	0	0_{-2}	6^{+2}	10 hours ago	(none)

Super-Build (master)

Site	Build Name	Update Files	Configure Error	Configure Warn	Build Error	Build Warn	Test Not Run	Test Fail	Test Pass	Build Time	Labels
kamino.kitware	Lion-nightlymaster-superbuild	1	0	0						10 hours ago	(none)
Blight.kitware	ubuntu-x64-nightlymaster-superbuild	1	0	0	0	34^{+15}_{-15}	0	0	1	11 hours ago	(none)
medea.kitware	WinXP64-VS9-nightlymaster-superbuild	1	0	0						12 hours ago	(none)
debian-x64.kitware	nightlymaster-superbuild									Expected build	(none)
debian-x32.kitware	nightlymaster-superbuild									Expected build	(none)
medea.kitware	WinXP32-VS9-nightlymaster-superbuild									Expected build	(none)

FIGURE 9.5
A portion of a CDash dashboard (from the ParaView open-source project). The dashboard color codes errors and warnings and is heavily hyperlinked to provide access into the input and output of the build process.

9.6.2.1 Overview

CMake is an open-source tool for building complex software systems across multiple computing platforms. As the platform consists of various combinations of operating system, hardware, and system libraries, CMake manages this complexity in a relatively transparent way. Using CMake requires specifying dependencies on third-party packages and selecting options to enable or disable certain features and behaviors of the software package in question. By embedding this information in CMake scripts, it is possible to standardize the process of configuration for many different platforms and to store such rules along with the source code.

CMake itself does not perform builds, but instead focuses on the configuration process that will produce standardized builds. In particular, CMake generates native project build files according to the platform, for example, Unix Makefiles, XCode, Visual Studio, Ninja, or Eclipse. In this way, the rules written in build system files are carried along with the project and are maintained and tracked in the same version control system that the project uses.

The use of CMake facilitates the sharing of software for scientific research by empowering developers to configure software to run on a variety of platforms that range from embedded systems and laptops through to supercomputers. Examples of packages that use CMake include KDE, LAPACK, CLAPACK, ParaView, Trilinos, VTK, and ITK, which are a few of the thousands of software projects using CMake [21].

CTest is a companion tool to CMake; it is also open source and is distributed as part of the CMake package. The goal of CTest is to facilitate the process of running tests and reporting their outcomes to centralized sites. The daily use of CTest is quite simple. It is reduced to scripting the command-line instructions that one would have used to run the test manually. However, in the process of scripting it, the developer must face the following questions:

- Where is the input data?
- Where to generate the output data?
- What parameters are necessary to execute programs?

The fact that open-source developers confront these questions on a daily basis forces them to be quite organized and methodical. They must figure out how to refer to data regardless of the particular computer system, that is, being used to run the test.

A key entry in the driving CMakeLists.txt file are the commands that describe tests that will be run later with CTest. A typical entry looks like

```
add_test( executable input1 input2 output
          parameter1 parameter2)
```

This includes the location and identification of the input data and the fully defined set of parameters required to run the test. It turns out that this also provides documentation for the test itself, at a level of granularity that is rarely found in scientific publications.

CDash is a web tool that collects and summarizes the results of the CTest testing process across multiple platforms. The project dashboard (Figure 9.5) provides a rich set of hyperlinks that supports rapid navigation through the output of the build process, and even into the source code if necessary. Hence, compile errors, or test failures, are easy to find and analyze. There are also many filtering options that make it possible to, for example, determine exactly when a test started failing, which, in combination with the information provided by the revision control system to track changes in the code, is invaluable when determining what change caused a failure displayed on the dashboard.

Finally, CPack is used to automatically package and distribute software releases across multiple operating systems. This greatly simplifies the release process and enables frequent, rapid releases of software. This supports the open-source tenet of *"release early, release often"* by which software is often released on a daily basis.

9.6.2.2 Unit Testing

Unit testing is a software engineering practice that focuses on creating tests for the smallest possible functional units of the software being developed. This makes it possible to locate errors with a high granularity when they are introduced into the software.

In the particular case of ITK and VTK, which are object-oriented C++ libraries, the practice of unit testing is tightly coupled with the design and implementation of classes and their methods. In ITK, we start by writing an empty C++ class with something similar to the following pseudocode:

```
class itkNewImageFilter
  {
  public:
  };
```

and a test for it in the simple form

```
int main(int argc, char *argv[])
  {
  itkNewImageFilter filter;

  return 0;
  }
```

Then we would add a piece to the test:

```
int main(int argc, char *argv[])
  {
  itkImage inputImage;
  itkNewImageFilter filter;
  filter->SetInput( inputImage );
  return 0;
  }
```

and then proceed to implement such method in the class:

```
class itkNewImageFilter
  {
  public:
    void SetInput(itkImage image)
  { this->SetInternalImage = image; }
};
```

This may appear to be an agonizingly slow way to write software, but in practice, it is the fastest way to write software that *does not have to be rewritten*. It is a common mistake for developers to go in long stretches of writing hundreds or even thousands of lines of code and then, as an afterthought, attempt to write tests for them. The consequence is that by the time they start writing tests, they have already introduced many bugs and inconsistencies in their code. Such defects now have to be found and fixed through the much more expensive and laborious process of detective work. The average density of errors in the software industry is one bug for every thousand lines of code* [22].

This is with the caveat that during the debugging process, new bugs will possibly be introduced. It is known that about 50% of all bugs are introduced while the developer is trying to fix other bugs [23]. These *second-generation* bugs are the beginning of a nearly endless task, because, again, attempts to fix either one of those bugs will, half of the time, introduce *third-generation* bugs, and so on. The mathematically inclined readers would already be estimating that one original bug becomes $\sum 1 + \frac{1}{2} + \frac{1}{4} + \frac{1}{8} \cdots$ bugs.

The methodical process described is at the same level of rigor that one would expect from any well-trained experimental researcher. Therefore, there should not be any objections to the cultural adoption of these practices when developing software for research applications. To put it bluntly: if a research software developer does not have the discipline to write unit tests, then they are also likely to lack the discipline to be a

---

* The average bug density of open-source projects is 0.45 defects per thousand lines of code [22].

well-qualified experimental researcher. Once again, this is not a technical challenge but rather a cultural challenge. To incentivize reproducibility in scientific research, it is therefore necessary to work simultaneously on multiple fronts. In particular, providing technical tools, while at the same time ensuring specific behaviors are celebrated or condemned through the culture of a community.

9.6.2.3 Examples of Code Review

The ITK and VTK projects use Gerrit (which depends on the DVCS Git) along with a simple evaluation script to quickly check a proposed change for basic correctness. Meaning, the script enforces certain guidelines such as style and naming conventions including inappropriate white space, appropriate line length and termination delimiters, and hard-coded path names. In addition to these checks, the events generated by Gerrit are monitored by another system, which submits a build request to an automated build farm if a developer is in the core group of developers. This initial build request may be a subset of the entire test suite in order to enable a quick turnaround. This build test utilizes a system called CDash@Home to request a build of the proposed change on Linux, Windows, and Mac OS X [24,25] systems. Hence, the automated check-in evaluation process not only verifies that the project successfully builds on these common computing platforms but also runs some quick tests and submits the results (Figure 9.6).

As a result of this initial smoke test, reviewers can view the build on the set of core-supported platforms and compilers to check for any serious regressions, freeing them to concentrate on reviewing the substance of the change. Once merged into the main development branch, a larger number of machines download the new version of the code and proceed to perform more comprehensive tests. This practice of reviewing and testing of patches before they are even merged into the main code base enables us to maintain much higher stability on the main development branch than was previously possible and also better engages the community in the maintenance effort. The result is to significantly blur the once sharp line between committer (a developer with commit rights) and budding contributor (someone who is just beginning to learn and contribute to a project).

9.6.3 Insight Journal

The *Insight Journal* is an open technical journal built on the principles of open access, open data, and open source [26]. This online journal focuses on the domain of medical image computing and enforces the verification of reproducibility for all contributed articles. The *Insight Journal* went online in 2005, thanks to generous funding from the National Library of Medicine at the National Institutes of Health. The journal began as an effort to facilitate the sharing of image analysis algorithms in support of the ITK community.

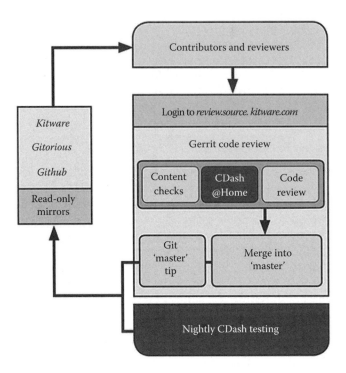

FIGURE 9.6
Graphical overview of the software process that incorporates Gerrit for code review, CDash@Home for pretesting, and CDash for nightly testing. Note the difference between those with write access and those without is reduced.

Today, there are several derivatives of the *Insight Journal*, such as the *VTK Journal*, the *Midas Journal*, and the *OSEHRA Technical Journal*, in use by other communities.

The creation of the *Insight Journal* was a response of the ITK developer community to the finding that a large number of papers published in the medical imaging field were not reproducible. While this is unfortunately common across other disciplines too, it was particularly frustrating to the development team of the ITK open-source software library. Initially, the team naively believed that published papers would have an associated open implementation necessary to produce useful results. Unfortunately, the culture of openness and verification in the open-source world collided with the failure of reproducibility that too often occurs in scientific research. The ITK development team found that for many algorithms, their publication in journals were too often just general guidelines to the overall flow of the algorithm and that the authors failed to provide a reproducible implementation covering all facets of the technique necessary to implement a working implementation.

From this experience, the community decided to create a journal of the type that would have been useful to the initial development of ITK. Such an ideal journal would require article submissions that included functioning source code, as well as tests and examples demonstrating the use of the code. These tests and examples further required the inclusion of all input data; and to support comparisons, the inclusion of the output data generated by running the contributed code. Finally, for each run of a test or an example, the article would include a full specification of the parameters necessary for it to run. The whole submission package, including the article, source code, tests, and data, would be available in its entirety to reviewers and readers of the journal under permissive licenses enabling them to download, use, modify, and redistribute the materials from the journal without having to involve the legal departments of their respective institutions. Based on these requirements, the *Insight Journal* was created.

The *Insight Journal*, and other similar open-access journals, fills a gap in the practice of scientific research by providing a venue where practitioners can share working versions of research code in a usable way. Despite the fact that the journal does not fit the traditional academic publishing model, which is mostly oriented to support career evaluations, it has become a key element of the ITK ecosystem. Running continuously for 7 years, it has (at the time of this writing)

- 3904 registered subscribers
- 540 published articles
- 821 reviews

The usefulness of the journal, as a vehicle for sharing contributions with peers, has been found to be extraordinary, although it currently does little to help academics score points essential for progression and tenure. However, since it enforces the verification of reproducibility, it is a real scientific journal that successfully facilitates communication across the research community, accelerating scientific progress by minimizing publication delays and providing an environment necessary for subscribers to use it in their own research. It is quite common for rapid dialogues to emerge between researchers and for members of the community to express appreciation at finding leading-edge computational tools, with associated data and documentation, which address their current challenges.

One of the major features of the journal is that it takes advantage of the near-zero costs to store and transmit data in today's networked world. In particular, it has eliminated most of the publishing restrictions that many traditional journals have inherited from the age of the printing press, including page limits, restrictions on number and type of figures, problems updating revisions, use of color, limitations on supplemental materials, and long turnaround cycles.

9.6.3.1 Practical Details

The journal follows well-established practices of open-source communities that are rooted in continuous openness and transparency and in particular heeds the mantra *release early, release often*. As a result, papers are published within 24 h of submission, allowing time only to remove spam submissions, followed by an open review process that is publicly visible to the entire community. This public process elevates the civility of the review dialogue while greatly accelerating access to the material contributed by the authors.

When we began designing the *Insight Journal* in 2005, one of the first concerns we had can be described this way: *We are inviting people on the Internet to send us arbitrary source code that we are going to compile and run on our machines.* It did not take long before we realized that an encapsulated environment was required to run these source code contributions in a secure way. The solution was implemented using the Xen virtualization platform [27], along with a process to launch a virtual machine on demand whenever an article was received by the journal. Thus, a web-based frontend triggers a request to launch a preconfigured virtual machine with the installed software tools and platforms required to run the code accompanying the submission. For example, the preconfigured VM has several recent versions of ITK, VTK, and CMake installed. A mechanism is provided to authors to specify the versions of ITK, VTK, and CMake required to build their submitted code. Automated scripts then take the source code from the submitted article package, copy it into the virtual machine, expand it, configure it, build it, and run the tests submitted by the authors. The results of the submitted tests are then posted as an initial, automatic review to the journal. In this way, readers are primed with the initial information as to whether the journal infrastructure was able to build and replicate the results that the original authors described in the submission.

Given that the authors can also submit revisions to their articles, along with modifications to the code and data (without having to go through editorial hurdles), the entire process unfolds in a rapid and agile manner. As soon as an article is submitted, notification is sent to all subscribers (about 3900 people), and to the ITK community mailing list (over 2300 people). The article is then made available for download, including the PDF document, all source code, test code, examples, and input data required to verify the content of the submission, as well as the output data resulting from executing the software on the input data. The goal is to facilitate reproducibility in a very pragmatic way, empowering any reader of the *Insight Journal* with the ability to rerun the experiments described in the article, with minimal effort, and verify or build upon the research described.

Any subscriber to the journal is able to contribute reviews to the article. The reviews are nonanonymous and are posted publicly. This spurs an open conversation in which the reviewer and the author(s) exchange views and ideas, point out errors, and suggest areas that could be improved. The entire community benefits from being exposed to the conversation captured within the journal website. By encouraging all members of the scientific community to participate, we cultivate an engaged and participatory community where we all share the responsibility and the opportunities for moving the science forward.

9.6.3.2 Community Involvement

From its inception, the *Insight Journal* encouraged authors to submit revisions of their papers, with corrections and ongoing improvements. Being free of the limitations of publishing on physical paper, we had the ability to correct any errors by simply allowing and encouraging authors to submit subsequent modified versions of their PDF documents and/or their source code, data, and configuration. This created a working environment suitable for spurring collaboration across the community.

The process was quite successful, and as the journal became more popular, readers and authors started to have conversations that led to improvements in the source code contributions. As this happened, it rapidly became evident that the process of uploading modified versions of the articles and source code, even though it was far more flexible than the traditional paper-based publishing venues, was too cumbersome when compared to other well-known agile open-source processes. More specifically, open-source projects routinely make modifications to their source code using version control systems (see Section 9.3.1). To honor this tradition, a second generation of the *Insight Journal* was put in place,* where every code contribution submitted to the Journal was automatically inserted into a back-end Git repository.†

The *Insight Journal* has dramatically collapsed the time from submission to publication, which has been enthusiastically embraced by the community. What used to be an arduous publication cycle of 2–5 years, now takes minutes with full disclosure. Not only is the code available almost immediately, but it is stored in an infrastructure that permits further development, improvement, and maintenance of the data, publication, and code.

9.6.3.3 Data Concerns

As the *Insight Journal* was adopted by the ITK community, it became clear that sharing data was more challenging than first thought and required subsequent modifications to the sharing process. Initially, we made it clear that licensing restrictions were to be minimal. We prefer that data are licensed using CC0 and similar nonreciprocal licensing models. One of our goals is

---

* http://www.kitware.com/blog/home/post/167.
† https://github.com/midas-journal.

to open up whole new fields of data reuse and meta-analyses. For example, we envision large analyses traversing hundreds or thousands of papers (and their data) to spot wider trends that the original researchers may have overlooked. This requires published work that uses open-access licenses and enables data mining with semantic meaning encoded and provides open APIs using open standards such as REST and XML/JSON to encode the results.

In addition, there were challenges managing data. Some of the issues we addressed include

- Large data set size
- Collections with a large number of data sets
- Limitations on access

To address those issues, the ITK community created a data access solution based on the Midas platform (see Section 9.6.1). Some of the more important features include

- Data revision control, based on content
- An API for downloading data on demand
- A mechanism for uniquely identifying data sets to be downloaded
- A mechanism for sharing data stored on local disks (for performance)

This collection of features enabled complex computational research scenarios such as the following:

- A research group gathers a data collection and uploads it to a database.
- The upload process generates unique identifiers for every data set, based on its content.
- A second research group decides to use this collection as input for a data analysis task.
- CMake scripts are written, which refer to the specific data sets to be used as input, and assign them to the specific executables of a computational experiment.
- An experiment is run by this second research group, which automatically downloads the data shared by the first research group and uses it as input to its computation.
- Finally, a third research group takes the source code and configuration provided by the second research group and replicates their experiment by building executables from the source code and downloading the original data shared by the first group.

With this infrastructure, it is possible to take a set of algorithms and run them rapidly on multiple data collections, a task that could have conceivably taken years of effort in the absence of such a computational platform employing the principles of open source, open access, open data, and open standards.

9.6.4 Scalable Computing

Modern science relies heavily on computation. Analytical processes can be used to tease relationships from data. Often, theories are simulated on a computer and compared with experimental results. Even the process of acquiring measurements relies heavily on computers: consider the image and signal processing that goes into observing stellar phenomena.

In our practice of open science, we do not rely on any single approach to perform computation. Typically, we employ open-source systems like Midas, VTK, and ITK to build custom applications. However, there are certain systems that we use when data becomes large and complex, or we need extra computing resources. We describe these systems in the following section.

9.6.4.1 High-Performance Computing

Throughout the world, researchers are increasingly turning to high-performance computing (HPC) to conduct their work. More often than not, this means making use of dedicated HPC resources that often have unusual computing environments. One of the major challenges is developing cross platform software that can run on these often specialized platforms. For example, some supercomputers do not provide graphics hardware, which means that applications that depend on OpenGL have to use software rendering. Further, the individual nodes of an HPC resource often run a limited version of the operating system (typically Linux), and as a result, significant work is necessary to port code designed for off-the-shelf desktop operating systems to work on such resources. These and other challenges are only going to become more pronounced as HPC moves toward exascale computing [28], in an environment where computational FLOPS are cheap (e.g., millions of computing cores) and I/O and data transfer are expensive (in terms of performance and energy costs).

The inherent complexity and challenges of HPC means that open-source software is essential to advancing the state of the art. There are several reasons for this:

- Open-source software can be more easily adapted to HPC platforms. There are minimal licensing issues and software engineers have full access to the code, which they can modify to fit to the platform.
- Problems can be more easily discovered and corrected since the code is not hidden. Debugging tends to be much easier. This is

particularly important as advanced parallel-computing algorithms use complex distributed and shared-memory techniques to maximize the performance of HPC resources.

- Computing time on these machines is typically limited and expensive, therefore carefully controlled, which makes verification of the correct operation of the code more important than ever.
- Commercial code is often licensed on a per CPU-core basis (or similar). With an explosion in the number of computing cores, the pricing model of commercial software causes dramatic increases in cost. In contrast, open-source software does not carry this burden.

In the following, we describe some of the HPC software that we use in our practice of open science. All of the open-source systems listed in the following use permissive, nonreciprocal BSD licenses.

VTK is a C++ toolkit (wrapped in Python, Java, and Tcl languages) developed by a large community of international contributors. It originated as companion software to a book on 3D visualization [19]. Now, nearly 20 years old (development started in 1993) with millions of lines of code, it has served as a foundational computing tool for 3D graphics, scientific and data visualization, computational geometry, human–computer interaction, informatics, image and volume analysis, engineering simulation, and more. The system is inherently portable and has been run on systems ranging from the Raspberry Pi to some of the largest supercomputers (at the scale of hundreds of thousands of processors).

ParaView is an open-source, large-scale parallel visualization application leveraging VTK to provide visual data analysis for many data sources, including computational fluid dynamics, medical computing, engineering simulation, combustion, point clouds (from LIDAR or other imaging sources), climate simulation, and video processing [29]. ParaView employs an advanced client–server computing architecture that enables lightweight clients to connect with computing and/or graphics servers residing on an HPC platform. Typically run using distributed, parallel computing model, it can also leverage large shared-memory parallel systems.

ParaViewWeb is a client application providing a collaborative, remote web interface for 3D visualization using the ParaView server [30]. It also provides a JavaScript API for ParaView scripting, features, and capabilities. ParaViewWeb has been designed so that advanced visualization tools can be easily integrated into a web page, and multiple viewers can simultaneously view, interact with, and collaborate around data (Figure 9.7).

Catalyst is a data analysis and visualization library designed to be tightly coupled with simulation codes [31]. It was created in response to the unfortunate reality that HPC systems produce too much data to be fully captured (due to IO and disk limitations); thus, co-processing systems like Catalyst are embedded into the computing process to analyze and extract only

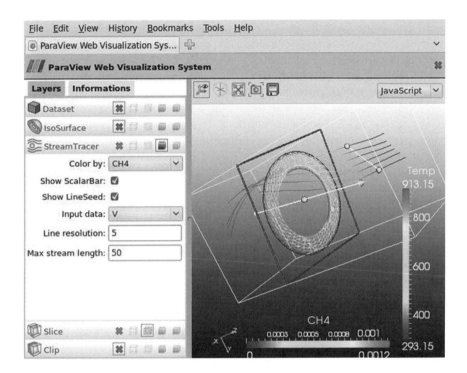

FIGURE 9.7
ParaViewWeb enables advanced, large data visualization capabilities through web clients, which in turn coordinate with a ParaView server that may reside on an HPC system. It also supports collaboration across multiple, simultaneous viewers.

essential data during computation. This also makes it possible to monitor long running analyses and control them during execution.

MoleQueue is an auxiliary application used to launch, monitor, and control HPC resources. Managing HPC systems is still a challenging task and MoleQueue makes it much easier by abstracting many of the differences between remote resources [32] and providing a simple API for client applications on the local desktop machine.

9.6.4.2 Science as a Service

In recent years, cloud computing has become an increasingly important part of scientific computing. Compared with HPC, cloud computing systems are quite similar in that the resources are time limited, and they often run a lean operating system. Thus, to create appropriate computational resources, distributed memory approaches must generally be employed. Indeed, one popular open-source package for scientific computing on the Amazon EC2 offering is StarCluster, which simplifies the process of configuring and deploying a Sun Grid Engine cluster on the Amazon platform that closely resembles a typical batch-scheduled HPC platform.

Once deployed, similar approaches to HPC can be used to schedule jobs and communicate between nodes. The communication between nodes is typically slower than purpose-built supercomputers, but time can be more easily purchased, and in some cases, the elasticity of the resource can be an enormous asset. Going forward, it is clear that the cloud will be a major part of the market for reproducible science, offering some unique opportunities.

Cloud providers, such as Amazon and Rackspace, make it relatively easy to customize the operating system running on one cloud instance and then deploy clones of it on one or more instances. These images can be shared publicly with others, enabling developers to produce pristine reference images of a full operating system where correct operation of the code has been tested. This means that others are able to verify results by purchasing time on the platform, deploying an instance with the reference image, and duplicating the reported results. When coupled with open-source codes and the Linux operating system, there is no restriction on distribution and even nonexperts have the opportunity to use complex codes where compilation, configuration, and deployment can take significant amounts of time and have now been made available to them ready to use.

The current widespread availability of cloud computing resources provides the opportunity to implement a reproducibility verification computational platform in a highly scalable way, without having to own and maintain the resources. By taking advantage of the network effects and the economies of scale, a full-fledged scientific computational platform is available at a cost that is very close the marginal cost of using the raw computational resources. A new scale of scientific research is made possible by these platforms, which will empower the computational research community to ask ambitious questions without having to add to their budget the full cost of large-scale resources. Large computational experiments are no longer the exclusive privilege of institutions that can afford the acquisition, installation, and maintenance of large computational resources such as clusters and supercomputers. Instead, it is now available to anyone, for the cost of the resources that are actually used during a given experiment.

As a result, another interesting development enabled by open science on the cloud is the development of a commercial marketplace for science [33]. The computational platform can be offered as a service to verify the reproducibility of reported results provided by a neutral third party. Such an approach is a complement to the [34], where a market-based system has been put in place to enable interested parties to contract services to replicate experiments from a set of trusted service providers. By delegating the experimental verification to organizations that have the suitable infrastructure, and that have a good reputation of being neutral and objective, opens up new possibilities in the practice of verification of experimental results, at a lower cost, thanks to a better allocation of resources and an open marketplace.

A typical scenario is for a pharmaceutical company to contract a *reproducibility verification service provider* to reproduce a set of bioinformatics experiments that they may have come across in a scientific publication. The original authors would have made available, as part of their publication, all the materials required to replicate their computational experiments, including the source code, data, and configuration parameters. The *verification provider*, who has a preconfigured and scalable computational platform, can then proceed to rerun those experiments and report back to its customer on the outcomes of these experiments. In terms of computational costs, the customer would only have to cover the cost of cloud resource usage incurred during the execution of the experiments, with no need to own and maintain the full computational platform.

At Kitware, we are currently experimenting with these and a variety of other open science practices in the cloud. We envision providing our advanced, open-source software tools to host data, support data-centric computing, and facilitate the sharing of scientific knowledge.

9.7 Challenges

The advent of the scientific method in the seventeenth century has enriched society in profound ways, from improving health to providing a multitude of goods and services to offering fundamental insights into the workings of the physical universe. Open science ensures that this legacy of innovation and understanding continues to address the challenges facing us in the twenty-first century and beyond. However, supporting open science comes at a significant cost; ensuring reproducibility requires resources for sharing and to nurture communities. Additionally, guaranteeing reproducibility in an ever-changing computing environment is difficult. And finally, with human reputation, recognition, and achievement on the line, we need to rethink the ways that scientists are evaluated and rewarded. These topics are addressed in the following paragraphs.

As described earlier, the size of data produced in science is growing at an enormous rate. Billions of dollars are spent to acquire or compute it; hence, it represents a scarce resource that cannot be easily replaced. Once it is collected, large data is expensive to store, provide access to, move, and analyze. In the past, data were often tabulated in paper publications and stored in a library; now, sophisticated data centers (including in some cases HPC support to process it) are required. This poses a problem in that very few institutions have the computing resources, or the wherewithal, to support such large-scale, data-intensive science. Fortunately, computing solutions are emerging (such as Amazon's EC2, EBS, and Glacier), but it remains to be seen whether commercial vendors are committed in the long-term to supporting scientific data. Supporting aging data is problematic as the justification for maintaining it wanes with the perception of declining value.

Another insidious problem is the shifting sands of the computing environment. Whether it is old software written in a programming language that has evolved or become obsolete or the computing platform (on which the software executes) that includes the operating system, software libraries, and hardware, computing environments change rapidly and play a major role in the reproducibility challenge. It is conceivable that software and data written in a certain era may no longer execute on future platforms. While open standards and commitment to backward compatibility do much to address this problem, computing environments have become so complex that it is hard to imagine indefinitely maintaining a reproducible configuration. Proposed solutions go so far as to propose virtual machines that are stored along with scientific software; however, there is no guarantee that future platforms will support existing VMs, and future computing architectures may be drastically different including high degrees of parallelism and based on distributed web resources.

Despite these technical obstacles, the biggest challenge may be addressing the entrenched scientific institution, social norms, and the way its various members and organizations interoperate. As described previously, the scientific publishing community is under siege due to their out-of-touch business model in the era of the Internet. Yet bigger issues remain including the tendency of some scientist's to be overly protective of their work (mostly due to the way they are evaluated for career rewards), which interferes with collaboration and community formation. This, despite the fact that the scale of scientific problems demands broader, collaborative expertise, and with strong evidence suggesting that sharing can be beneficial. Consider the open arXiv preprint server and open access PLoS journals that are examples of influential and successful scientific communities and Steve Lawrence's recent article in *Nature* that shows strong correlation between open access and the number of citations [2].

The scientific method has evolved over centuries of practice and has the enviable feature that it is self-correcting, with committed and passionate practitioners. Thus, despite these challenges, we are optimistic about the future of open science and the likelihood that its practice will be more open and collaborative than ever before. This optimism must also be tempered with the realization that changes of this magnitude can be generational, with new researchers quickly seeing the value of sharing if appropriate credit can be obtained when seeking career progression.

References

1. J. Gray. Chapter E-Science: A transformed scientific method. In *The Fourth Paradigm: Data-Intensive Scientific Discovery*. Microsoft Research, Redmond, WA, 2009.

2. S. Lawrence. Free online availability substantially increases a paper's impact. http://www.nature.com/nature/debates/e-access/Articles/lawrence.html (Accessed December 2, 2013).

3. A. Donald, In science today, a genius never works alone. http://www.guardian.co.uk/commentisfree/2013/feb/03/teamwork-science-transforming-the-world

4. K. Popper. *Conjectures and Refutations: The Growth of Scientific Knowledge*, 2nd edn. Routledge, London, UK, 2002.

5. NIH Data Sharing Policy and Implementation Guidance. http://grants.nih.gov/grants/policy/data_sharing/data_sharing_guidance.htm

6. E. Ostrom. *Governing the Commons: The Evolution of Institutions for Collective Action*. Cambridge University Press, 1990.

7. Open Access Overview. http://www.earlham.edu/~peters/fos/overview.htm

8. Directory of Open Access Journals. http://www.doaj.org/doaj?func=home&uiLanguage=en (Accessed December 2, 2013).

9. Wikipedia Editorial Oversight and Control. http://en.wikipedia.org/wiki/Wikipedia:Editorial_oversight_and_control

10. History of the Royal Society. http://royalsociety.org/about-us/history (Accessed December 2, 2013).

11. Open Standards. http://en.wikipedia.org/wiki/Open_standard

12. B. Perens. Is opendocument an open standard? Yes! http://www.dwheeler.com/essays/opendocument-open.html

13. J. Jomier, S.R. Aylward, C. Marion, J. Lee, and M. Styner. A digital archiving system and distributed server-side processing of large datasets. *Proc. SPIE*, 7264, 18, 2009.

14. J. Jomier, S. Jourdain, U. Ayachit, and C. Marion. A digital archiving system and distributed server-side processing of large datasets. *Proceedings of SPIE*. 7264, 726413, 2009.

15. OSA Interactive Science Publication. http://www.opticsinfobase.org/isp.cfm

16. National Alliance of Medical Image Computing NA-MIC. http://na-mic.org

17. The Publication Database hosted by SPL http://www.na-mic.org/publications

18. K. Martin and W. Hoffman. *Mastering CMake*. Kitware, Inc., New York, 2009.

19. The Visualization Toolkit VTK. http://www.vtk.org

20. Insight Segmentation and Registration Toolkit ITK. http://www.itk.org
21. CMake Programming Language Statistics. http://www.ohloh.net/languages/cmake (Accessed December 2, 2013).
22. Coverity scan: 2011 open source integrity report. http://softwareintegrity.coverity.com/coverity-scan-2011-open-source-integrity-report-registration.html (Accessed February 10, 2013).
23. Capers J. Quality quest. *CIO*, February 1995.
24. The CDash@Home Cloud. http://www.kitware.com/source/home/post/21
25. Code Review, Topic Branches and VTK. http://kitware.com/source/home/post/ 62.
26. The *Insight Journal*. http://www.insight-journal.org/
27. Xen virtualization platform. The Xen Project. http://www.xen.org/
28. DOE Office of Science. The opportunities and challenges of exascale computing. http://science.energy.gov/~/media/ascr/ascac/pdf/reports/exascale_subcommittee_report.pdf,2010 (Accessed December 2, 2013).
29. J. Ahrens, B. Geveci, and C. Law. Paraview: An end-user tool for large data visualization. In C.D. Hansen and C.R. Johnson, eds., *Visualization Handbook*. Elsevier, San Diego, CA, 2004.
30. ParaViewWeb. http://www.paraview.org/Wiki/ParaViewWeb
31. Catalyst: Scalable in-situ analysis. http://catalyst.paraview.org/ (Accessed December 2, 2013).
32. MoleQueue HPC Resource Manager. http://wiki.openchemistry.org/MoleQueue
33. Renee DiResta. Science as a service. http://radar.oreilly.com/2013/01/science-as-a-service.html
34. Exchange Reproducibility Initiative. https://www.scienceexchange.com/reproducibility

10

Reproducibility, Virtual Appliances, and Cloud Computing

Bill Howe

CONTENTS

In many contexts, virtualization and cloud computing can mitigate the challenges of computational reproducibility without significant overhead to the researcher.

10.1 Introduction

Science in every discipline increasingly relies on computational and data-driven methods. Perhaps paradoxically, these in silico experiments are often *more* difficult to reproduce than traditional laboratory techniques. Software pipelines designed to acquire and process data have complex version-sensitive interdependencies, their interfaces are often complex and underdocumented, and the datasets on which they operate are frequently too large to efficiently transport.

At the University of Washington eScience Institute, we are exploring the role of cloud computing in mitigating these challenges. A virtual machine (VM) can snapshot a researcher's entire working environment, including data, software, dependencies, notes, logs, and scripts. Snapshots of these images can be saved, hosted publicly, and cited in publications. This approach not only facilitates reproducibility, but incurs very little overhead for the researcher. Coupled with cloud computing, either commercial or taxpayer-funded [17], experimenters can avoid allocating local resources to host the VM, large datasets and long-running computations can be managed efficiently, and costs can be partially shared between producer and consumer.

In many cases, the application of virtualization to support reproducible research does not require any significant change to the researchers' workflow: the same experiments can be conducted in the virtual environment as in the physical environment, using the same code, data, and environment. When the experiments are complete, the experimenter will save a snapshot of the VM, make it publicly available, and cite it in all appropriate papers. Readers of the paper who wish to reproduce the results can launch their own instance of the author's virtual appliance, incurring no additional cost to the author (and, we will argue, only minimal cost to the reproducer), and reexecute the experiments. Further, new experiments, modifications, and extensions implemented by the reproducer can be saved in a new VM image and reshared as desired. The consequences and benefits of this basic model, as well as discussion of adaptations to support more complicated reproducibility scenarios, is the subject of this chapter. We provide examples from the literature of how this approach can significantly improve reproducibility with minimal additional effort and that this approach is largely complementary to other technologies and best practices designed to improve reproducibility. We will conclude with the remaining challenges to be overcome to fully realize this model.

10.2 Background on Cloud Computing and Virtualization

A VM provides a complete interface to a physical computer in software. This interface allows a complete operating system and any other software to run within a managed "virtual" software environment without requiring any direct access to the underlying "host" computer. A VM along with an operating system and other software can be stored, transported, and managed as a single file called an "image." The use of virtualization helps solve problems in a variety of areas technique can be applied to a huge number of problems: software can be tested on N different platforms without having to purchase and maintain N different computers. In a data center, if a physical machine fails, the VM can be migrated elsewhere, in some cases without interruption. Software developed for Windows can be used in a Linux environment, without complicated dual-boot scenarios. In the enterprise, IT departments can upgrade thousands of desktop environments by distributing a new VM image nching new VMs rather than physically installing software on each machine individually. In this chapter, we focus on using VMs to distribute software. This mechanism allows users to skip the installation step entirely—an attractive option for software with many complex dependencies.

Computing resources offered on demand, elastically, over the Internet—cloud computing—affords even more use cases for virtualization. Amazon Web Services (AWS), for example, allows VMs to be shared among users and launched on Amazon's hardware—users rent not only the software but the hardware on which to run it. This model has been wildly successful, allowing customers to get out of the business of administering computing resources and focus entirely on their business or their research.

10.3 Related Approaches and Examples

Lincoln Stein cogently argued why cloud computing and virtualization will be transformative for genome informatics [32]:

> Cloud computing ... creates a new niche in the ecosystem for genome software developers to package their work in the form of virtual machines. For example, many genome annotation groups have

developed pipelines for identifying and classifying genes and other functional elements. Although many of these pipelines are open source, packaging and distributing them for use by other groups has been challenging given their many software dependencies and site-specific configuration options. In a cloud computing environment these pipelines can be packaged into virtual machine images and stored in a way that lets anyone copy them, run them and customize them for their own needs, thus avoiding the software installation and configuration complexities.

Stein's argument focuses on the cloud's role for distributing *software* rather than distributing and reproducing specific experimental results, but the mechanisms are closely related. Dudley and Butte articulate the connection between cloud computing and reproducible research [11] and argue that the need for specialized, nonstandard software motivates the need to share complete operating environments ("whole system snapshot exchange") as opposed to packaged tools. Dudley and Butte consider many of the issues we discuss in this chapter: reproducibility in the context of large datasets and specialized computational environments and the economics of long-term preservation.

As Stein and others predicted, the use of VMs for the purposes of software dissemination is becoming commonplace in the life sciences [1,4,9,18]. For example, the CloVR project provides a set of metagenomic analysis tools and a browser-based graphical dashboard for interacting with them [14]. Both the tools and the dashboard are distributed as a VM, reducing installation effort for the client, eliminating the need for installation documentation by the providers, and trivially providing cross-platform support. Examples in other fields are emerging as well: the CernVM [31] simplifies the process to set up and run high-energy physics codes and now supports all experiments associated with the Large Hadron Collider [19], as well as many other experiments.

These applications of virtualization and cloud computing for tool delivery can help improve reproducibility by encouraging standardization on particular tools. However, new experiments tend to require new software that has not yet been packaged into clean reusable components. In this chapter, we consider the role of VMs and cloud computing even in these "early stage" software situations: *ad hoc* software to demonstrate experimental results as opposed to *engineered* software intended for long-term reuse.

In these situations, we can put current approaches to reproducibility into four categories: researchers can simply post their *raw code and data* on the web and rely on the published paper for documentation. This approach requires very little investment from the researcher, although they still need to find a place to host their materials. The effort required by the reproducer is significant, however, and this approach is generally not seen as sufficient (although it is still arguably an improvement over common current practices, where the code is simply not made available at all).

Beyond just posting the code and data, experiments and software can be equipped with *extensive documentation* for recreating the original experimental environment. This approach requires significant up-front effort by the experimenter but can improve long-term reusability and will typical improve near-term reproducibility as well. Brown et al. emphasize the use of source control repositories and explicit documentation for reproducibility and include instructions for reproducing results with every paper [7]. Their argument is that complete instructions for reproducing the environment can be more robust than providing a single instance of a working environment. However, reconstruction of the software environment typically requires a different set of skills than simply reproducing the experiments: installation of software using Linux package managers, the use of version control software, the configuration of environment variables, and sometimes familiarity with relatively advanced Linux tools such as *screen* [8]. As a result, this approach restricts reproducibility to those who have similar technical expertise to the authors. We conjecture that of those who wish to reproduce the results published in this way, some nonzero fraction will be unable or unwilling to do so due to the additional overhead of configuring the environment. But perhaps more significantly, Brown et al. are exemplars in providing thorough, clear, and accurate instructions for reproducing their results. For authors who are unable to justify the cost of this effort given that long-term reusability may or may not be one of their goals, virtualization offers a low-overhead "minimum bar" for reproducibility. Further, virtualization is entirely complementary to this documentation-oriented approach: a working example environment disseminated as a VM provides a means of checking that you have followed the authors, instructions properly and provides redundancy in case the instructions are incomplete.

A third approach is to adopt some kind of *controlled environment* in which to conduct your experiment that simplifies the metadata capture, provenance, logging, and dissemination process. These environments may be scientific workflow systems [16,35,36,38] or an augmented programming environments [10,28]. But in each case, the environment restricts the language, programming style, or libraries to which the researcher has access. Projects with specialized needs (large datasets, combinations of languages and libraries) may not be able to tolerate these restrictions, or the cost of reengineering the experiment to conform to the provided environment may make the value proposition unclear. We will discuss these solutions in more detail later in the chapter.

The fourth approach, described in this chapter, is to use VMs to capture and publish the code, data, and experimental environment.

Figure 10.1 illustrates how these four approaches compare in relative effort for the experimenter, those who wish to reproduce the experiments (left-hand plot) and those who wish to reuse and extend the experimental software (right-hand plot). Posting raw code and data requires little effort from the experimenter (lower half of each plot) but requires significant effort

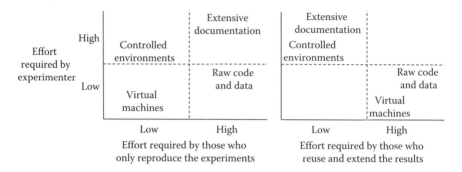

FIGURE 10.1
These four approaches to disseminating science software vary in the effort required by the original experimenter, those who wish to directly reproduce the results and those who wish to reuse and extend the software for other purposes. VMs incure very little overhead for the original experimenter and support direct reproducibility but are not sufficient for long-term extensibility. For extensibility, complete documentation is generally required, though some scientific workflow systems and other controlled environments offer a possible solution.

from both reproducers and extenders. Augmenting the code with documentation requires more up-front effort from the experimenter, and reproducers are required to reestablish the original environment from scratch, a task they may or may not possess the skills to do. However, this approach is critical to support those who wish to reuse and extend the software and adapt it for their own projects—there is no "shortcut" for software reuse. For the extenders, this documentation significantly reduces the effort required. Controlled environments also require some up-front effort but can significantly reduce the effort required by both reproducers and extenders. Finally, VMs impose very little overhead on the experimenter, and direct reproducibility of results is straightforward, but an undocumented VM with all software preinstalled does very little to support long-term reusability and extensibility, perhaps offering only a small improvement over providing the raw code and data. These approaches are not mutually exclusive; releasing a VM demonstrating particular results along with complete documentation for the requisite software is an appropriate strategy [8].

10.3.1 Other Uses of Virtual Machines

Beside reproducibility, the creation and exchange of VMs has other benefits for scientific knowledge sharing. First, VMs are also increasingly used to distribute software for educational purposes. Sorin Mitran at the University of Washington uses VMs in classes ranging from nontechnical first-year seminars to graduate classes to package the software environment for teaching scientific computing.* He finds that "using VMs allows a class to

---

* http://mitran.web.unc.edu/teaching/.

concentrate on the math and programming as opposed to installing all the utilities that come together to solve a problem." Second, VMs can be used to deliver custom prototypes and proofs of concept. Paradigm 4, the company that develops and distributes the SciDB database engine [33], routinely uses Amazon Machine Images (AMI) for customer projects. They develop a prototype on behalf of a customer and deliver it as an AMI, allowing the customer to reproduce results by running the scripts developed by P4. This approach provides a "try before you buy" mechanism that allows customers to experiment with the system without investing IT resources to install the software locally. Another facet of reproducibility exercised by the SciDB system is to adopt a "no overwrite" philosophy for operation of the system—all results are derived from previous results, affording complete reproducibility and provenance.

10.3.2 Toward Services Instead of Artifacts

A VM avoids the need to install unfamiliar software on a potentially new platform but still presumes that the reproducer can navigate your experimental environment and operate your code. Increasingly, we see bioinformatics tools exposed as a service, where users can interact with a web interface instead of the "raw" scripts and files. In the future, we can imagine publishing scripts as web-based interfaces directly, without going through an engineering project to do so. This kind of capability is among the goals of the HUBZero project [22].* Instances of the HUBZero framework for specific domains, for example, NanoHUB in the area of nanotechnology, allow analysis routines to be uploaded to a server, attached to simple graphical interfaces for passing parameters and viewing results, and executed remotely. We are evolving toward a "standard pipeline" of computational science that can expose an experimental result as a reusable and reproducible tool in a matter of hours.

10.4 How Cloud Computing Can Improve Reproducibility

10.4.1 Capturing More Variables

VMs allow researchers to share the entire context of their environment— data, code, logs, usage history, intermediate results, figures, notes, failed experiments, operating system configuration details, and more. Sharing at this level of abstraction mitigates most portability issues encountered when trying to install, configure, and run someone else's software.

---

* http://hubzero.org/.

The experimenter need not repackage their code for multiple platforms, and the reproducer need not install additional software or debug portability problems.

The VM provides an exact replica of the original "laboratory" complete with all variables—controlled and uncontrolled—intact. The success of reproducibility is not contingent on the experimenter's awareness and explicit control of every variable that might have influenced the outcome (library versions, operating system versions, subtle bugs). The analogy is a "crime scene": keep everything pristine so investigators can reconstruct the events that led to your paper.

10.4.2 Fewer Constraints on Research Methods

In many cases, no changes are required to your research methodology to use a VM (except for a one-time cost of establishing a virtual environment in which to work). You are free to use whatever operating system, languages, libraries, tools, conventions, and practices you see fit. Except for experiments requiring specialized hardware or large external datasets (cases we will consider later), any experiments that can be run on a local machine can also be run on a VM in the cloud. Other proposals to enhance reproducibility rely on the experimenter adopting some form of managed environment: language extensions and packages [27], technology-assisted metadata and documentation conventions [13,23,30], or scientific workflow systems with visual programming environments and advanced provenance features [3,20]. These systems offer enormous benefits in certain contexts, but they put significant constraints on the experimenter's research methodology: a particular language must be used, a particular programming style must be adopted, existing code must be ported to a workflow environment, or a particular documentation convention must be adopted.

In those contexts where these approaches are feasible, we advocate their use—they are generally compatible with (and complementary to) virtualization. Also, virtualization alone provides no support for managing provenance, typechecking workflows, generating documentation, or most other features provided by, say, scientific workflow systems. However, we contend that there will always be experiments performed (and data stored) outside the jurisdiction of any managed environment. Rather than ignore this data or rely on fiat, our approach is to cast a wide net to capture all data, all dependencies, and all possible variables.

The freedom to mix and match a variety of tools, to throw out one solution and rebuild another from scratch, and to reexecute one's experiment over and over at essentially zero cost are strengths of computational science that are not shared by most laboratory techniques. We should be conservative about sacrificing these properties by artificially constraining which tools can be used and how they may be combined.

10.4.3 On-Demand Backups

Snapshots of VMs, saved at regular intervals or on demand, offer a comprehensive (though noisy) laboratory notebook with minimal overhead. The entire state of the experiment, including controlled and uncontrolled variables, is saved and is immediately accessible. Returning to a previous state in the project involves identifying the appropriate snapshot and launching the VM. The overhead of saving many copies of nearly identical VMs is increasingly mitigated by deduplication and delta techniques [34]. Search and management services for a large set of related VMs are also beginning to emerge [2].

10.4.4 Virtual Machines as Citable Publications

VMs hosted on AWS, unlike those created and managed locally, are given a unique and permanent identifier that can be referenced in papers with no additional work by the experimenter. Concerns about longevity and preservation of public cloud resources can be addressed over time and need not be considered a limitation in the near term (we will discuss preservation and longevity issues in more detail in the next section).

10.4.5 Code, Data, Environment, Plus Resources

So far, all the reasons we have described apply to virtualization alone, whether or not the cloud is involved. Virtualization certainly predated cloud computing and has been used regularly in a variety of computing contexts for many years. However, only a public cloud computing environment (whether commercial or taxpayer-funded) provides not only the VM, but a host in which to run it that is identical to the original authors. An experiment with any significant resource requirements cannot be fully reproduced by simply downloading the VM to run on one's laptop. For example, *de novo* assembly tasks for analyzing short read *omics sequences typically use graph-based methods that require significant memory resources [39]. Such techniques are difficult to replicate in local environments, especially by small labs or individual researchers who cannot absorb significant investments in hardware. Cloud computing provides a common platform for anyone to use to reproduce experiments, at a cost that scales elastically with the resources required to run it. Sharing VMs via the cloud provides access to not only the code, data, and environment used by the experimenter but a carbon copy of the computing resources to run them.

10.4.6 Automatic Upgrades

Data and code hosted in the cloud automatically benefits from technology advancements with no additional effort from the experimenter or those who wish to reproduce their results. For example, VM images can be launched using new instance types—with, say, more memory, more cores, or special hardware*—as they become available. Additionally, cloud providers such as Amazon routinely release new capabilities. Consider Elastic MapReduce,[†] a new parallel processing framework natively deployed in AWS that can be applied to any data stored in their Simple Storage Service (S3). This software need not be installed or configured prior to use.

10.4.7 Competitive, Elastic Pricing

The prices of one EC2 compute unit, one gigabyte of RAM, and one terabyte of storage have all dropped over 50% in most usage scenarios since AWS was first released [15]. These price drops are automatically applied for all users, and in some cases, retroactively. The price drops generally reflect the falling cost of hardware and new economies of scale realized in the design of AWS.

10.4.8 Reproducibility for Complex Architectures

Computational experiments increasingly involve complex architectures, consisting of multiple servers interacting in application-specific ways: database servers [6], many-core architectures [12], and specialized resources such as GPGPUs [21]. In these cases, the code and the data are not enough. Reproducers need access to the specific hardware platforms or application architectures used in the experiment. The only choices are to document the platform carefully and hope those who wish to reproduce your results can build an appropriate environment from scratch or to provide outside access to one's own environment. AWS and other public cloud providers offer native facilities for precisely this purpose. Amazon's CloudFormation service allows configurations of related VMs to be saved and deployed as a single unit, making such complex architectures significantly easier to reproduce.[‡] Moreover, specialized hardware including GPGPUs and clusters with fast interconnects are now available on AWS.

10.4.9 Unfettered Collaborative Experiments

Two researchers at different institutions cannot typically work in the same development environment without one institution provisioning accounts, a process that undermines security and often takes weeks. A shared instance launched in the cloud provides "neutral territory" for developers to work in a common environment while maintaining local security. Reproducible

---

* http://aws.amazon.com/ec2/.
† http://aws.amazon.com/elasticmapreduce/.
‡ http://aws.amazon.com/cloudformation/.

experiments can therefore be shared among multiple researchers, potentially reducing the number of independent verifications of the same result.

10.4.10 Data-Intensive Computing

As science becomes increasingly data-intensive, reproducibility requires shared access to large datasets. Downloading large datasets to one's local environment to reproduce experiments simply cannot scale as datasets grow beyond a few terabytes. The only viable solution is to bring the computation to the data rather than bring the data to the computation. Unless each experimenter is equipped to open one's own environment to outside usage, the public cloud becomes the only platform that can both host the data and host the computation.

10.4.11 Cost Sharing

Reproducibility necessarily involves consideration of costs. By hosting data and code in the public cloud, the costs are shared by both the experimenter and those who wish to reproduce their results. In the simplest case, the experimenter incurs only the minimal costs of a single VM image: a 30 GB image will cost less than US$3.00 a month under most usage scenarios. Those wishing to reproduce the results launch their own instances from this image, incurring all relevant costs themselves. In more complex circumstances, the experimenter may need to pay storage costs to host large datasets. In some circumstances, Amazon and other cloud providers offer free hosting of public datasets in the interest of attracting traffic to their services, and online data markets are emerging that can more effectively share data storage costs between producers and consumers [5].

10.4.12 Foundation for Single-Payer Funding

Federal funding agencies require a data management plan to accompany all proposals [25]. Hosting one's research output in the public cloud lays a foundation for a single-payer system where NSF, NIH, and other agencies work directly with cloud providers to pay the costs of hosting scientific data. Individual investigators can put their research grants fully into science rather than having to plan, design, and implement a sustainable data management strategy. This approach also neatly solves a current obstacle to the uptake of cloud computing: universities inadvertently subsidize local deployments of hardware by charging large indirect cost rates on every dollar spent on "services," including cloud computing. Capital expenditures, including computing infrastructure, are ironically *not* subject to this overhead, despite incurring significant ongoing costs to the university in the form of power, cooling, space, and maintenance. By passing these funds directly from the funding agencies to the cloud providers, no overhead is charged.

10.4.13 Compatibility with Other Approaches

The approach we advocate—performing one's computational experiments inside VMs hosted in the public cloud—is compatible with and complementary to other approaches. If researchers begin to converge on a particular language, framework, workflow engine, file format, or coding convention, then these shared a VMs will become increasingly easier to (re)use. But it is unlikely that the VMs will become obsolete. We argue that there will always be exceptional circumstances that require an unconstrained programming environment, and VMs provide a "catch all" solution for such exceptions.

10.5 Remaining Challenges

Cloud computing offers a compelling approach to improving reproducibility in computational research, but there are risks and obstacles.

10.5.1 Cost

The economics favor a shift toward cloud computing in many situations, but not all. Storage in particular is still too expensive for science use cases, having been designed for a nearly opposite set of requirements: high availability, low latency, high-frequency access by a large number of concurrent users. In contrast, most science use cases need to store large amounts of data for use by relatively few individuals who can typically tolerate delays in accessing data or even occasional outages. In some cases, even lost data can be tolerated (e.g., simulations can be reexecuted). In return for this tolerance, they expect prices to approximate the cost of the raw disks, which is dropping precipitously.

There are three mitigating factors to this problem: First, there are programs to host public data at no charge.* Second, centralization in the cloud lays a foundation for a single-payer system to pay directly for publicly funded research, as we argued in the previous section. Third, the requirements for scientific data storage are not dissimilar from those of archival applications, suggesting that the cloud providers will soon offer lower-cost services with the performance characteristics described [14].

10.5.2 Culture

A more difficult problem to solve is one of culture. Researchers are accustomed to the "ownership" of laboratory equipment, and computing infrastructure is considered just a new form of such equipment. There is

------

* http://aws.amazon.com/publicdatasets/.

skepticism (and sometimes outright misunderstanding) of the cost, reliability, security, and longevity of cloud systems. This difficulty will diminish over time as computing is increasingly considered a utility (akin to power, telephone, or Internet) and less of a specialized solution designed expressly for one's unique requirements.

The implicit subsidies for purchasing and maintaining local hardware in the university context, as we considered earlier, must also be eliminated before cloud computing will be fully competitive.

10.5.3 Provenance

There is no *de facto* method of storing and querying the history of activity within a VM, apart from primitive methods provided by the guest operating system. Reasoning about the sequence of steps that led to a particular result is the real goal of reproducing another's work, and tools to manipulate the provenance of results are key enabler [24].

10.5.4 Reuse

A VM alone offers no assistance in reusing or extending software for new purposes. A preinstalled, preconfigured environment simplifies the direct reproducibility of specific commands but is a relatively opaque representation of the underlying technique and implementation. As illustrated in Figure 10.1, other techniques make different trade-offs in attempting to minimize effort for the experimenter and the consumers of their work, but we find that reproducibility can and should be separable from the far more general software engineering problem of reuse. In fact, demanding that all experimental results also deliver effective reusability may be harmful: researchers will tend to over-rely on existing, standard tools and services if they know that new tools will be difficult to properly publish.

10.6 Nonchallenges

Not all concerns about the application of cloud computing for scientific research are warranted in practice.

10.6.1 Security

Perceived security limitations of cloud-based systems are largely untenable. At the physical layer, it is not controversial to claim that the security of the data centers owned and managed by Microsoft, Amazon, and Google is more secure than the server room in a university lab. At the virtual layer, the system in the cloud is no less and no more vulnerable to attacks from the external Internet than local systems: firewalls are still enabled. Other potential vulnerabilities, such as those arising from the hypervisor itself, have had no significant impact on the uptake of cloud computing in the enterprise, and it is difficult to argue that science data—mandated to be made public by funding agencies—are substantially more sensitive.

For sensitive data (HIPAA, ITAR, etc.), the cloud may or may not be appropriate for a given institution or application. Exemplar applications demonstrating the feasibility of cloud computing for applications involving sensitive data do exist, however,* and the federal government is a significant customer.[†]

10.6.2 Licensing

Licensing issues may seem to complicate the use of public cloud infrastructures, but the problems appear to be transient. In some cases, licensing issues may actually be simpler in the cloud due to the rigorous administrative control of the underlying infrastructure, aggregate buying power of large cloud vendors, and the payment infrastructure already integrated. For example, one can rent fully licensed VMs equipped with Mathematica,[‡] and MathWorks offers a variety of demonstrations of using MATLAB® with AWS.[§]

10.6.3 Vendor Lock-In and Long-Term Preservation

Dependencies on commercial providers of goods and services are ubiquitous in all areas of science, but dependencies on vendors of computing infrastructure receive significantly more scrutiny. We have encountered enough instances of inaccessible data locked in university-owned resources that the service level agreements (SLAs) and relative mature infrastructure offered by cloud providers appear significantly more reliable for long-term access. In addition to commercial providers, open-source and taxpayer-funded efforts, sometimes modeled after AWS but designed expressly for research, are also emerging [17,26,29]. Moreover, high-performance computing facilities are increasingly interested in supplying cloud-like facilities, especially virtualization and on-demand provisioning [37].

The long-term role of cloud computing in the sciences is still evolving, but there appear to be immediate benefits for reproducibility in using and sharing virtual environments for computational experiments. Informed by

---

* http://aws.amazon.com/about-aws/whats-new/2009/04/06/whitepaper-hipaa/.
[†] http://fedcloud.gov.
[‡] http://www.wolfram.com/news/cloudcomputing.html.
[§] http://aws.typepad.com/aws/2008/11/parallel-comput.html.

these benefits, the University of Washington eScience Institute emphasizes cloud computing as a key strategy in enabling the next generation of rigorous and reproducible computational science.

References

1. E. Afgan, D. Baker, N. Coraor, B. Chapman, A. Nekrutenko, and J. Taylor. Galaxy cloudman: Delivering cloud compute clusters. *BMC Bioinformatics*, 11(Suppl. 12):S4, 2010.
2. G. Ammons, V. Bala, T. Mummert, D. Reimer, and X. Zhang. Virtual machine images as structured data: The mirage image library. In *Proceedings of the 3rd USENIX Conference on Hot Topics in Cloud Computing, HotCloud'11*, USENIX Association, Berkeley, CA, 2011, pp. 22–22.
3. E. Andersen, S. P. Callahan, D. A. Koop, E. Santos, C. E. Scheidegger, H. T. Vo, J. Freire, and C. T. Silva. Vistrails: Using provenance to streamline data exploration. In *Poster Proceedings of the International Workshop on Data Integration in the Life Sciences (DILS)*, Philadelphia, PA, 2007, p. 8. Invited for oral presentation.
4. S. Angiuoli, M. Matalka, A. Gussman, K. Galens, M. Vangala, D. Riley, C. Arze, J. White, O. White, and W. F. Fricke. Clovr: A virtual machine for automated and portable sequence analysis from the desktop using cloud computing. *BMC Bioinformatics*, 12(1):356, 2011.
5. M. Balazinska, B. Howe, and D. Suciu. Data markets in the cloud: An opportunity for the database community. *PVLDB*, 4(12):1482–1485, 2011.
6. BioSQL. 2011, http://biosql.org (Accessed on May, 2013.)
7. C. T. Brown. Our approach to replication in computational science. http://ivory.idyll.org/blog/replication-i.html (Accessed on May, 2013.)
8. C. T. Brown, A. Howe, Q. Zhang, A. B. Pyrkosz, and T. H. Brom. Running the diginorm paper script pipeline. http://ged.msu.edu/angus/diginorm-2012/pipeline-notes.html (Accessed on May, 2013.)
9. J. G. Caporaso, J. Kuczynski, J. Stombaugh, K. Bittinger, F. D. Bushman, E. K. Costello, N. Fierer et al. QIIME allows analysis of high-throughput community sequencing data. *Nature Methods*, 7(5):335–336, 2010.
10. A. Davison. Automated capture of experiment context for easier reproducibility in computational research. *Computing in Science and Engineering*, 14(4):48–56, 2012.
11. J. T. Dudley and A. J. Butte. In silico research in the era of cloud computing. *Nature Biotechnology*, 28(11):1181–1185, 2010.
12. F. J. Esteban, D. Daz, P. Hernndez, J. A. Caballero, G. Dorado, and S. Glvez. Direct approaches to exploit many-core architecture in bioinformatics. *Future Generation Computer System*, 29(1):15–26, 2013.

13. M. Gavish and D. Donoho. Three dream applications of verifiable computational results. *Computing in Science and Engineering*, 14(4):26–31, 2012.
14. J. Hamilton. Internet scale storage. Keynote presentation, *SIGMOD 2011*, Athens, Greece, 2011.
15. B. Howe. Cloud economics: Visualizing AWS prices over time. http://escience.washington.edu/blog/cloud-economics-visualizing-aws-prices-over-time, 2010. (Accessed on May, 2013.)
16. The Kepler Project. http://kepler-project.org
17. J. Klinginsmith, M. Mahoui, and Y. M. Wu. Towards reproducible escience in the cloud. In *CloudCom2011*, IEEE, Athens, Greece, December 2011.
18. K. Krampis, T. Booth, B. Chapman, B. Tiwari, M. Bicak, D. Field, and K. E. Nelson. Cloud biolinux: Pre-configured and on-demand bioinformatics computing for the genomics community. *BMC Bioinformatics*, 13:42, 2012.
19. Large Hadron Collider (LHC). 2013, http://lhc.web.cern.ch (Accessed on May, 2013.)
20. B. Ludäscher, I. Altintas, C. Berkley, D. Higgins, E. Jaeger-Frank, M. Jones, E. Lee et al. Scientific workflow management and the Kepler system. *Concurrency and Computation: Practice and Experience*, 18(10): 1039–1065, 2006.
21. S. F. Mahmood and H. Rangwala. GPU-Euler: Sequence assembly using GPGPU. In *HPCC*, Banff, Alberta, Canada, 2011, pp. 153–160.
22. M. McLennan and R. Kennell. Hubzero: A platform for dissemination and collaboration in computational science and engineering. *Computing in Science and Engineering*, 12(2):48–53, 2010.
23. L. Moreau, B. Clifford, J. Freire, J. Futrelle, Y. Gil, P. Groth, N. Kwasnikowska et al. The open provenance model core specification (v1.1). *Future Generation Computer Systems*, 27(6):743–756, June 2011.
24. K.-K. Muniswamy-Reddy, P. Macko, and M. Seltzer. Provenance for the cloud. In *Proceedings of the 8th USENIX Conference on File and Storage Technologies, FAST'10*, USENIX Association, Berkeley, CA, 2010, pp. 15–14.
25. NSF data management plan requirements. http://www.nsf.gov/eng/general/dmp.jsp
26. D. Nurmi, R. Wolski, C. Grzegorczyk, G. Obertelli, S. Soman, L. Youseff, and D. Zagorodnov. The eucalyptus open-source cloud-computing system. In *Cloud Computing and Its Applications (CCA '08)*, Chicago, IL, 2008.
27. R. Peng. Caching and distributing statistical analyses in R. *Journal of Statistical Software*, 26(7):1–24, 2008.
28. R. D. Peng. Reproducible research and biostatistics. *Biostatistics*, 10(3):405–408, 2009.

29. San Diego supercomputing center cloud storage services. https://cloud.sdsc.edu

30. E. Schulte, D. Davison, T. Dye, and C. Dominik. A multi-language computing environment for literate programming and reproducible research. *Journal of Statistical Software*, 46(3):1–24, 2012.

31. B. Segal, P. Buncic, C. Aguado Sanchez, J. Blomer, D. Garcia Quintas, A. Harutyunyan, P. Mato et al. LHC cloud computing with CernVM. In *Proceedings of the 13th International Workshop on Advanced Computing and Analysis Techniques in Physics Research*, Jaipur, India, February 22–27, 2010.

32. L. Stein. The case for cloud computing in genome informatics. *Genome Biology*, 11(5):207, 2010.

33. M. Stonebraker. Scidb: An open-source dbms for scientific data. *ERCIM News*, 2012(89), 2012.

34. P. Svärd, B. Hudzia, J. Tordsson, and E. Elmroth. Evaluation of delta compression techniques for efficient live migration of large virtual machines. *Virtual Execution Environments*, 46:111–120, 2011.

35. The Taverna Project. 2009. http://taverna.sourceforge.net (Accessed on May, 2013.)

36. The Triana Project. 2012. http://www.trianacode.org (Accessed on May, 2013.)

37. U.S. Department of Energy Office of Science Office of Advanced Scientific Computing Research (ASCR), December 2011.

38. The VisTrails Project. 2013. http://www.vistrails.org (Accessed on May, 2013.)

39. D. R. Zerbino and E. Birney. Velvet: Algorithms for de novo short read assembly using de bruijn graphs. *Genome Research*, 18(5):821–829, 2008.

11

The Reproducibility Project: A Model of Large-Scale Collaboration for Empirical Research on Reproducibility

Open Science Collaboration[*]

CONTENTS

---

[*] See Endnote at the end of the chapter for a listing of authors.

The goal of science is to accumulate knowledge that answers questions such as "How do things work?" and "Why do they work that way?" Scientists use a variety of methodologies to describe, predict, and explain natural phenomena. These methods are so diverse that it is difficult to define a unique scientific method, although all scientific methodologies share the assumption of reproducibility (Hempel and Oppenheim, 1948; Kuhn, 1962; Popper, 1934/1992; Salmon, 1989).

In the abstract, reproducibility refers to the fact that scientific findings are not singular events or historical facts. In concrete terms, reproducibility—and the related terms repeatability and replicability—refers to whether research findings recur. "Research findings" can be understood narrowly or broadly. Most narrowly, reproducibility is the repetition of a simulation or data analysis of existing data by reexecuting a program (Belding, 2000). More broadly, reproducibility refers to *direct replication*, an attempt to replicate the original observation using the same methods of a previous investigation but collecting unique observations. Direct replication provides information about the reliability of the original results across samples, settings, measures, occasions, or instrumentation. Most broadly, reproducibility refers to *conceptual replication*, an attempt to validate the *interpretation* of the original observation by manipulating or measuring the same conceptual variables using different techniques. Conceptual replication provides evidence about the validity of a hypothesized theoretical relationship. As such, direct replication provides evidence that a finding can be obtained, and conceptual replication provides evidence about what it means (Schmidt, 2009).

These features of reproducibility are nested. The likelihood of direct replication is constrained by whether the original analysis or simulation can be repeated. Likewise, the likelihood that a finding is valid is constrained by whether it is reliable (Campbell et al., 1963). All of these components of reproducibility are vitally important for accumulating knowledge in science, with each directly answering its own specific questions about the predictive value of the observation. The focus of the present chapter is on direct replication.

An important contribution of direct replication is to identify false-positives. False-positives are observed effects that were inferred to have occurred because of features of the research design but actually occurred by chance. Scientific knowledge is often gained by drawing inferences about a population based on data collected from a sample of individuals to make inferences about the population as a whole. Since this represents an example of induction, the knowledge gained in this way is always uncertain. The best a researcher can do is estimate the likelihood that the research findings are not a product of ordinary random sampling variability and provide a probabilistic measure of the confidence they have in the result. Independently reproducing the results reduces the probability that the original finding occurred by chance alone and, therefore, increases confidence in the inference. In contrast, false-positive findings are unlikely to be replicated.

Given the benefits of direct replication to knowledge building, one might expect that evidence of such reproducibility would be published frequently. Surprisingly, this is not the case. Publishing replications of research procedures is rare (Amir and Sharon, 1990; Makel et al., 2012; Morrell and Lucas, 2012; Open Science Collaboration, 2012). One recent review of psychological science estimated that only 0.15% of published studies were attempts to directly replicate a previous finding (Makel et al., 2012). As a consequence, there is a proliferation of scientific findings, but little systematic effort to verify their validity, possibly leading to a proliferation of irreproducible results (Begley and Ellis, 2012; Prinz et al., 2011). Despite the low occurrence of published replication studies, there is evidence that scientists believe in the value of replication and support its inclusion as part of the public record. For example, a survey of almost 1300 psychologists found support for reserving at least 20% of journal space to direct replications (Fuchs et al., 2012).

In this chapter, we first briefly review why replications are highly valued but rarely published. Then we describe a collaborative effort—the *Reproducibility Project*—to estimate the rate and predictors of reproducibility in psychological science. We emphasize that, while a goal of direct replication is to identify false-positive results, it does not do so unambiguously. Direct replication always includes differences in sample, setting, or materials that could be theoretically consequential boundary conditions for obtaining the original result. Finally, we detail how we are conducting this project as a large-scale, distributed, open collaboration. A description of the procedures and challenges may assist and inspire other teams to conduct similar projects in other areas of science.

11.1 Current Incentive Structures Discourage Replication

The ultimate purpose of science is the accumulation of knowledge. The most exciting science takes place on the periphery of knowledge, where researchers suggest novel ideas, consider new possibilities, and delve into the unknown. As a consequence, innovation is a highly prized scientific contribution, and the generation of new theories, new methods, and new evidence is highly rewarded. Direct replication, in contrast, does not attempt to break new ground; it instead assesses whether previous innovations are accurate. As a result, there are currently few incentives for conducting and publishing direct replications of previously published research (Nosek et al., 2012).

Current journal publication practices discourage replications (Collins, 1985; Mahoney, 1985; Schmidt, 2009). Journal editors hope to maximize the impact of their journals and are inclined to encourage contributions that are associated with the greatest prestige. As a consequence, all journals encourage innovative research, and few actively solicit replications, whether successful or unsuccessful (Neuliep and Crandall, 1990). An obvious response to these publication practices is to create journals devoted to publishing replications or null results. Of multiple attempts to start such a journal over the last 30 years, success is fleeting. Several versions exist today (e.g., http://www.jasnh.com/; http://www.jnr-eeb.org/; http://www.journalofnullresults.com/), but challenges remain: journals that publish what no other journal will publish ensures their low status (Nosek et al., 2012). It is not in a scientist's interest to publish in low-status journals.

Because prestigious journals do not provide incentives to publish replications, researchers do not have a strong incentive to conduct them (Hartshorne and Schachner, 2012a; Koole and Lakens, 2012). Scientists make reasonable assessments of how they should spend their time. Publication is the central means of career advancement for scientists. Given the choice between replication and pursuing novelty, career researchers can easily conclude that their time should be spent pursuing novel research. This may be especially true for researchers that do not yet have academic tenure.

Complicating matters is the presence of additional forces rewarding positive over negative results. A common belief is that it is easier to obtain a negative result erroneously than it is to obtain a positive result erroneously. This is true when using statistical techniques and sample sizes designed to detect differences (Nickerson, 2000) and when designs are underpowered (Cohen, 1962; Lipsey and Wilson, 1993; Sedlmeier and Gigerenzer, 1989). Although both of these features are common, researchers can design studies so that they will be informative no matter the outcome (Greenwald, 1975). There are many reasons why a null result may be observed erroneously such as imprecise measurement, poor experimental design, or other forms of random error (Greenwald, 1975; Nickerson, 2000). There are also many reasons why a positive result may be observed erroneously such as introducing artifacts into the research design (Rosenthal and Rosnow, 1960), experimenter bias, demand characteristics, systematic apparatus malfunction, or other forms of systematic error (Greenwald, 1975). Further, false-positives can be inflated through selective reporting and adventurous data analytic strategies (Simmons et al., 2011). There is presently little basis other than power of research designs to systematically prefer positive results compared to negative results. Decisions about whether to take a positive or negative result seriously are based on evaluation of the research design, not the research outcome.

Layered on top of legitimate epistemological considerations are cultural forces that favor significant (Fanelli, 2010, 2012; Greenwald, 1975; Sterling, 1959) and consistent (Giner-Sorolla, 2012) results over inconsistent or ambiguous results. These incentives encourage researchers to obtain and publish positive, significant results and to suppress or ignore inconsistencies that disrupt the aesthetic appeal of the findings. As examples, researchers might decide to stop data collection if preliminary analyses suggest that the findings will be unlikely to reach conventional significance, examine multiple variables or conditions and report only the subset that "worked," accept those studies that confirm the hypothesis as effective designs, and dismiss those that do not confirm the hypothesis as pilots or methodologically flawed because they fail to support the hypothesis (LeBel and Peters, 2011). These practices, and others, can inflate the likelihood that the results are false-positives (Giner-Sorolla, 2012; Ioannidis, 2005; John et al., 2012; Nosek et al., 2012; Schimmack, 2012; Simmons et al., 2011). This is not to say that researchers engage in these practices with deliberate intent to deceive or manufacture false effects. Rather, these are natural consequences of motivated reasoning (Kunda, 1990). When a particular outcome is better for the self, then decision making can be influenced by factors that maximize the likelihood of that outcome. Researchers may tend to carry out novel scientific studies with a confirmatory bias such that they—without conscious intent—guide themselves to find support for their hypotheses (Bauer, 1992; Nickerson, 1998).

11.2 Publishing Incentives Combined with a Lack of Replication Incentives May Reduce Reproducibility

The strong incentives to publish novel, positive, and clean results may lead to problems for knowledge accumulation. For one, the presence of these incentives leads to a larger proportion of false-positives, which produces a misleading literature and makes it more likely that future research will be based on claims that are actually false. Any individual result is ambiguous; but because the truth value of a claim is based on the aggregate of individual observations, ignoring particular results undermines the accuracy of a field's collective knowledge. This occurs both by inflating the true size of the effect and by concealing potential limitations to the effect's generalizability. Knowing the rate of false-positives in the published literature would clarify the magnitude of the problem and indicate whether significant intervention is needed. However, there is very little empirical evidence on the rate of false-positives. Simulations, surveys, and reasoned arguments provide some evidence that the false-positive rate could be very high (Greenwald, 1975; Hartshorne and Schachner, 2012a; Ioannidis, 2005). For example, asking psychologists about the proportion of research findings that would be reproduced from their journals in a direct replication yielded an estimate of 53% (Fuchs et al., 2012). The two known empirical estimates of nonrandom samples of studies in biomedicine provide disturbing reproducibility estimates of 25% or less (Begley and Ellis, 2012; Prinz et al., 2011). There are few other existing attempts to estimate the rate of false-positives in any field of science.

The theme of this chapter is reproducibility, and the focus of this section is on the primary concern of irreproducibility: that the original results are false. Note, however, that the reproducibility rate is not necessarily equivalent to the false-positive rate. The *maximum* reproducibility rate is 1 minus the rate of false-positives tolerated by a field. The ubiquitous alpha level of 0.05 implies a false-positive tolerance of 5%, meaning a reproducibility rate of 95%. However, in practice, there are many reasons why a true effect may fail to replicate. A low-powered replication, one with an insufficient number of data points to observe a difference between conditions, can fail for mathematical rather than empirical reasons.

The reproducibility rate can be lowered further for other reasons. Imprecise reporting practices can inadvertently omit crucial details necessary to make research designs reproducible. Description of the methodology—a core feature of scientific practice—may become more illustrative than substantive. This could be exacerbated by editorial trends encouraging short-report formats (Ledgerwood and Sherman, 2012). Even when the chance to offer additional online material about methods occurs, it may not be taken. For example, a Google Scholar search on articles published in *Psychological Science*—a short-report format journal—for the year 2011 revealed that only 16.8% of articles included the phrase "supplemental material" denoting additional material available online, even without considering whether or not that material gave a full accounting of methods. As a consequence, when replication does occur, the replicating researchers may find reproduction of the original procedure difficult because key elements of the methodology were not published. This makes it difficult both to clarify the conditions under which an effect can be observed and to accumulate knowledge.

In sum, both false-positives and weak methodological specification are challenges for reproducibility. The current system of incentives in science does not reward researchers for conducting or reporting replications. As a consequence, there is little opportunity to estimate the reproducibility rate, to filter out those initial effects that were false-positives, and to improve specification of those initial effects that are true but specified inadequately. The *Reproducibility Project* examines these issues by generating an empirical estimate of reproducibility and identifying the predictors of reproducibility.

11.3 Reproducibility Project

The Reproducibility Project began in November 2011 with the goal of empirically estimating the reproducibility of psychological science. The concept was simple: Take a sample of findings from the published literature in psychology and see how many of them could be replicated. The implementation, however, is more difficult than the conception. Replicating a large number of findings to produce an estimate of reproducibility is a mammoth undertaking, requiring much time and diverse skills. Given the incentive structures for publishing, only a person who does not mind stifling their own career success would take on such an effort on their own even if they valued the goal. Our solution was to minimize the costs for any one researcher by making it a massively collaborative project.

The Reproducibility Project is an open collaboration to which anyone can contribute according to their skills and available resources. Project tasks are distributed among the research team, minimizing the demand on each individual contributor but still allowing for a large-scale research design. As of this writing (March 2013), 118 researchers have joined the project, a complete research protocol has been established, and more than 50 replication studies are underway or completed. The project, though incomplete, has already provided important lessons about conducting such large-scale, distributed projects. The remainder of this chapter describes the design of the project, what can be learned from the results, and the lessons for conducting a large-scale collaboration that could be translated to similar efforts in other disciplines.

11.3.1 Project Design

To estimate the rate and predictors of reproducibility in the psychological sciences, we selected a quasi-random sample of studies from three prominent psychological journals (*Journal of Personality and Social Psychology*; *Journal of Experimental Psychology: Learning, Memory, and Cognition*; and *Psychological Science*) from the 2008 publication year—a year far enough in the past that there is evidence for variation in impact of the studies and variability in independent replication attempts and not so far in the past that original materials would not be available. Studies were selected for replication as follows: Beginning with the first issue of 2008, the first 30 articles that appeared in each journal made up the initial sample. As project members started attempting to replicate studies, additional articles were added to the eligible pool in groups of 10. This strategy minimized selection biases by having only a small group of articles available for selection at any one time while maintaining a sufficient number of articles so that interested replication teams could find tasks that match their resources and expertise.

Each article in the sampling frame was reviewed with a standard coding procedure*. The coding procedure documented (1) the essential descriptors of the article such as authors, topic, and main idea; (2) the key finding from one of the studies and key statistics associated with that finding such

---

* Linked resources are also available via the Reproducibility Project's page on the Open Science Framework website: http://openscienceframework.org/project/VMRGu/wiki/home.

as sample size and effect size; (3) features of the design requiring specialized samples, procedures, or instrumentation; and (4) any other unusual or notable features of the study. This coding provided the basis for researchers to rapidly review and identify a study that they could potentially replicate. Also, coding all articles from the sampling frame will allow systematic comparison of the articles replicated with those that were available but not replicated.

Most articles contain more than one study. Since the Reproducibility Project is concerned with the state of replicability in general, a single key finding was sampled from a single study. By default, the last study reported in a given article was the target of replication. If a replication of that study was not feasible, then the second to the last study was considered. If no studies were feasible to replicate, then the article was excluded from the replication sample. A study was considered feasible for replication if its primary result could be evaluated with a single inference test and if a replication team on the project had sufficient access to the study's population of interest, materials, procedure, and expertise. Although every effort is made to make the sample representative, study designs that are difficult to reproduce for practical reasons are less likely to be included. In psychology, for example, studies with children and clinical samples tend to be more resource intensive than others. Likewise, it is infeasible to replicate some study designs with large samples, many measurements over time, a focus on one-time historical events, or expensive instrumentation. It is not obvious whether studies with significant resource challenges would have more or less reproducible findings as compared to those that have fewer resource challenges.

11.3.2 Maximizing Replication Quality

A central concern for the Reproducibility Project was the quality of replication attempts. Sloppy, nonidentical, or underpowered replications would be unlikely to replicate the original finding, even if that original finding was true. While these are potential predictors of reproducibility, they are not particularly interesting ones. As a consequence, the study protocol involved many features to maximize quality of the replications. As a first step, each replication attempt was conducted with a sufficient number of observations so that replications of true findings would be likely. For each eligible study, a power analysis was performed on the effect of interest from the original study. The power analysis determined the samples necessary for 80%, 90%, and 95% power to detect a statistically significant effect the same size as the prior result using the same analytic procedures. Replication teams planned their sample size aiming for the highest feasible power. All studies were designed to achieve at least 80% power, and about three-fourths of the studies conducted to date have an anticipated power of 90% or higher.

In another step to maximize replication quality, replication teams contacted the original authors of each study to request copies of project materials and clarify any important procedures that did not appear in the original report (http://bit.ly/rpemailauthors). As of this writing, authors of every original article have shared their materials to assist in the replication efforts, with one exception. In the exceptional case, the original authors declined to share all materials that they had created and declined to disclose the source of materials that they did not own so that the replication team could seek permission for their use. Even so, a replication attempt of that study is underway with the replication team using its own judgment on how to best implement the study.

Next, for all studies, the replication team developed a research methodology that reproduced the original design as faithfully as possible. Methodologies were written following a standard template and included measurement instruments, a detailed project procedure, and a data analysis plan. Prior to finalizing the procedure, one or two Reproducibility Project contributors who were not a part of the replication team reviewed this proposed methodology. The methodology was also sent to the original authors for their review. If the original authors raised concerns about the design quality, the replication teams attempted to address them. If the design concerns could not be addressed, those concerns were documented as a priori concerns raised by the original authors. The evaluations of the original authors were documented as endorsing the methods of the replication, raising concerns based on informed judgment or speculation (which are not part of the published record as constraints on the design), raising concerns that are based on published empirical evidence of the constraints on the effect, or no response. This review process minimized design deficiencies in advance of conducting the study and also obtained explicit ratings of the design quality in advance. These steps should make it easier to detect post hoc rationalization if the replication results violate researchers' expectations.

Some studies that were originally conducted in a laboratory were amenable to replication via the Internet. Using the web is an excellent method for recruiting additional power for human research, but it could also alter the likelihood of observing the original effects. Thus, we label such studies "secondary replications." These studies remained eligible to be claimed for "primary replications"—doing the study in the laboratory following the original demonstration. As of this writing, there were more than 10 secondary web replications underway in addition to the more than 50 primary replications. This provides an opportunity to evaluate systematically whether the change in setting affects reproducibility.

Upon finalization, the replication methodology was registered and added to an online repository. At this point, data collection could start. After data collection, the replication teams conducted confirmatory analyses following the registered methodology. The results and interpretation were documented and submitted to a team member (who was not part of the

replication team) for review. In most cases, an additional attempt was made
to contact the authors of the original study in order to share the results of
the replication attempt and to consult with them as to whether any part
of the data collection or data analysis process may have deviated from that of
the original study. Finally, the results of the replication attempt were written
into a final manuscript, which was logged in the central project repository.
As additional replication attempts are completed, the repository is updated
and a more complete picture of the reproducibility of the sample emerges
(http://openscienceframework.org/project/EZcUj/).

The project is ongoing. In principle, there need not be an end date. Just
as ordinary science accumulates evidence about the truth value of claims
continuously, the Reproducibility Project could accumulate evidence about
the reproducibility, and ultimately truth value, of its particular sample of
claims continuously. Also, new resources provide opportunities to improve
and enlarge the sample of replication studies. For example, in February 2013,
the project received a grant of more than $200,000 to support replication
projects. The project team formed a committee and grant application process
to encourage more researchers to join the project and strengthen the study.
Eventually, the collaborative team will establish a closing date for replication
projects to be included in an initial aggregate report. That aggregate report
will provide an estimate of the reproducibility rate of psychological science
and examine predictors of reproducibility such as the publishing journal, the
precision of the original estimate, and the existence of other replications in
the published literature.

11.4 What Can and Cannot Be Learned from the Reproducibility Project

The Reproducibility Project will produce an estimate of the reproducibility rate of psychological science. In fact, it will produce multiple estimates,
as there are multiple ways to conceive of evaluating replication (Open Science Collaboration, 2012). For example, a standard frequentist solution is to
test whether the effect reaches statistical significance with the same ordinal
pattern of means as the original study. An alternative approach is to evaluate whether the meta-analytic combination of the original observation and
replication produces a significant effect. A third possibility is to test whether
the replication effect is significantly different from the original effect size
estimate. Each of these will reveal distinct reproducibility rates, and each
offers a distinct interpretation. Notably, none of the possible interpretations
will answer the question that is ultimately of interest: At what rate are the
conclusions of published research true?

11.4.1 Of the Studies Investigated, Which of Their Conclusions Are True?

The relationship between the validity of a study's results and the validity of the conclusions derived from those results is, at best, indirect. Replication only addresses the validity of the results. If the original authors used flawed inferential statistics, then replicating the result may say nothing of the accuracy of the conclusion (e.g., Jaeger, 2008). Similarly, if the study used a confounded manipulation, and that confound explains the reported results rather than the original interpretation, then the interpretation is incorrect regardless of whether the result is reproducible. More generally, replication cannot help with misinterpretation Piaget's (1952, 1954) demonstrations of object permanence and other developmental phenomena are among the most replicable findings in psychology. Simultaneously, many of his interpretations of these results appear to have been incorrect (e.g., Baillargeon et al., 1985).

Reinterpretation of old results is the ordinary process of scientific progress. That progress is facilitated by having valid results to reinterpret. Piaget's conclusions may have been overthrown, but his empirical results still provide the foundation for much of developmental psychology. The experimental paradigms he designed were so fruitful, in part, because the results they generate are so easily replicated. In this sense, reproducibility is essential for theoretical generativity. The Reproducibility Project offers the same contribution as other replications toward increasing confidence in the truth of conclusions. Findings that replicate in the Reproducibility Project are ones that are more likely to replicate in the future. The aggregate results will provide greater confidence in the validity of the findings, whether or not the conclusions are correct.

11.4.2 Of All Published Studies, What Is the Rate of True Findings?

It is of great importance to know the rate of valid findings in a given field. Even under the best of circumstances, at least some findings will be false due to random chance or simple human error. While there is a concern that science may be far from the ideal (e.g., Ioannidis et al., 2001), there are little systematic data in any field and hardly any in psychology. There are at least two barriers to obtaining empirical data on the rate of true findings. The first is that accumulating such data across a large sample of findings requires a range of expertise and a supply of labor that is difficult to assemble. In that respect, one of the contributions of the Reproducibility Project is to show how this can be accomplished. The second is that, as discussed earlier, failure to replicate a result is not synonymous with the result being a false-positive.

The Reproducibility Project attempts to minimize the other factors that are knowable and undesirable (e.g., low power and poor replication design) and to estimate the influence of others. There are three possible interpretations of a failure to replicate the results of an original study:

Interpretation 1: The original effect was false. The original result could have occurred by chance (e.g., setting alpha = 0.05 anticipates a 5% false-positive rate), by fraud, or unintentionally by exploiting flexible research practices in design, analysis, or reporting (Greenwald, 1975; John et al., 2012; Simmons et al., 2011).

Interpretation 2: The replication was not sufficiently powered to detect the true effect (i.e., the replication is false). Just as positive results occur by chance when there is no result to detect (alpha = 0.05), negative results occur by chance when there is a result to detect (beta or power). Most studies are very underpowered (Lipsey and Wilson, 1993; Sedlmeier and Gigerenzer, 1989; see Cohen, 1962, 1992). Adequate power is a necessary feature of fair replication attempts. The Reproducibility Project sets 80% as the baseline standard power for replication attempts (Cohen, 1988) and encourages higher levels of power whenever possible. The actual power of our replications can be used as a predictor of reproducibility in the analytic models and as a way to estimate the false-negative rate among replications. For example, an average power of 85% across replications would lead us to expect a false-negative rate of 15% on chance alone.

Interpretation 3: The replication methodology differed from the original methodology on unconsidered features that were critical for obtaining the true effect. There is no such thing as an exact replication. A replication necessarily differs somehow, or else it would not be a replication. For example, in behavioral research, even if the same participants are used, their state and experience differ. Likewise, even if the same location, procedures, and apparatus are used, the history and social context have changed. There are infinite dimensions of sample, setting, procedure, materials, and instrumentation that could be conditions for obtaining an effect. Keeping with the principle of Occam's razor, these variables are assumed irrelevant until proven otherwise. Indeed, if an effect is interpreted as existing only for the original circumstances, with no explanatory value outside of that lone occasion, its usefulness for future research and application is severely limited. Consequently, authors almost never exhaustively report procedural details when writing about effects.

Part of standard research practice is to understand the conditions necessary to elicit an effect. Does it depend on the color of the walls? The hardness of the pencils used? The characteristics of the sample? The context of measurement? How the materials are administered? There is an infinite number of possible conditions, and a smaller number of plausible conditions, that could be necessary for obtaining an effect.

A replication attempt will necessarily differ in many ways from the original demonstration. The key question is whether a failure to replicate could plausibly be attributed to any of these differences. The answer may rest upon what aspect of the original effect each difference violates:

1. *Published constraints on the effect*: Does the original interpretation of the effect suggest necessary conditions that are not part of the replication attempt? If the original interpretation is that the effect will only occur for women, and the replication attempt includes men, then it is not a fair replication. The existing interpretation (and perhaps empirical evidence) already imposes that constraint. Replication is not expected. Replication teams avoid violating these constraints as much as possible in the Reproducibility Project. Offering original authors an opportunity to review the design provides another opportunity to identify and address these constraints. When the constraints cannot be addressed completely, they are documented as potential predictors of reproducibility.

2. *Constraints on the effect, identified a priori*: An infinitely precise description requires infinite journal space, and thus every method section is necessarily an abridged summary. Thus, there may be design choices that are known (to the original experimenters, if to no one else) to be crucial to obtaining the reported results, but not described in print. By contacting the original authors prior to conducting the replication attempt, the Reproducibility Project minimizes this flaw in the published record.

3. *Constraints on the effect, identified post hoc*: Constraints identified beforehand are distinct from the reasoning or speculation that occurs after a failed replication attempt. There are many differences between any replication and its original, and subsequent investigation may determine that one of these differences, in fact, was crucial to obtaining the original results. That is, the original effect is not reproducible as originally interpreted but is reproducible with the newly discovered constraints. The Reproducibility Project only initiates this process: For studies that do not replicate, interested researchers may search for potential reasons why. This might include additional studies that manipulate the factors identified as possible causes of the replication failure. Such research will produce a better understanding of the phenomenon.

4. *Errors in implementation or analysis for the original study, replication study, or both*: Errors happen. What researchers think and report that they did might not be what they actually did. Discrepancies in results can occur because of mistakes. There is no obvious difference between "original" or "replication" studies in the likelihood of errors occurring. The Reproducibility Project cannot control errors in original studies, but it can make every effort to minimize their occurrence in the replication studies. For example, it is conceivable

that the Reproducibility Project will fail to replicate studies because some team members are incompetent in the design and execution of the replication projects. While this possibility cannot be ruled out entirely, procedures including carefully detailed experimental protocols minimize its impact and maximize the likelihood of identifying whether competence is playing a role. Moreover, features of the replication team (e.g., relevant experience, degrees, publishing record) can be used as predictors of reproducibility.

The key lesson from this section is that failure to replicate does not unambiguously suggest that the original effect is false. The Reproducibility Project examines all of the possibilities described earlier in its evaluation of reproducibility. Some can be addressed effectively with design. For example, all studies will have at least 80% power to detect the original effect, and the power of the test will be evaluated as a predictor for likelihood of replication. Also, differences between original and replication methods will be minimized by obtaining original materials whenever possible and by collaborating with original authors to identify and resolve all possible published or a priori identifiable design constraints. Finally, original authors and other members of the collaborative team review and evaluate the methodology and analysis to minimize the likelihood of errors in the replications, and the designs, materials, and data are made available publicly in order to improve the likelihood of identifying errors. Notwithstanding the ambiguity surrounding the interpretation of a replication failure, the key value of replication remains: as data accumulate, the precision of the effect estimate increases.

11.4.3 What Practices Lead to More Replicable Findings?

Perhaps the most promising possible contribution of the Reproducibility Project will be to provide empirical evidence of the correlates of reproducibility or to make a more informed assessment of the reproducibility of existing results. Researchers have no shortage of hypotheses as to what research practices would lead to higher replicability rates (e.g., LeBel and Paunonen, 2011; LeBel and Peters, 2011; Nosek and Bar-Anan, 2012; Nosek et al., 2012; Vul et al., 2009). Without systematic data, there is no way to test these hypotheses (for discussion, see Hartshorne and Schachner, 2012a,b). Note that this is a correlational study, so it is possible that some third factor, such as the authors' conscientiousness, is the joint cause of both the adoption of a particular research practice and high replicability. However, the lack of a correlation between certain practices and higher replicability rates is—assuming sufficient statistical power and variability—more directly interpretable, suggesting that researchers should look elsewhere for methods that will meaningfully increase the validity of published findings.

11.4.4 Summary

Like any research effort, the most important factor for success of the Reproducibility Project is the quality and execution of its design. The quality of the design, execution of replications, and ultimate interpretations of the findings will define the extent to which the Reproducibility Project can provide information about the reproducibility of psychological science. As with all research, that responsibility rests with the team conducting the research. The last section of this chapter summarizes the strategies we are pursuing to conduct an open, large-scale, collaborative project with the highest-quality standards that we can achieve (Open Science Collaboration, 2012).

11.5 Coordinating the Reproducibility Project

The success of the Reproducibility Project hinges on effective collaboration among a large number of contributors. In business and science, large-scale efforts are often necessary to provide important contributions. Sending an astronaut to the moon, creating a feature film, and sequencing the human genome are testaments to the power of collaboration and social coordination. However, most large-scale projects are highly resourced with money, staff, and administration in order to assure success. Further, most large-scale efforts are backed by leadership that has direct control over the contributors through employment or other strong incentives, giving contributors compelling reasons to do their part for the project.

The Reproducibility Project differs from the modal large-scale project because it started light on resources and light on leadership. Most contributors are donating their time and drawing on whatever resources they have available to conduct replications. Project leaders cannot require action because the contributors are volunteers. How can such a project succeed? Why would any individual contributor choose to participate?

The Reproducibility Project team draws its project-design principles from open-source software communities that developed important software such as the Linux operating system and the Firefox web browser. These communities achieved remarkable success under similar conditions. In this section, we describe the strategies used for coordinating the Reproducibility Project so that other groups can draw on the project design to pursue similar scientific projects. An insightful treatment of these project principles and strategies is provided in Michael Nielsen's (2011) book *Reinventing Discovery*.

The challenges to solve are the following: (1) recruiting contributor, (2) defining tasks so that contributors know what they need to do and can do it, (3) ensuring high-quality contributions, (4) coordinating effectively so that contributions can be aggregated, and (5) getting contributors to follow through on their commitments. The next sections describe the variety of strategies the project uses to address these challenges.

11.5.1 Clear Articulation of the Project Goals and Approach

Defining project goals is so obvious that it is easy to overlook. Prospective contributors must know what the project will accomplish (and how) to decide whether they want to contribute. The Reproducibility Project's primary goal is to estimate the reproducibility of psychological science. It aims to accomplish that goal by conducting replications of a sample of published studies from major journals in psychology. The extent to which prospective contributors find the goal and approach compelling will influence the likelihood that they volunteer their time and resources. Further, once the team is assembled, a clear statement of purpose and approach bonds the team and facilitates coordination. This goal and approach is included in every communication about the Reproducibility Project.

11.5.2 Modularity

Even though potential contributors may find the project goal compelling, they recognize that they could never conduct so many replications by themselves. The Reproducibility Project's goal of replicating dozens of studies is appealing because it has the potential to impact the field, but actually replicating those many studies is daunting. One solution is crowdsourcing (Estellés-Arolas and González-Ladrón-de-Guevara, 2012), in which work is decomposed into smaller, modular tasks that are distributed across volunteers.

Modularity is the extent to which a project can be separated into independent components and then recombined later. Also, if contributors are highly dependent on each other, then the time delay is multiplicative: delay by one affects all. The Reproducibility Project is highly modularized. Individuals or small teams conduct replications independently. Some replications are completed very rapidly, others over a longer time scale. Barriers to progress are isolated to the competing schedules and responsibilities of the small replication teams.

Besides accelerating progress, modularizing is attractive to volunteer contributors because they have complete control over the extent and nature of their participation. Modularization is useful, but it will provide limited value if there are only a few contributors. One way for crowdsourcing to overcome this problem is to have a low barrier to entry.

11.5.3 Low Barrier to Entry

Breaking up a large project into pieces reduces the amount of contribution required by any single contributor. For volunteers with busy lives, this is vital. The Reproducibility Project encourages small contributions so that contributors can volunteer their services incrementally without incurring inordinate costs to their other professional responsibilities or allowing unfulfilled commitments to impede workflow.

Even with effective modularization, prospective contributors may have difficulty in estimating the workload required when making the initial commitment to contribute. Uncertainty itself is a formidable barrier to entry. The Reproducibility Project provides specific documentation to reduce this barrier. In particular, prospective contributors can review studies available for replication in a summary spreadsheet, consult with a team member whose role is to connect available studies to new contributors with appropriate skills and resources, and review the replication protocol that provides instruction for every stage of the process. Effective supporting material and personnel simplify the process of joining the project.

11.5.4 Leverage Available Skills

Collaborations can be particularly effective when they incorporate researchers with distinct skill sets. A problem that is very difficult for a nonexpert may be trivial for an expert. Further, there are many potential contributors that do not have resources or skills to do the central task: conducting a replication. In any large-scale project, there are additional administrative, documentation, or consulting tasks that can be defined and modularized. The Reproducibility Project has administrative contributors with specified roles and contributors who assist by documenting and coding the studies available for replication. There are also consultants for common issues such as data analysis.

11.5.5 Collaborative Tools and Documentation

As a distributed project, the Reproducibility Project coordination must embrace asynchronous schedules. Communication among the entire team occurs via an e-mail LISTSERV (https://groups.google.com/group/openscienceframework?hl=en) that maintains a record of all communications. New ideas, procedural issues, project plans, and task assignments are discussed on the LISTSERV. Decisions resulting from team discussion are codified in project documentation that is managed with Google Docs and the Open Science Framework (OSF; http://openscienceframework.org/).

Print documentation is extensive, as it is the primary means of providing individual contributors with knowledge of (1) what is happening in the project, (2) their role in the project, and (3) what they must do to fulfill their role. The project documentation defines the overall objective of the project, tables of subgoals and actions necessary to achieve them, protocols for conducting a replication project, and templates for communicating results. This workflow is designed to maximize the quality of the replication,

make explicit the standards and expectations of each replication, and minimize the workload for the individual contributors. With a full specification
of the workflow, templates for report writing, and material support for cor-
respondence with original study authors, the replicating teams can smoothly
implement the project's standard procedures and focus their energies on
the unique elements of the replication study design and data collection to
conduct the highest-quality replication possible.

Unlike modular replications, administrative tasks require frequent and
timely upkeep and can impact the workflow of other team members. Thus,
although initially run by volunteers, dedicated administrative support was
needed as the project increased in scale. Together, documentation and
dedicated administrators provide continuity in the projects' objectives and
methods across time and individual replication teams.

The highly defined workflow also makes it easy to track progress of one's
own replication—and those of others. Each stage of the project has explicitly
defined milestones, described in the project's researcher guide, and team
members denote on the project tracksheet when each stage is completed. At
a glance, viewers of the tracksheet can see the status of all projects. Besides its
information value, tracking progress provides normative information for the
research teams regarding whether they are keeping up with the progress of
other teams. Without that information, individual contributors would have
little basis for social comparison and also little sense of whether the project
as a whole is making progress.

11.5.6 Light Leadership with Strong Communication

Large-scale, distributed projects flounder without leadership. However,
leadership cannot be overly directive when volunteers staff the project.
Project leaders are responsible for facilitating communication and discussion
and then guiding the team to decisions and action. Without someone taking
responsibility for the latter, projects will stall with endless discussion and no
resolution.

To maximize project investment, individual contributors should have
the experience that their opinions about the project design matter and can
impact the direction of the project. Simultaneously, there must be sufficient
leadership to avoid having each contributor feel like they shoulder inordinate responsibility for decision making. Contributors vary in the extent
to which they desire to shape different aspects of the project. Some have
strong opinions about the standard format of the replication report; others
would rather step on a nail than spend time on that. To balance this, the
Reproducibility Project leadership promotes open discussion without requiring contribution. Simultaneously, leadership defines a timeline for decision
making, takes responsibility for reviewing and integrating opinions, and
makes recommendations for action steps.

11.5.7 Open Practices

The Reproducibility Project is an open project. This means that anyone can join, that expectations of contributors are defined explicitly in advance, and that the project discussion, design, materials, and data are available publicly. Openness promotes accountability among the team. Individuals have made public commitments to project activities. This transparency minimizes free-riding and other common conflicts that emerge in collaborative research. Openness also promotes accountability to the public. Replication teams are trying to reproduce research designs and results published by others. The value of the evidence accumulated by the Reproducibility Project relies on these replications being completed to a high standard. Making all project materials available provides a strong incentive for the replication teams to do an excellent job. Further, openness increases the likelihood that errors will be identified and addressed. In addition to public accessibility, the Reproducibility Project builds in error checking by requiring each replication team to contact original authors to invite critique of their study design prior to data collection and by having members review and critique each others' project reports.

11.5.8 Participation Incentives

Why participate in a large-scale project? What is in it for the individual contributor? The best designed and coordinated project will still fail if contributors have no reason to participate voluntarily. The Reproducibility Project has a variety of incentives that may each have differential impact on individual contributors. For one, many contributors have an intrinsic interest in the research questions the project has set out to answer or, more generally, view the project as an important service to the field.

Another class of incentives is experiential. Some contributors want to belong to a large-scale collaboration, try open science practices, or conduct a direct replication. For some, this may be for the pleasure of working with a group or trying something new. For others, this may be conceived as a training opportunity. Other incentives are the more traditional academic rewards. The most obvious is publication. Publication is the basis of reward, advancement, and reputation building (Collins, 1985). Contributors to the Reproducibility Project earn coauthorship on publication about the project and its findings. The relative impact for each individual contributor is most certainly reduced by the fact that there are many contributors. However, the nature of the research question, the scale of the project, and (in our humble brag opinion) quality of the endeavor mean that the project may have a high impact on psychology and science more generally. While no contributor will establish a research career using publications with the Open Science Collaboration exclusively, authorship on an important, high-profile project provides an added bonus for the more intrinsic factors that motivate contributions to the Reproducibility Project.

11.6 Conclusion

The Reproducibility Project is the first attempt to systematically and empirically estimate the reproducibility of a subdiscipline of science. It draws on the lessons of open-source projects in software development: leveraging individuals' opinions about how things should be done while providing strong coordination to enable progress. What will be learned from the Reproducibility Project is still undetermined. But if the current progress is any indicator, the high investment of its contributors and the substantial interest and attention by observers suggest that the Reproducibility Project could provide a useful initial estimate of the reproducibility of psychological science and perhaps inspire other disciplines to pursue similar efforts.

Systematic data on replicability do not exist. The Reproducibility Project addresses this shortcoming. If large numbers of findings fail to replicate, that will strengthen the hand of the reform movements and lead to a significant reevaluation of the literature. If most findings replicate satisfactorily—as many as would be expected given our statistical power estimates—then that will suggest a different course of action. More likely, perhaps, is that the results will be somewhere in between and will help generate hypotheses about particular practices that could improve or damage reproducibility.

We close by noting that even in the best of circumstances, the results of any study—including the Reproducibility Project—should be approached with a certain amount of skepticism. While we attempt to conduct replication attempts that are as similar as possible to the original study, it is always possible that "small" differences in method may turn out to be crucial. Thus, while a failure to replicate should decrease confidence in a finding, one does not want to make too much out of a single failure (Francis, 2012). Rather, the results of the Reproducibility Project should be understood as an opportunity to learn whether current practices require attention or revision. Can we do science better? If so, how? Ultimately, we hope that we will contribute to answering these questions.

Endnote

1. Alexander A. Aarts, Nuenen, the Netherlands; Anita Alexander, University of Virginia; Peter Attridge, Georgia Gwinnett College;

Štìpán Bahník, Institute of Physiology, Academy of Sciences of the Czech Republic; Michael Barnett-Cowan, Western University; Elizabeth Bartmess, University of California, San Francisco; Frank A. Bosco, Marshall University; Benjamin Brown, Georgia Gwinnett College; Kristina Brown, Georgia Gwinnett College; Jesse J. Chandler, PRIME Research; Russ Clay, University of Richmond; Hayley Cleary, Virginia Commonwealth University; Michael Cohn, University of California, San Francisco; Giulio Costantini, University of Milan–Bicocca; Jan Crusius, University of Cologne; Jamie DeCoster, University of Virginia; Michelle DeGaetano, Georgia Gwinnett College; Ryan Donohue, Elmhurst College; Elizabeth Dunn, University of British Columbia; Casey Eggleston, University of Virginia; Vivien Estel, University of Erfurt; Frank J. Farach, University of Washington; Susann Fiedler, Max Planck Institute for Research on Collective Goods; James G. Field, Marshall University; Stanka Fitneva, Queens University; Joshua D. Foster, University of South Alabama; Rebecca S. Frazier, University of Virginia; Elisa Maria Galliani, University of Padova; Roger Giner-Sorolla, University of Kent; R. Justin Goss, University of Texas at San Antonio; Jesse Graham, University of Southern California; James A. Grange, Keele University; Joshua Hartshorne, M.I.T.; Timothy B. Hayes, University of Southern California; Grace Hicks, Georgia Gwinnett College; Denise Humphries, Georgia Gwinnett College; Georg Jahn, University of Greifswald; Kate Johnson, University of Southern California; Jennifer A. Joy-Gaba, Virginia Commonwealth University; Lars Goellner, University of Erfurt; Heather Barry Kappes, London School of Economics and Political Science; Calvin K. Lai, University of Virginia; Daniel Lakens, Eindhoven University of Technology; Kristin A. Lane, Bard College; Etienne P. LeBel, University of Western Ontario; Minha Lee, University of Virginia; Kristi Lemm, Western Washington University; Melissa Lewis, Reed College; Stephanie C. Lin, Stanford University; Sean Mackinnon, Dalhousie University; Heather Mainard, Georgia Gwinnett College; Nathaniel Mann, California State University, Northridge; Michael May, University of Bonn; Matt Motyl, University of Virginia; Katherine Moore, Elmhurst College; Stephanie M. Müller, University of Erfurt; Brian A. Nosek, University of Virginia; Catherine Olsson, M.I.T.; Marco Perugini, University of Milan–Bicocca; Michael Pitts, Reed College; Kate Ratliff, University of Florida; Frank Renkewitz, University of Erfurt; Abraham M. Rutchick, California State University, Northridge; Gillian Sandstrom, University of British Columbia; Dylan Selterman, University of Maryland; William Simpson, University of Virginia; Colin Tucker Smith, University of Florida; Jeffrey R. Spies, University of Virginia; Thomas Talhelm, University of Virginia; Anna van 't Veer, Tilburg University; Michelangelo Vianello, University of Padova.

References

Amir, Y. and Sharon, I. (1990). Replication research: A "must" for the scientific advancement of psychology. *Journal of Social Behavior and Personality,* 5, 51–69.

Baillargeon, R., Spelke, E. S., and Wasserman, S. (1985). Object permanence in five-month-old infants. *Cognition,* 20, 191–208.

Bauer, H. H. (1992). *Scientific Literacy and the Myth of the Scientific Method.* Chicago, IL: University of Illinois Press.

Begley, C. G. and Ellis, L. M. (2012). Raise standards for preclinical cancer research. *Nature,* 483, 531–533. doi:10.1038/483531a.

Belding, T. C. (2000). Numerical replication of computer simulations: Some pitfalls and how to avoid them. arXiv preprint nlin/0001057.

Campbell, D. T., Stanley, J. C., and Gage, N. L. (1963). *Experimental and Quasi-Experimental Designs for Research* (pp. 171–246). Boston, MA: Houghton Mifflin.

Cohen, J. (1962). The statistical power of abnormal-social psychological research: A review. *Journal of Abnormal and Social Psychology,* 65, 145–153.

Cohen, J. (1988). *Statistical Power Analysis for the Behavioral Sciences.* Hillsdale, NJ: Lawrence Erlbaum Associates.

Cohen, J. (1992). A power primer. *Psychological Bulletin,* 112, 155–159.

Collins, H. M. (1985). *Changing Order.* London, U.K. Sage.

Estellés-Arolas, E. and González-Ladrón-de-Guevara, F. (2012). Towards an integrated crowdsourcing definition. *Journal of Information Science,* 38(2), 189–200.

Fanelli, D. (2010). "Positive" results increase down the hierarchy of the sciences. *PLoS ONE,* 5(4), e10068. doi:10.1371/journal.pone.0010068.

Fanelli, D. (2012). Negative results are disappearing from most disciplines and countries. *Scientometrics,* 90, 891–904. doi:10.1007/s1192-011-0494-7.

Francis, G. (2012). Publication bias and the failure of replication in experimental psychology. *Psychonomic Bulletin and Review,* 19, 975–991.

Fuchs, H., Jenny, M., and Fiedler, S. (2012). Psychologists are open to change, yet wary of rules. *Perspectives on Psychological Science,* 7, 634–637. doi:10.1177/1745691612459521.

Giner-Sorolla, R. (2012). Science or art? How esthetic standards grease the way through the publication bottleneck but undermine science. *Perspectives on Psychological Science,* 7, 562–571. doi:10.1177/1745691612457576.

Greenwald, A. G. (1975). Consequences of prejudice against the null hypothesis. *Psychological Bulletin,* 82, 1–20.

Hartshorne, J. K. and Schachner, A. (2012a). Tracking replicability as a method of post-publication open evaluation. *Frontiers in Computational Neuroscience,* 6(8), 1–13. doi:10.3389/fncom.2012.0008.

Hartshorne, J. K. and Schachner, A. (2012b). Where's the data? *The Psychologist*, 25, 355.

Hempel, C. G. and Oppenheim, P. (1948). Studies in the logic of explanation. *Philosophy of Science*, 15, 135–175.

Ioannidis, J. P. A. (2005). Why most published research findings are false. *PLoS Medicine*, 2(8), e124. doi:10.1371/journal.pmed.0020124.

Ioannidis, J. P. A., Ntzani, E. E., Trikalinos, T. A., and Contopoulos-Ioannidis, D. G. (2001). Replication validity of genetic association studies. *Nature Genetics*, 29, 306–309.

Jaeger, T. F. (2008). Categorical data analysis: Away from ANOVAs (transformation or not) and towards logit mixed models. *Journal of Memory and Language*, 59, 434–446. doi: 10.1016/j.bbr.2011.03.031.

John, L., Loewenstein, G., and Prelec, D. (2012). Measuring the prevalence of questionable research practices with incentives for truth-telling. *Psychological Science*, 23, 524–532. doi: 10.1177/0956797611430953.

Koole, S. L. and Lakens, D. (2012). Rewarding replications: A sure and simple way to improve psychological science. *Perspectives on Psychological Science*, 7, 608–614. doi:10.1177/1745691612462586.

Kuhn, T. S. (1962). *The Structure of Scientific Revolutions*. Chicago, IL: University of Chicago Press.

Kunda, Z. (1990). The case for motivated reasoning. *Psychological Bulletin*, 108, 480–498. doi:10.1037/0033-2909.108.3.480.

LeBel, E. P. and Paunonen, S. V. (2011). Sexy but often unreliable: Impact of unreliability on the replicability of experimental findings involving implicit measures. *Personality and Social Psychology Bulletin*, 37, 570–583.

LeBel, E. P. and Peters, K. R. (2011). Fearing the future of empirical psychology: Bem's (2011) evidence of psi as a case study of deficiencies in modal research practice. *Review of General Psychology*, 15, 371–379. doi:10.1037/a0025172.

Ledgerwood, A. and Sherman, J. W. (2012). Short, sweet, and problematic? The rise of the short report in psychological science. *Perspectives on Psychological Science*, 7, 60–66. doi:10.1177/1745691611427304.

Lipsey, M. W. and Wilson, D. B. (1993). The efficacy of psychological, educational, and behavioral treatment: Confirmation from meta-analysis. *American Psychologist*, 48, 1181–1209. doi:10.1037/0003-066X.48.12.1181.

Mahoney, M. J. (1985). Open exchange and epistemic process. *American Psychologist*, 40, 29–39.

Makel, M. C., Plucker, J. A., and Hagerty, B. (2012). Replications in psychology research: How often do they really occur? *Perspectives on Psychological Science*, 7, 537–542. doi:10.1177/1745691612460688.

Morrell, K. and Lucas, J. W. (2012). The replication problem and its implications for policy studies. *Critical Policy Studies*, 6, 182–200. doi:10.1080/19460171.2012.689738.

Neuliep, J. W. and Crandall, R. (1990). Editorial bias against replication research. *Journal of Social Behavior and Personality*, 5, 85–90.

Nickerson, R. S. (1998). Confirmation bias: A ubiquitous phenomenon in many guises. *Review of General Psychology*, 2, 175–220.

Nickerson, R. S. (2000). Null hypothesis significance testing: A review of an old and continuing controversy. *Psychological Methods*, 5(2), 241–301. doi:10.1037/1082-989X.5.2.241.

Nielson, M. (2011). Reinventing discovery: The new era of networked science. Princeton University Press.

Nosek, B. A. and Bar-Anan, Y. (2012). Scientific utopia: I. Opening scientific communication. *Psychological Inquiry*, 23(3), 217–243. doi:1080/1047840X.2012.692215.

Nosek, B. A., Spies, J. R., and Motyl, M. (2012). Scientific utopia: II. Restructuring incentives and practices to promote truth over publishability. *Perspectives on Psychological Science*, 7, 615–631. doi:10.1177/1745691612459058.

Open Science Collaboration. (2012). An open, large-scale collaborative effort to estimate the reproducibility of psychological science. *Perspectives on Psychological Science*, 7, 657–660. doi:10.1177/1745691612462588.

Piaget, J. (1952). *The Origins of Intelligence in Children*. New York: International University Press.

Piaget, J. (1954). *The Construction of Reality in the Child*. New York: Basic.

Popper, K. (1934/1992). *The Logic of Scientific Discovery*. New York: Routledge.

Prinz, F., Schlange, T., and Asadullah, K. (2011). Believe it or not: How much can we rely on published data on potential drug targets? *Nature Reviews Drug Discovery*, 10, 712–713. doi:10.1038/nrd3439-c1.

Rosenthal, R. and Rosnow, R. L. (1960). *Artifact in Behavioral Research*. New York: Academic Press.

Salmon, W. (1989). *Four Decades of Scientific Explanation*. Minneapolis, MN: University of Minnesota Press.

Schimmack, U. (2012). The ironic effect of significant results on the credibility of multiple-study articles. *Psychological Methods*, 17, 551–566.

Schmidt, S. (2009). Shall we really do it again? The powerful concept of replication is neglected in the social sciences. *Review of General Psychology*, 13, 90–100. doi:10.1037/a0015108.

Sedlmeier, P. and Gigerenzer, G. (1989). Do studies of statistical power have an effect on the power of studies? *Psychological Bulletin*, 105, 309–316.

Simmons, J. P., Nelson, L. D., and Simonsohn, U. (2011). False-positive psychology: Undisclosed flexibility in data collection and analysis allows presenting anything as significant. *Psychological Science*, 22, 1359–1366. doi:10.1177/0956797611417632.

Sterling, T. D. (1959). Publication decisions and their possible effects on inferences drawn from tests of significance—Or vice versa. *Journal of the American Statistical Association*, 54, 30–34.

Vul, E., Harris, C., Winkielman, P., and Pashler, H. (2009). Puzzlingly high correlations in fMRI studies of emotion, personality, and social cognition. *Perspectives in Psychological Science*, 4, 274–90. doi:10.1111/j.1745-6924.2009.01125.x.

12

What Computational Scientists Need to Know about Intellectual Property Law: A Primer

Victoria Stodden*

CONTENTS

12.1 Introduction

Data and code are becoming as important to research dissemination as the traditional manuscript. For computational science, the evidence is clear: it is typically impossible to verify scientific claims without access to the code and data that generated published findings. Gentleman and Lang [1] introduced the notion of the "research compendium" as the unit of scholarly communication, a triple including the explanatory narrative, the code, and the data used in deriving the results. One of the reasons for including the code and data is to facilitate the production of *really reproducible research*, a phrase coined by Jon Claerbout in 1991[†] to mean research results that can

---

* Victoria thanks an anonymous reviewer for many extremely helpful comments. This research was supported by Alfred P. Sloan Foundation award number PG004545, Facilitating Transparency in Scientific Publishing and NSF award number 1153384, EAGER: Policy Design for Reproducibility and Data Sharing in Computational Science.
[†] See http://sepwww.stanford.edu/doku.php?id=sep:research:reproducible for the Stanford Exploration Project's pioneering recommendations for reproducible research.

be regenerated from the available code and data. Claerbout's approach was paraphrased by Donoho and Buckheit [2] as follows:

> The idea is: An article about computational science in a scientific publication is **not** the scholarship itself, it is merely **advertising** of the scholarship. The actual scholarship is the complete software development environment and the complete set of instructions which generated the figures.

Enabling computational replication typically means supplying the data, software, and scripts, including all parameter settings, which produced the results [3,4]. This approach runs headlong and unavoidably into current intellectual property law, which creates a stumbling block rather than an impassable barrier to the dissemination of really reproducible research. In this chapter, I describe these intellectual property stumbling blocks to the open sharing of computational scientific knowledge and present solutions that coincide with long-standing scientific norms. In Section 12.2, I motivate scientific communication as a narrative with a twofold purpose: to communicate the importance of the findings within the larger scientific context and to provide sufficient information that the results may be verified by others in the field. Sections 12.3 and 12.4 then discuss intellectual property barriers and solutions that enable code and data sharing, respectively. Each of these three research outputs, the research article, the code, and the data, requires different legal analyses and action in the scientific context as described in the following. The final section discusses citation for digital scholarly output, focusing on code and data.

A widely accepted scientific norm, as labeled by Robert K. Merton, is *communism* or *communalism* [5]. By this, Merton meant that property rights in scientific research extend only to the naming of scientific discoveries (Arrow's impossibility theorem, e.g., named for its originator Kenneth Arrow), and *all other intellectual property rights* are given up in exchange for recognition and esteem. This notion, at least in the abstract, underpins the current system of publication and citation that forms the basis for academic promotion and reward.

Computational science today is facing a credibility crisis: without access to the code and data that underlie scientific discoveries, published findings are all but impossible to verify [4]. Reproducible computational science has attracted attention since Claerbout wrote some of the first really reproducible manuscripts in 1992.* More recently, a number of researchers have adopted reproducible methods [2,6,7] or introduced them in their role as journal editors [8–10]. This chapter discusses how intellectual property law applies to data in the context of communicating scientific research.

---

* See http://sepwww.stanford.edu/doku.php?id=sep:research:reproducible.

12.2 Publishing the Research Article

Scientific publication has taken the well-recognized form of the research article since 1665 with the first issue of the *Philosophical Transactions of the Royal Society of London*.* This section motivates the sharing of the research paper and discusses the clash that has arisen between the need for scientific dissemination and modern intellectual property law in the United States.

Scientific results are described in the research manuscript, including their derivation and context, and this manuscript is typically published in an established academic journal. It is of primary importance that the body of scientific knowledge, today comprised primarily of journal publications, has as little error as possible. This is in part accomplished through peer review and in part through the very act of publication and permitting a wide audience access to the work. The recognition that the scientific research process is error prone, that error can creep in anywhere and from any source, is central to the scientific method, and wider access to the findings increases the chances that errors will be caught.

The second reason property rights have been deemphasized in scientific research is the idea that scientific knowledge about our world, such as physical laws, mathematical theorems, or the nature of biological functions, is to be discovered, rather than invented or created, and these discoveries belong to all of humanity. This is not to say that scientific discovery is not a creative act, quite the contrary, but that the underlying scientific fact is a public good, a facet of our world not subject to ownership. This is the underlying rationale behind US federal government grants of over $40 bil- lion dollars for scientific research in 2012 [11]. This vision is also reflected both in the widespread understanding of scientific facts as "discoveries" and not "inventions" and in current intellectual property law, which does not recognize a scientific discovery as rising to the level of individual ownership, unlike an invention or other contribution. We will see this notion rise again in the discussion on scientific data:

Copyright law in the United States originated in the Constitution, stating that "The Congress shall have Power … To promote the Progress of Science and useful Arts, by securing for limited Times to Authors and Inventors the exclusive Right to their respective Writings and Discoveries."[†] Through a series of subsequent laws, copyright has come to assign a specific set of rights to authors of original expressions of ideas *by default*. In the context of scientific research, this means that the written description of a finding is

---

* For a brief history, see http://rstl.royalsocietypublishing.org/ including an image of the first issue with the endearing title "Philosophical Transactions Giving Some Account of the Present Undertakings, Studies, and Labours of the Ingenious in Many Considerable Parts of the World."

[†] US Const. art. I, §8, cl. 8.

automatically copyrighted by the author(s) (how copyright applies to data and code is discussed in the following two sections). Copyright secures exclusive rights vested in the author to both reproduce the work and prepare derivative works based upon the original. There are exceptions and limitations to this power, such as fair use, but these do not provide for an intellectual property framework for scientific knowledge that matches long-standing scientific norms of openness, access, and transparency.

Intellectual property law, and its interpretation by academic and research institutions, means that authors have copyright over their research manuscripts. Copyright can be transferred to others, and the copyright holders can grant permissions for use to others as they see fit. In a system established many decades ago, journals typically request that copyright be assigned to the publisher for free, rather than remain with the authors, as a condition of publication. Many journals have a second option for authors if they request it, where copyright remains with the author, but permission is granted to the journal to publish the article.* If copyright was transferred, access to the published article usually involves paying a fee to the publisher. Typically, scientific journal articles are available only to the privileged few affiliated with a university library that pays the journal subscription fees, and articles are otherwise offered for a surcharge of about $30 each. Authors of scientific articles, and the owners of copyright, typically transfer copyright to publishers as a condition of publication.

Publishing scientists today have other options. A transformation is underway that has the potential to make scientific knowledge openly and freely available. The *open-access movement* has established ways of publishing that secure long-term public access to the research article. This may still involve the journal requesting a transfer of copyright to them, and it usually involves an upfront fee to compensate the journal for the loss of revenue from library subscriptions and article purchases.

This transformation started in 1991 when Paul Ginsparg, professor of Physics at Cornell University, set up an open repository called arXiv.org (pronounced "archive") for physics articles awaiting journal publication. In the biosciences, the *Public Library of Science (PLoS)* was launched 2000.† They publish under a new model, open-access publishing, which publishes scientific articles by charging the authors the costs upfront, typically about $2000 per article, and making the published papers freely available online.‡

On balance openly available articles appear to be cited at higher rates than those behind subscription paywalls [12,13]. There are steps a researcher

---

* See, for example, *Science* Magazine's alternative license at http://www.sciencemag.org/site/feature/contribinfo/prep/lic_info.pdf (last accessed January 29, 2013).
† See http://blogs.plos.org/plos/2011/11/plos-open-access-collection-%E2%80%93-resources-to-educate-and-advocate/ for a collection of articles on Open Access.
‡ See http://www.plos.org/publish/pricing-policy/publication-fees/ for up-to-date pricing information.

can take when publishing a manuscript, to help maximize the future access to their article. First, a researcher can request the alternative copyright agreement that gives the journal permission to publish the article but leaves copyright with the author. Another approach is to use the SPARC addendum, to retain rights to post the article on the author's webpage, in scholarly repositories, or more widely on the Internet.* The SPARC addendum, for example, ensures the right of the author to retain the following:

1. The rights to reproduce, to distribute, and to publicly display the article in any medium for noncommercial purposes.
2. The right to prepare derivative works from the article.
3. The right to authorize others to make any noncommercial use of the article so long as the author receives credit as author and the journal in which the article has been published is cited as the source of first publication of the article. For example, the author may make and distribute copies in the course of teaching and research and may post the article on personal or institutional websites and in other open-access digital repositories.

These are valuable rights authors likely wish to retain so they can reuse their own work and share with others, and this can be accomplished by using the The Scholarly Publishing and Academic Resources Coalition (SPARC) addendum with the traditional publisher's agreement.

A second option is choosing to publish in open-access journals. This is a personal decision for the authors as journal impact factor is often tied to career advancement, but open-access journals like *PLoS ONE* have been gaining in prestige.†

When publishing in an open-access journal, authors are sometimes asked to designate a Creative Commons license for their article. Authors can also find themselves confronted with this choice when depositing to a repository, or even when posting the article on their own webpage, depending on the downstream use they wish to permit. Creative Commons licenses are very useful for researchers, and I will discuss their various licensing options. In the Creative Commons sense, "license" is the term used to mean that an owner gives advance permission for use of his or her copyrighted works. Although related, this is a different sense of the term than, say, a software license or patent license that is paid for and permits use of the software or patent for a period of time. In this case, license refers to the granting of certain uses by the copyright holder in advance—without charge to anyone—so there is no need to contact the copyright holder to request permission.

---

* See http://www.arl.org/sparc/author/addendum.shtml.
† See http://scholarlykitchen.sspnet.org/2011/06/28/plos-ones-2010-impact-factor/ for recent impact factor information.

Creative Commons has provided documents (licenses) that encode certain terms of use in formal legal language, making it easy for researchers and others to grant permission for use of their work if they happen to want what these licenses provide. The most basic Create Commons license is "CC-BY," and essentially, it permits unrestricted downstream use so long as attribution is given to the original author. Note that in this case, the author is also the copyright holder. Licensing options that grant permission for use can only be applied by the copyright holder (or with the copyright holder's permission), so think carefully before signing your copyright over to other entities, such as journals.

CC-BY is the closest permission structure to that which scientists and researchers are used to—essentially saying, use my work however you wish, but make sure you credit me.* Creative Commons, however, designed licenses with a broader community in mind and offers other licensing options. For certain specialized scientific research, these may be useful, so I touch on them here for completeness, but each option adds further restrictions over CC-BY that I believe should be outweighed by their benefits over CC-BY. Creative Commons has licenses that restrict downstream use to noncommercial purposes only (NC), which forbid the creation of derivative works (ND), and direct downstream users as to what license they must use on their work (SA). The simplest choice that matches scientific community norms is CC-BY.

With broader sharing of publications, scientific knowledge could be spread more widely, more mistakes caught, and the rate of scientific progress improved. In addition, more downstream activity would be encouraged, such as technological development, industry growth, and further scientific discoveries [14,15]. Open archiving is mandated by the National Institutes for Health (NIH), where published articles arising from NIH-funded research must be deposited in PubMed Central† within 12 months of publication. On February 22, 2013, this was extended to all federal funding agencies through an executive memorandum released by the Office of Science and Technology Policy in the Whitehouse.‡ To maximize access, we need a streamlined and uniform way of managing copyright over scientific publications, and also copyright on data and code, as discussed in the next section.

12.3 Publishing Scientific Software, Code, and Tools

The computational steps taken to arrive at a result are often complex enough that their complete communication is prohibitive in a typical scientific

---

* See http://creativecommons.org/licenses/by/3.0/.
† PubMed Central is located at http://www.ncbi.nlm.nih.gov/pmc/.
‡ See   http://www.whitehouse.gov/blog/2013/02/22/expanding-public-access-results-federally-funded-research.

publication. This is a key reason for releasing the code that contains all the steps, instructions, data calls, and parameter settings that generated the published findings. Of the three digital scholarly objects discussed in this chapter, the code has the most complex interactions with intellectual property law since it is both subject to copyright and patent.

Software is considered an original expression of an underlying idea, and therefore it is subject to copyright. As discussed in the previous section, copyright adheres by default—a programmer who does nothing other than write software will produce code copyrighted to herself.* The algorithm or methods that the code implements are not subject to copyright themselves, but copyright adheres to the code that implements the algorithm or methods. The effect of copyright in this case is the prohibition on others to reproduce or modify the code[†] (see Box 12.1).

BOX 12.1   INSET: COPYRIGHT IN A NUTSHELL

The original expression of ideas falls under copyright by default (text, code, figures, tables, original selection, and arrangement of data)

Subject to some exceptions and limitations, copyright secures exclusive rights vested in the author to

1. Reproduce the work
2. Prepare derivative works based upon the original

Copyright is of limited but long duration, generally life of the author plus 70 years, and is subject to exceptions and limitations such as fair use.

Copyright works counter to long-standing scientific norms that encourage reuse and verification of results. This means running the code on a different system (reproducing) or adapting the code to a new problem (reusing). Authors must grant permission to others to use their code in these ways. The Creative Commons licenses discussed in the previous section were created for digital artistic works, and they are not suitable for code and so cannot solve our problem. There are, however, a great number of open licenses for software that permit authors to give permission for replication and reuse. Software exists primarily in two forms: source and compiled, and the transmission of the complied form alone is not sufficient for scientific purposes.

---

* Although the exception in academic research, the copyright can initially go to an employer or commissioning party under the "work made for hire" doctrine.

[†] There are exceptions and limitations to copyright, such as fair use, but these do not extend to scientific scholarly objects and how researchers would typically use them. From a computational researcher's perspective, these exceptions and limitations should not be relied on to provide sufficient access and affirmative steps such as licensing should be taken. For more on fair use, see http://www.copyright.gov/fls/fl102.html and [16,17].

Communication of the source code, whether intended to be compiled or not, is essential to understanding and reusing scientific code. In the context of scientific research, source code is often in the form of scripts, for example, in MATLAB® or Python, which execute in association with an installed package and are not compiled.

There are several open licenses for code that place few restrictions on reuse beyond attribution, creating an intellectual property framework resembling conventional scientific norms. The (Modified) Berkeley Software Distribution (BSD) license, for example, permits the downstream use, copying, and distribution of either unmodified or modified source code, as long as the license accompanies any distributed code and the previous authors' names are not used to promote any modified downstream software. The license is brief enough it can be included here:

```
Copyright (c) ⟨YEAR⟩, ⟨OWNER⟩
All rights reserved.
```

Redistribution and use in source and binary forms, with or without modification, are permitted provided that the following conditions are met:

- Redistributions of source code must retain the previous copyright notice, this list of conditions, and the following disclaimer.
- Redistributions in binary form must reproduce the previous copyright notice, this list of conditions, and the following disclaimer in the documentation and/or other materials provided with the distribution.
- Neither the name of the ⟨ORGANIZATION⟩ nor the names of its contributors may be used to endorse or promote products derived from this software without specific prior written permission.

This text is followed by a disclaimer releasing the author from liability for use of the code. The Modified BSD license is very similar to the MIT license, with the exception that the MIT license does not include a clause forbidding endorsement. The Apache 2.0 license is also commonly used to specify terms of use on software. Like the Modified BSD and MIT licenses, the Apache license requires attribution, but it differs in that it permits users to exercise patent rights that would otherwise only extend to the original author, so that a patent license is granted for any patents needed for use of the code (probably a fairly obscure situation for academic research). The Apache 2.0 license further stipulates that the right to use the software without patent infringement will be lost if the downstream user sues the licensor for patent infringement. Attribution under Apache 2.0 requires that any modified code carries a copy of the license, with notice of any modified files and all copyright, trademark, and patent notices that pertain to the work be included. Attribution can also be done in the notice file. The *Reproducible Research Standard* [18,19] recommends using one of these three licenses or a similar attribution license for scripts and software released as part of a scientific research compendium.

Patents are a second form of intellectual property that can create a barrier to the open sharing of scientific codes. Columbia University, for example, states in its *Faculty Handbook* that

> ... the University and a member of the faculty may expect and require of one another cooperation in the development and exploitation of conceptions ... In particular, the University will advise a faculty member about securing a patent, and will participate with him or her in seeking patent protection, in every way compatible with their several capacities and common interests. ... The obligations of a faculty member include the execution of an assignment or a patent, and of rights thereunder, in appropriate circumstances.*

There are exceptions, but this expectation of patenting is typical in academic research institutions.

Patenting is often viewed as a method of enabling access, especially by institutional technology transfer offices, to technology that would otherwise remain inaccessible in academic institutions and research journals. In the case of software, patents add a layer of complexity, and possible fees, to the scientific notion of reproducibility of results. Reproducibility implies the open availability of the software that permits replication along with the published results (Gentleman and Lang's *research compendium* introduced previously in this chapter). Researchers seeking a patent appear to be reluctant to release their code publicly, possibly for fear of creating "prior art" and thus creating a barrier to patent granting, or a perceived loss of revenue from researchers who would like to use their software for research purposes [20].

Neither of these reasons should prevent a patent-seeking researcher from making his or her code publicly and openly available. Under US law, an inventor or rights holder can apply for a patent on a published invention, so long as it is within 1 year of disclosure.† A dual system of patent licensing for industry application can coexist with openly downloadable software for academic research purposes. If a researcher feels inclined to pursue a patent on software, he or she should ensure that academic researchers are able to openly and easily download the software, without going through a patent

---

* See http://www.columbia.edu/cu/vpaa/handbook/appendixd.html (last accessed February 12, 2013).
† This is known as a "statutory bar" to an otherwise valid patent.

licensing process (even one without a fee) in accordance with the Principle of Scientific Licensing, which states [18]:

> **Principle of Scientific Licensing**: Legal encumbrances to the dissemination, sharing, use, and re-use of scientific research compendia should be minimized, and require a strong and compelling rationale before application.

Code can be made available in a dedicated code repository such as GitHub, BitBucket, SourceForge, or RunMyCode [7].* All will provide links to the stored code, permitting it to be associated with the manuscript and data. This theme of accessibility of research compendia continues in the next section with a discussion on publishing the data associated with scientific findings.

12.4 Publishing Datasets and "Raw Facts"

Data are understood as integral in the communication of computational findings, part of the *research compendium* introduced earlier in the chapter. Data can refer to an input into scientific analysis, such as a publicly available dataset like those at Data.gov[†] or those gathered by researchers in the course of the research, or it can refer to the output of computational research, as is the case in computational simulations. In short, it is typically an array of numbers or descriptions, to which analysis and interpretation is applied. It does not include computer code, discussed in the previous section.

In 1991, the US Supreme Court held in Feist v. Rural Telephone Service Co. that raw facts are not copyrightable but the original "selection and arrangement" of these raw facts may be.[‡§] The Supreme Court has not made a ruling concerning intellectual property in data since, and modern computational research may create a residual copyright in a particular dataset, if original selection and arrangement of facts takes place. Collecting, cleaning, and readying data for analysis is often a significant part of scientific research and arguably could be considered "original selection and arrangement" in the sense of Feist.

The *Reproducible Research Standard* recommends therefore releasing data under a Creative Commons CC0, or "no rights reserved" publication, in part

---

* See https://github.com/, https://bitbucket.org/, http://sourceforge.net/, and http://www.runmycode.org/.
† See https://explore.data.gov/.
‡ Copyright does extend to databases under European intellectual property law. This is a key distinction between European and US intellectual property systems in the context of scientific research.
§ See Feist Publications v. Rural Telephone Service Co., 499 U.S. 360 (1991).

because of the possibility of such a residual copyright existing in the dataset.* The public domain certification means that as the dataset author, and potential copyright holder, you will not exercise any rights you may have in the dataset that may derive from copyright (or any other ownership rights). A public domain certification also means that as the author, you are relying on downstream users to cite and attribute your work appropriately. For this reason, a specific citation recommendation should be included with the dataset, suggesting to downstream users that they cite any use of the dataset itself.

Datasets may have barriers to reuse and sharing that do not stem from intellectual property law, such as confidentiality of records, privacy concerns, and proprietary interests from industry or other external collaborators that may assert ownership over the data. Good practice suggests planning for maximal data release at the time of publication at the beginning of a research collaboration, whether it might be with industrial partners who may foresee different uses for the data than supporting reproducible research or with scientists subject to a different intellectual property framework for data, such as those in Europe.

Datasets should be made available in recognized repositories for the field, if they exist, and conform to any established standards for formats, metadata, or exposition. If recognized repositories don't exist, both the DataVerse Network and Dryad will host datasets from any field, for example, and provide association with the manuscript and code through persistent links.[†] They are able to accommodate access restriction on the datasets, due to privacy concerns or other constraints. A number of federal funding agencies have data sharing requirements in their grant guidelines. The National Science Foundation grant guidelines state that

> Investigators are expected to share with other researchers, at no more than incremental cost and within a reasonable time, the primary data, samples, physical collections and other supporting materials created or gathered in the course of work under NSF grants.[‡]

Similarly, the NIH grant guidelines state that

> The NIH expects and supports the timely [no later than the acceptance for publication of the main findings from the final data set] release and sharing of final research data from NIH-supported studies for use by other researchers.[§]

---

* See http://creativecommons.org/about/cc0 for further details on the CC0 license.
[†] See http://thedata.org/ and http://datadryad.org/.
[‡] See http://www.nsf.gov/pubs/policydocs/pappguide/nsf11001/aag_6.jsp.
[§] The NIH data sharing guidelines apply to grants greater than $500,000. See http://grants.nih.gov/grants/guide/notice-files/NOT-OD-03-032.html.

These guidelines have been minimally enforced, but this may change. The February 22, 2013, Executive Memorandum mentioned previously requires federal funding agencies to develop enforceable open data plans.

12.5 Citation

The research article, code, and data are shared with the hope that they will be used by other researchers. Citation of data and software use is not standard in the computational sciences and must become so. Aside from being a plagiarism violation [21], using uncited code and data is poor scientific practice, and it impedes both transparency in research and rewards for scientific contributions [22]. When sharing code or data, it is helpful to provide citation information both to guide downstream users and to remind users that citation is expected.

Throughout this chapter, the use of open attribution-only licensing has been recommended, but it is worth commenting on the relationship between this legal concept and traditional academic citation. They are not identical. In the case of open software licensing as discussed in this chapter, attribution generally refers to listing contributions and authors in a file that accompanies the software. This is important for provenance and transparency, but doesn't satisfy citations standards used in academic rewards. Some open licenses require this type of attribution, but it must be noted that additional, not legal, citation should take place to satisfy scientific norms. Any software use should receive a scientific citation in the list of references, on a par with referenced publications. A footnote mentioning the software use is not adequate. The content of this citation should include, at minimum, the author(s), the software version, the location of the code on the Internet, the date of software release, and the data of software access. If the authors suggest further citation information, for example, a report describing the software, this should be cited.

In the case of Creative Commons attribution licensing, the two concepts lie slightly closer. Section 4(b) of the CC-BY 3.0 license states that

> If You Distribute ... the Work or any Adaptations ..., You must ... keep intact all copyright notices for the Work and provide, reasonable to the medium or means You are utilizing: (i) the name of the Original Author ... (ii) the title of the Work ... (iii) to the extent reasonably practicable, the URI, if any... and (iv) ... in the case of an Adaptation, a credit identifying the use of the Work in the Adaptation. ... The credit required by this Section 4 (b) may be implemented in any reasonable manner...*

---

* Note that CC-BY 4.0 has now been publicly released for comment. http://creativecommons.org/weblog/entry/36713.

Arguably, what is "reasonable to the medium" in the research context is scientific citation. The CC-BY license is most likely to be applied to the research paper itself, for which citation practices exist, but if applied to text describing data selection and arrangement, for example, it could be interpreted as requiring standard scientific citation. Hopefully the research community quickly adopts practices that include code and data citation as standard, and legal requirements remain a last resort.

12.6 Conclusion

The current set of scientific norms evolved over hundreds of years to maximize the integrity of our stock of scientific knowledge. They espouse standards of independent verification and transparency and publication of research findings to disseminate the knowledge widely. Current scientific practice has not kept up with technological advancement, meaning much of the published computational findings are unreplicable since the source code and data are not made conveniently and routinely available. To make reproducibility possible in today's computational research environment, the communication of new types of scholarly objects, for example, a digital research paper, code, or data, requires engaging intellectual property law. In this chapter, I have traced how intellectual property law interacts with digital scholarly communication, through both the relevant aspects of the copyright and patent systems, for scholars sharing really reproducible computational research.

For broad reuse, sharing, and archiving of code to be a commonly accepted practice in computational science, it is important that open licenses be used that minimize encumbrances to access and reuse, such as attribution-only licenses like the MIT license or the Modified BSD license or the Creative Commons attribution license. A collection of code with an open licensing structure permits archiving, persistence of the code, and research on the code base itself, just as is the case for collections of research articles. For these reasons, as well as the integrity of our body of scholarly knowledge, it is essential to address the barriers created by current intellectual property law in such a way that access and reuse are promoted and preserved and future research encouraged.

References

1. R. Gentleman and D. T. Lang. Statistical analyses and reproducible research, 2004. http://biostats.bepress.com/bioconductor/paper2/ (Accessed January 23, 2014.)

2. D. Donoho and J. Buckheit. WaveLab and reproducible research, Stanford University, Department of Statistics Technical Report 474, 1995.
3. G. King. Replication, Replication. *PS: Political Science and Politics* 28: 443–499, 1995. Copy at http://j.mp/jCyfF1 (Accessed January 23, 2014.)
4. D. Donoho, A. Maleki, M. Shahram, I. Ur Rahman, and V. Stodden. Reproducible research in computational harmonic analysis. *Computing in Science and Engineering* 11: 8–18, January 2009.
5. R. K. Merton. The normative structure of science, in: R. K. Merton, ed., *The Sociology of Science: Theoretical and Empirical Investigations*, Chicago, IL: University of Chicago Press, OCLC 755754, p. 267, 1973.
6. D. Donoho, V. Stodden, and Y. Tsaig. About SparseLab, 2007. See http://sparselab.stanford.edu.. (Accessed January 23, 2014.)
7. V. Stodden, C. Hurlin, and C. Perignon. RunMyCode.Org: A novel dissemination and collaboration platform for executing published computational results. *eSoN IEEE eScience Conference*, Chicago, IL, 2012. Available at http://papers.ssrn.com/sol3/papers.cfm?abstract_id=2147710 (Accessed January 23, 2014.)
8. *Journal of Experimental Linguistics*, Linguistic Society of America, http://elanguage.net/journals/jel (Accessed January 23, 2014.)
9. *Biostatistics.* Oxford Press. http://biostatistics.oxfordjournals.org/ (Accessed January 23, 2014.)
10. R. Trivers. Fraud, Disclosure, and Degrees of Freedom in Science, May 20, 2012. Available at http://www.psychologytoday.com/blog/the-folly-fools/201205/fraud-disclosure-and-degrees-freedom-in-science (Accessed January 23, 2014.)
11. Budget of the U.S. Government, Budget of the United States Government, Fiscal Year 2012. Available at http://www.gpo.gov/fdsys/search/pagedetails.action?packageId=BUDGET-2013-BUD (Accessed January 23, 2014.)
12. Y. Gargouri, C. Hajjem, V. Larivière, Y. Gingras, L. Carr et al. Self-selected or mandated, open access increases citation impact for higher quality research. *PLoS ONE* 5(10): e13636. doi:10.1371/journal.pone.0013636, 2010.
13. M. McCabe. Online access and the scientific journal market: An economist's perspective, national academies of science report. Available at http://sites.nationalacademies.org/PGA/step/PGA_058712 (Accessed January 23, 2014.)
14. V. Stodden. Innovation and growth through open access to scientific research: Three ideas for high-impact rule changes, in: *Rules for Growth: Promoting Innovation and Growth Through Legal Reform*, Kansas City, MO: The Kauffman Task Force on Law, Innovation, and Growth, ed. February, 2011.
15. V. Stodden. Open science: Policy implications for the growing phenomenon of user-led scientific innovation. *Journal of Science Communication* 9(1), 2010.

16. W. Fisher III. Reconstructing the fair use doctrine. *Harvard Law Review*, 101(8): 1659–1795, June 1988.

17. P. A. David. The economic logic of 'Open Science' and the balance between private property rights and the public domain in scientific data and information: A primer. SIEPR Discussion Paper No 02-30, 2005. http://ideas.rhttp://www.nap.edu/openbook.php?record_id=10785& page=19 (Accessed January 23, 2014.)

18. V. Stodden. Enabling reproducible research: Licensing for scientific innovation. *International Journal of Communications Law and Policy* 13: 1–25, 2009.

19. V. Stodden. The legal framework for reproducible research in the sciences: Licensing and copyright. *IEEE Computing in Science and Engineering* 11(1): 35–40, January 2009.

20. V. Stodden. The scientific method in practice: Reproducibility in the computational sciences, MIT Sloan Research Paper No. 4773-10. Available at http://papers.ssrn.com/sol3/papers.cfm?abstract_id=1550193.

21. National Research Council. *Responsible Science, Volume I: Ensuring the Integrity of the Research Process*. Washington, DC: The National Academies Press, 1992.

22. National Research Council. *For Attribution—Developing Data Attribution and Citation Practices and Standards: Summary of an International Workshop*. Washington, DC: The National Academies Press, 2012.

Part III

Platforms

13

Open Science in Machine Learning

Mikio L. Braun and Cheng Soon Ong

CONTENTS

13.1 Introduction

The advent of Big Data has resulted in an urgent need for flexible analysis tools. Machine learning addresses part of this need, providing a stable of potential computational models for extracting knowledge from the flood of data. In contrast to many areas of the natural sciences, such as physics, chemistry, and biology, machine learning can be studied in an algorithmic and computational fashion. In principle, machine learning experiments can be precisely defined, leading to perfectly reproducible research. In fact, there have been several efforts in the past of frameworks for reproducible experiments [3,13,15,30]. Since machine learning is a data-driven approach, we need to have access to carefully structured data [25], that would enable direct comparison of the results of statistical estimation procedures. Some headway has been seen in the statistics and bioinformatics community. The success of R and Bioconductor [8,9] as well as projects such as Sweave [15] and Org-mode [29] have resulted in the possibility to embed the code that produces the results of the paper in the paper itself. The idea is to have a unified computation and presentation, with the hope that it results in reproducible research [14,24].

Machine learning is an area of research that spans both theoretical and empirical results. For methodological advances, one key aspect of reproducible research is the ability to compare a proposed approach with the current state of the art. Such a comparison can be theoretical in nature, but often a detailed theoretical analysis is not possible or may not tell the whole story. In such cases, an empirical comparison is necessary. To produce reproducible machine learning research, there are three main requirements (paraphrased from [32]):

1. Software (possibly open source) that implements the method and produces the figures and tables of results in the paper
2. Easily available (open) data on which the results are computed
3. A paper (possibly open access) describing the method clearly and comprehensively

The approach taken by projects that embed computation into the description may not be suitable for the machine learning community, as the datasets may be large and computations may take significant amounts of time. Instead of a unified presentation and computation document, we propose to have independent interacting units of software, data, and documentation of the scientific result.

Motivated by the ideals of the free and open-source software movement and current trends for open access to research, we flavor our advocacy for reproducible research with that of open science. It has been shown that researchers who are not experts are likely to find solutions to scientific problems [7]. Corresponding to the three requirements for reproducible research earlier, we advocate open-source software, open data, and open access to research. Open-source software enables nonexperts to use the same tools that engineers in professional organizations use. Open data enable non-experimentalists access to measurements made under difficult and expensive experimental conditions. Open access to research publications enables nonspecialists to obtain the same information that scientists in well-funded institutions can discover. The long-term goal is to make (often publicly funded) results freely available to the general public to maximize the potential impact of the scientific discoveries.

In recent years, there has been a move in machine learning to open science. The theoretical contributions are freely available in an open access journal, the *Journal of Machine Learning Research*, and from the proceedings of several conferences. This journal also now accepts submissions to a special section on open-source software. Furthermore, there is an active community on mloss.org that provides a collection of open-source software projects in machine learning. For empirical data, there have been several recent projects such as mldata.org, mlcomp.org, tunedit.org, and kaggle.com, which provide a more computational focus than the earlier efforts such as the UCI machine learning database. We envision that in the future there will be interdependent units of software, data, and documentation that interact based on common protocols.

In the rest of this chapter, we will relay our design choices and experiences in organizing open-source software in machine learning and open access to machine learning data. We briefly introduce the area of machine learning and how to make it reproducible. After identifying similarities and differences between reproducible research and open science, we advocate open-source software and open data. Finally, we discuss several issues that have arisen, such as standards for interoperability, and the trade-off between automation and flexibility.

13.2 What Is Machine Learning?

Lying at the intersection of computation, mathematics and statistics, machine learning aims to build predictive models from data. It is the design and development of algorithms that allow computers to evolve behaviors based on empirical data.* A more precise definition is given by [18]: "A computer program is said to learn from experience E with respect to some class of tasks T and performance measure P, if its performance at tasks in T, as measured by P, improves with experience E." Refer to Figure 13.1 for an illustration for supervised learning. The generality of the approach has found application in various fields such as bioinformatics, social network analysis, and computer vision. It is particularly useful when there are relatively large amounts of data, and the desired outcomes are well defined. For example, in bioinformatics, given the genome sequence of a particular individual one might be interested to predict whether this person is likely to

---

* http://en.wikipedia.org/wiki/Machine_learning.

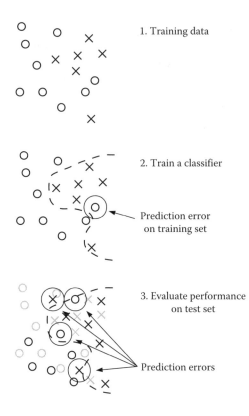

1. Training data

2. Train a classifier

Prediction error
on training set

3. Evaluate performance
on test set

Prediction errors

FIGURE 13.1
The different stages in a supervised learning application. The learning data are presented as training dataset. On this training set, a classifier is trained to separate the two classes. Already on the training set, the classifier might choose the predict different labels than the ones that are specified if it assumes that there is noise in the dataset. The classifier is eventually evaluated on an independent test set not available at training. This test set is used to estimate the expected test error on unseen examples. Again, depending on the level of noise, the test error might be nonzero for the optimal decision boundary. However, if the test error is much larger than the training error, one says that the learning method has *overfitted* to the training data.

get a certain disease. Instead of more traditional approaches in biochemistry of building mechanistic models of the process at hand, machine learning directly tries to infer the prediction rule (diseased or not) from the data using statistical methods. Naturally, expert domain knowledge is captured in this process by constructing features from the known contributing factors of the disease. There are numerous books describing machine learning in great depth [1,2,16,18,21,28], and this section focuses on how machine learning algorithms interact with experimental data, and how it is often used in the natural sciences.

While the techniques of machine learning are widely applicable to data-rich applications, machine learning researchers tend to focus on the methodological questions. Experiments in computational science generally have the following workflow [17]:

1. Preprocessing of data, using knowledge of the measurement process to remove artifacts.
2. Feature construction and selection, aiming to capture the contributing factors.
3. Model construction and parameter estimation from training data, resulting in a predictor. This is often called "training" or "learning."
4. Preparation of test data. This is often done in parallel to the preparation of the training data above.
5. Evaluation and validation of estimated predictor on the test data.

While each step in this workflow is crucial to the success of the data analysis approach, machine learning tends to focus on step 3: choosing the right computational representation of the problem and estimating the model parameters from available data. One key characteristic of machine learning approaches is the focus on "generalization error," which is the estimated performance of the trained method on future data. This emphasis on future data is important because many machine learning approaches are so flexible that they could exactly explain all the training data while capturing nothing about the underlying process. This behavior is called "overfitting." Hence, the importance of steps 4 and 5 in the aforementioned workflow, in checking the performance of the trained method on data that was not used during training. Steps 1 and 2 in the aforementioned process require close collaboration with domain experts, and it remains an open question whether this creative process of converting human intuition into precise representations on the computer can be fully automated.

13.2.1 Supervised, Unsupervised, and Reinforcement Learning

Machine learning problems can be broadly categorized into three approaches, depending on the type of problem that needs to be solved:

- Supervised learning
- Unsupervised learning
- Reinforcement learning

Supervised learning is used when the training data consists of examples of the input features along with their corresponding target values. For example, disease classification uses the genome sequence as input features, and the target value is a simple binary "yes/no" label. Depending on the type of label required by the problem, this results in classification, regression, or structured prediction. Note that it is often advantageous to consider a representation of input features that are more amenable to computation. In the aforementioned example, since there are many common sections between the genomes of different individuals, the whole genome sequence may be processed to identify relevant mutations and the mutations are used as features instead. In recent years, many robust and efficient methods have been developed for this well-specified problem.

In other applications, training data may not have corresponding labels. This more exploratory mode of learning is called *unsupervised learning* and is often used to discover structure within the data. The scientist may be interested in discovering groups of similar examples (called clustering), determining the distribution of the data (called density estimation), or finding low-dimensional representations (called principle component analysis or manifold learning). In contrast to supervised learning, unsupervised learning methods do not depend on manual human annotation to obtain the target outputs. This allows a higher degree of automation in the data generation process, but the final evaluation of the method and the results becomes considerably more difficult. Naturally, there has been a spectrum of approaches, collectively called semisupervised learning [5], which try to combine the benefits of both supervised and unsupervised methods.

The aforementioned approaches, including the computational science workflow, assume a "passive" application of machine learning. First, the data are collected, then the computational approach is applied to analyze the resulting data. However, data collection may be expensive or difficult, and furthermore the experiment may include choosing different conditions. Approaches called *active learning* and *reinforcement learning* are concerned with finding suitable actions to take in a given situation. For example, a classifier may actively choose which individuals it would like to obtain a label for during training. Or a robot may choose the action of moving to a new location before collecting more data. As will be discussed later, these more interactive data collection paradigms pose novel conceptual challenges to reproducible research.

13.2.2 Role of the Dataset

The dataset plays an important role in machine learning because it typically tackles problems where a formal explanation of the data analysis task is not possible. One could even say that this is one of the distinguishing features of the machine learning approach.

In theoretical computer science, for instance, problems are typically formally defined. An introductory problem taught in basic computer science courses is the problem of finding the shortest path in a graph. This can be precisely defined in mathematical terms, and the value of the shortest path can be formally verified. This means that for a given algorithm one can prove that it indeed solves the problem. In addition, one may also prove theorems about other aspects of the shortest path problem like the time it takes to compute a solution.

For many machine learning problems, such a definition may not be possible. Consider the problem of handwritten character recognition where the goal is to correctly classify images of handwritten digits. The problem here is that there is no formal specification of what exactly the handwritten, digitized image of a character looks like. The performance of the machine learning algorithm can only be measured relative to the collected dataset, which is in the aforementioned example the set of images of handwritten numbers.

13.2.3 Applied Statistics, Data Mining, and Artificial Intelligence

As with any human defined separation between fields, what is labeled as machine learning [1,2,16,18,21,28] instead of applied statistics, data mining or artificial intelligence, tends to have more to do with the community that a researcher belongs to than any particular technical difference. Nevertheless, we shall attempt to outline some general trends in the different communities in this section. It also serves as a brief guide to further literature for interested readers.

Applied statistics [4,6,10,34] tends to focus more on theoretical understanding of the statistical properties, and traditionally has been focused on regression-type estimation problems. Data mining [22,36] has a more business-oriented origin, and has historically been focused on finding relationships in data in an unsupervised learning fashion. One example is *association rule mining*, which discovers interesting relations between items in databases, and has been popular in market basket analysis. In contrast to machine learning, data mining aims to discover structure in a given dataset, which makes evaluation of the discovery more challenging as it is hard to know the true structure in a given dataset.

Machine learning is often considered to be a subfield of artificial intelligence [12,27] where artificial intelligence is concerned with the study and construction of computer algorithms that exhibit intelligent behavior. In addition to learning, there is significant research on reasoning and planning, which has traditionally been based on mathematical logic. In many real-world problems, such as commuting to work, the solution involves multiple steps that have to be in a particular order.

13.3 Machine Learning and Reproducibility

When it comes to reproducibility, an interesting aspect in machine learning is that reproducibility can be achieved to a higher degree by automation than in other sciences. The reason is that all components of the research are available on a computer. Unlike, say, experimental biology, where one has to physically construct an experiment, machine learning is mostly about data.

As explained in the previous section, machine learning is concerned with creating learning methods that perform well on certain application problems, and that can be verified independently by third parties. Therefore, a research result consists of the method found, the dataset it has been evaluated on, and a full description of the experimental setup, including feature extraction and estimation of free model parameters. Requiring reproducibility therefore implies requiring publication of the method, the data, and the experimental set up.

In statistics and for easily computable problems, there has been significant progress in reproducible research that follows the vein of literate programming [13]. The proposal is to embed the code for calculating the results of the paper directly in the paper itself, hence simplifying management of code and data, and significantly improving reproducibility of the paper's results. In recent years, there has been a trend in machine learning to analyze large scale data, where a significant investment in time and computational infrastructure is required to produce the results reported in the paper. The model of embedding code and data into the generation of a PDF paper does not scale to such issues, and a restructuring of what is means to have reproducible results is required.

Since data mining has traditionally been applied to business intelligence problems, it has been difficult to obtain access to such private and sensitive data. In artificial intelligence, similar to the challenges faced when trying to reproduce research in reinforcement learning, there are open questions on how to compare and validate solutions. We will discuss this further in Section 13.7.

For machine learning, we believe that by adopting the procedures and concepts of open source, open data, and open access, one can create an environment that supports reproducibility, encourages collaboration between researchers, and removes many of the limitations and delays inherent in the current scientific environment (Figure 13.2).

13.3.1 Openness

The idea of open-source software, which emerged in the 1980s, has interesting connections to the problem of reproducibility, although its motivations and goals are ultimately different. Open source is less about reproducibility, but more an attempt to create a process for collaborative software creation, which is not so different from the way science is organized.

Open-source software was the first in a whole series of movements about openness to ease collaboration. Originally, open-source software emerged as a countermovement to the increasingly commercial nature of writing software. Software essentially became trade secrets. While open-source software was first restricted to academia, it eventually became a widely accepted

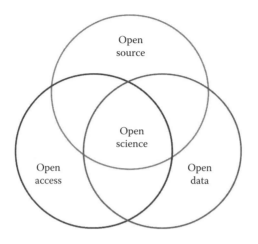

FIGURE 13.2
Open science = Open-source software + Open access papers + Open data.

alternative to commercial closed source software. The Linux operating system has helped a lot in this respect as it was one of the first large-scale pieces of software completed in this way.

The open-source model for collaboration has since been copied in other areas as well, of which the open data, open access, and open research movements are the most relevant for our current discussion. As we will discuss in much more detail in the context of open source in the next chapter, the main aspect of these approaches is to create the legal and organizational foundations for collaborative research. The licenses are just one facet of this approach.

So openness is not mainly about reproducibility, but about collaboration. To achieve reproducibility, it would suffice to publish the relevant pieces of information under a classical copyright license, which would allow others to reproduce the results, but would not allow them to directly reuse the code and data.

One could argue whether openness is required for scientific progress, but we believe that the community processes that come with openness significantly simplify scientific collaboration and therefore help to speed up scientific progress as a whole.

13.4 Open-Source Software

The basic idea of open-source software is very simple, programmers or users can read, modify, and redistribute the source code of a piece of software.

TABLE 13.1

Attributes of Open-Source Software

1.	Free redistribution
2.	Source code
3.	Derived works
4.	Integrity of the author's source code
5.	No discrimination against persons or groups
6.	No discrimination against fields of endeavor
7.	Distribution of license
8.	License must not be specific to a product
9.	License must not restrict other software
10.	License must be technology-neutral

Source: Open Source Initiative. http://www.opensource.org/ docs/osd.

The underlying idea is both to make software freely available and to establish a collaborative community where people contribute to software they find interesting, weeding out bugs if they can, without relying on a software company to take care of this.

The Open Source Initiative (OSI) has compiled a definition of open source according to the criteria listed in Table 13.1. Note that this includes not discriminating against certain persons or groups (e.g., by restricting the use to certain countries) or uses (e.g., to include nonacademic uses). Software that violates any of these requirements is not considered open source. For example, software projects that restrict usage to "noncommercial use only" or "research only" violates OSI definitions and should not be labeled open source.

13.4.1 Open-Source Licenses

Since traditional copyright is the default in most country jurisdictions, open-source software has to come with an explicit copyright license that gives permissions for others to exercise the exclusive rights of copyright. This permission is sometimes given under certain terms and conditions. Since individual licenses might be hard and costly to enforce, people have quickly begun to use a number of standard licenses. Organizations like the Free Software Foundation (www.fsf.org) have also stepped in to defend the GPL, LGPL, and AGPL licenses to set a legal precedent.

Table 13.2 collects key features of these licenses. A detailed comparison of the different styles of licenses is beyond the scope of this chapter, and the interested reader is referred to other resources [20]. The main differences to consider is whether one wants to ensure that derived work is again open source or not. Also note that there is always the option of releasing the software under a different license by the authors themselves. That way,

TABLE 13.2

The Rights of the Developer to Redistribute a Modified Product

License	Apache	BSD/MIT	GPL	LGPL	MPL/CDDL	CPL/EPL
Reciprocity	No	No	Yes	Maybe	No	No
Modification release	No	No	Yes	Yes	Yes	Yes
Patent	Yes	No	No	No	Yes	Yes
Jurisdiction	Silent	Silent	Silent	Silent	California	New York
Freedom	PR	Free	PR	PR	Free	PR

A comparison of open-source software licenses listed as "with strong communities" on http://opensource.org/licenses/category. The reciprocity term of GPL states that *if* derivative works from a GPLed licensed software are distributed in binary form, *then* the recipient of the binary form must also be given the source code of the derivative work licensed under the same GPL license. Other important questions are whether the source code to modifications must be released (Modification release); whether it provides an explicit license of patents covering the code (Patent); the legal jurisdiction the license falls under (Jurisdiction); freedom to adapt license terms (Freedom) (PR = Permission Required from license drafter). Apache: License used by the Apache web server; BSD: License under which the BSD Unix variant is released; MIT: developed by the MIT; GPL/LGPL: (lesser) GNU General Public License; MPL: License used by the Mozilla web browser; CDDL: Common Development and Distribution License developed by Sun Microsystems based on the MPL; CPL: Common Public License published by IBM; EPL: Eclipse Public License used by the Eclipse Foundation, derived from the CPL.

for example, companies can buy software from the authors by paying for an alternative software license.

Another complication in choosing the right license is that some licenses are not compatible with one another in the sense that one cannot combine two pieces of software that have conflicting licenses because it would then be impossible to satisfy both licenses at the same time.

Generally speaking, the BSD/MIT style licenses are the most admissible. They basically state that you are free to reuse the software as long as the original copyright notices and the license stay intact. The GNU Public License (GPL) requires that any derivative work is also published under the GPL. Variants of this exist like the Lesser GPL (LGPL), which just linking by a non-LGPLed work against a LGPLed library without modifying the library is not defined as a derived work, or the Affero GPL, which even extends the notion of derived work to include software that uses the original software over some sort of network interface communication.

13.4.2 Open-Source Collaboration Model

As discussed in Section 13.3.1, while openness and reproducibility are not equivalent, the open-source software movement has developed a number of standards and processes that are also relevant for reproducible research.

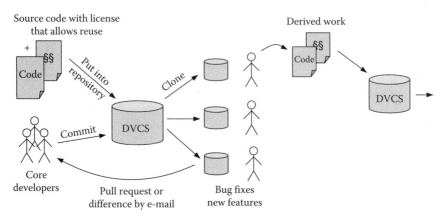

FIGURE 13.3

An overview of the open-source collaboration model. The source code is released with a license that permits collaboration. Typically, the source code is put into a distributed version control system (DVCS) that tracks changes and lets people access the source code more easily. A group of core developers (also known as "committers") control which changes are incorporated in the main code base. Other users can clone the source code, for example, to fix bugs. Changes are offered to the core developers in the form of so-called pull requests, or by e-mail. The source code can be incorporated into other projects that may again be published under an open-source model.

One important component of open source, which is often overlooked, is that open source is not simply a publishing model but comes with a community that emphasizes collaborative work. The book by Raymond [26] gives a good overview over this matter. Figure 13.3 illustrates this process.

Since the source code is freely available and may be modified by others, in principle anyone with the necessary skill and motivation is able to contribute to a project. The level of involvement ranges from fixing minor bugs, to proposing new features, and finally joining the project as a main developer. In order to control the overall direction and consistency and quality of the source code, projects are typically organized into different layers of users. The way this typically works is that anyone may suggest minor fixes or additions, but only a selected number of people are able to actually push changes to the main source tree. Such users are called "committers" from the technical term for adding a revision to a source control system. Within the inner circle, any kind of organization is possible from strictly hierarchical organizations to more or less egalitarian organizations.

Source code version control systems play an important role to facilitate collaboration in open-source projects. Two major categories of them are available today: centralized and distributed. A source code version control system is a database that tracks the changes to source code such that one can always revert changes or go back in time to earlier versions easily.

Typical examples are CVS (www.nongnu.org/cvs) or subversion (subversion. apache.org).

One restriction of these systems is that there is only one central database or repository to track the changes that makes coordination in large multimodule projects difficult. Distributed version control systems (for example, git git-scm.com, mercurial mercurial.selenic.com, or bazaar bazaar. canonical.com) remove this restriction by making it easy to set up local copies of existing repositories (this act is often referred to as "cloning"). People can locally work on the code and offer their changes for integration with the main repository (a process often called "pull request"). In addition, a number of web services exist to give a more user-friendly web interface to these version control systems for managing collaborators, controlling access rights, and organizing software projects. Examples include sourceforge (sourceforge.net), github (github.com), Google code (code.google. com), bitbucket (bitbucket.org), and many others.

Open-source software goes beyond simply making source code available. By publishing the full source codes used in a scientific study, reproducibility can be supported in significant ways. But there are more benefits from the open-source approach, which we will discuss in the context of machine learning.

13.4.3 Machine Learning and Open-Source Software

By making machine learning methods available to others by source code, fair comparison of methods is much easier. Instead of having to reconstruct algorithms from the descriptions in the papers, one can simply reuse existing software. Just as in open source, exposing all the details of the computation also helps to uncover problems in the methods much more quickly. Adopting an open-source approach can help to transfer research results to distant academic disciplines or even the industry much faster because people can build on existing software and integrate them in their own products.

The open-source approach has the potential to further transform the way scientists collaborate because it allows people to share their work much sooner. Traditionally, researchers keep their results private until they are published, based on the misconceived fear that others may steal their results. This often leads to significant delays because the review process takes months, sometimes years until a paper is published. Adopting an open-source approach, researchers could put their work in progress into a version control system that would track individual contributions of researchers and also provide timestamps to resolve precedence between different research groups. Using the open-source collaboration model brings us eventually to the open research approach where all work in progress is made public to invite collaboration already at an early stage.

A number of different initiatives have been started in recent years to support the use of open-source software in machine learning. These efforts are a first step toward making it easier and more rewarding for researchers to publish their code under an open-source license.

Originally started as the *Machine Learning Tools Satellite Workshop* in December 2005, researchers in the machine learning community first came together a day before the annual *Neural Information Processing Systems Conference* (nips.cc) to discuss possible ways to support machine learning. A year later, a second workshop took place at the NIPS conference, with a closing discussion that led to the position paper "The need for open source software in machine Learning" [31]. Among the obstacles identified against machine learning open-source software was that writing software is not considered a scientific contribution in academic circles, and hence there is no incentive for researchers to publish source code. Furthermore, researchers may not be good programmers, and there is entrenched behavior of reviewers accepting papers that may not be reproducible, where the sloppiness may hide more subtle problems. For more industrial laboratories, there is a common misconception among management that open-source software conflicts with commercial interests. In fact, open-source software *is* commercial software [35], particularly in terms of the federal regulations in the United States.

To recognize the contribution of good software in academic currency, a special machine learning open-source software track at the *Journal of Machine Learning Research (JMLR)*, and a community website mloss.org where people can register their machine learning open-source software projects, were created. The main motivation for the special track at the *JMLR* was to give people a way to publish software. Otherwise, it would not seem wise to spend a significant time on publishing software because this effort is not captured in the usual metric used to measure scientific productivity.

As of November 2013, 501 projects are listed on mloss.org, and the *JMLR* has published 55 papers on the special MLOSS track, which demonstrates that the initiative has been very well received by the community. So far, the initiative has been highly successful, but has focused mostly on the "method" side of the problem to make machine learning research more reproducible. Unfortunately, there has been little interaction in terms of common standards and interfaces that would make it much easier to exchange pieces of software.

One problem is also that an open-source project has a much different life cycle than a scientific publication. The main difference is that once a paper is published, although one usually continues to research on the topic, the publication stays fixed. A successful open source software project, on the other hand, lives on and ideally attracts an active user and developer base. The problem is that an open-source project can require a significant amount of work to keep running, which is then not reflected in the aforementioned publication model.

13.5 Open Access

In recent years, it has become accepted that open access is a desirable and viable publication model for papers. Open access benefits researchers, institutions, nations, and society as a whole. For researchers, it brings increased visibility, usage and impact for their work. Institutions enjoy the same benefits as researchers but in aggregated form. Countries also benefit because open access increases the impact of the research in which they invest public money and therefore there is a better return on investment. Society as a whole benefits because research is more efficient and more effective, delivering better and faster outcomes for us all (www.openoasis.org).

Not only is open access a desirable avenue for research output, but it is in fact practical and economically viable. Enabled by low-cost distribution on the Internet, open access literature is digital, online, free of charge, and free of most copyright and licensing restrictions. For example, Creative Commons (creativecommons.org) lays out a flexible range of protections and freedoms for authors, artists, and educators. Many journals (more than 8000 according to www.doaj.org) have adopted the open access model. In fact, NIH supports open access to research funded via its grants, but the publishers are fighting back.

As mentioned in the introduction, machine learning has multiple open access publication venues, including its flagship journal the *Journal of Machine Learning Research*, and the proceedings of conferences such as the *International Conference on Machine Learning (ICML)*, *Neural Information Processing Systems (NIPS)*, the *Conference on Uncertainty in Artificial Intelligence (UAI)*, and the *International Conference on Artificial Intelligence and Statistics (AISTATS)*.

13.6 Open Data

Based on the model provided by open-source software and open access papers, the approach has been extended to other areas, most notably Open Data (opendatacommons.org). As mentioned earlier, datasets are very important in machine learning because they define learning problems that cannot be defined formally. A new well-designed dataset has the potential to spark a completely new line of research.

Historically, machine learning publications mostly focused on new data analysis methods, therefore datasets were often compiled or used for publications that presented new methods. Another typical way to publish datasets consists in organizing a challenge. Here, the challenge organizer puts together a dataset, keeping part of the data private and inviting others to develop methods for their datasets during a given challenge runtime. Afterward, the methods are ranked on the private data based on the published performance measure. The top performing methods are often invited to publish in a special issue of a journal, or in a workshop.

Over time, such datasets are often collected in dataset repositories with the goal of making it easier to find relevant datasets and existing results. However, there still is not an open exchange in the same way as there is with source code.

Part of this problem might be that while authorship is usually clear with source code, the number of people involved in data acquisition is often much larger, and often more interdisciplinary. Privacy and legal considerations may be much more complex for data related to people such as medical information.

13.6.1 Machine Learning Dataset Repositories

Recall that open science consists of three components: open-source software, open access papers, and open data. While mloss.org provides the "method" software, the actual experimental protocols for a particular paper are not available, and neither is the data used for producing the results and figures.

Several repositories focusing on machine learning datasets exist, for example, the UCI machine learning repository (archive.ics.uci.edu/ml), or the DELVE repository (www.cs.toronto.edu/~delve). These sites have quite a long history, the DELVE site being run since 1995. Both sites host a number of standard benchmark sets which have been used in hundreds of publications.

Still, both sites do not allow for the level of interactivity that would be require to become a main repository for open data exchange. Both sites are rather static, one cannot simply add a dataset. The sites contain mostly datasets that are generally considered to be too easy, with the focus lying mostly on regression and binary classification. DELVE in particular has been mostly unmaintained in the last few years.

When designing mldata.org, we had the goals in mind to create a community run website where people can publish datasets. The website has mechanisms to stimulate interaction between users, such as tagging, discussions, and ratings. The whole dataset can be edited in a wiki-like fashion, such that the community can continually improve the archived data.

Another goal was to provide standardized means for benchmarking. The DELVE repository has been rather ambitious in this respect, but to our knowledge, it is currently referred to only for the datasets. As we will discuss in more depth in Section 13.7, building and establishing a standard framework for benchmarking datasets is a difficult task, but this problem has to be solved ultimately to make machine learning research reproducible.

Our website mldata.org supports four kinds of information: raw datasets, learning tasks, learning methods, and challenges. A raw dataset is just some data, while the learning task also specifies the input and output variables and the cost function used in evaluation. A learning method is the description of a full learning pipeline, including feature extraction and learner. One can upload predicted labels for a dataset and a task to create a solution entry that automatically evaluates the error on the predicted labels. Finally, a number of learning tasks can be grouped to create a challenge.

Most of these data are text. We did not attempt a full formal specification of the learning method, but as a first step, we defined a general file exchange format for supervised learning based on HDF5, a structured compressed file format. It is similar to an archive of files but has additional structure on the level of the files, such that users can directly store and access matrices, or numerical arrays. Using the specified file format is not mandatory, but using it unlocks a number of additional features like a summary of the dataset and converting the dataset into a number of other formats.

The website went live in 2007. To jump start the community, we uploaded hundreds of freely available datasets. So far, our experience with the website is mixed. As we will discuss in the final section of this chapter, achieving an acceptable level of interoperability has to be balanced against the complexity of the system. Here, we are still in a process to find the optimal mix.

13.6.2 Business Models around Machine Learning Datasets

In recent years, other approaches to disseminating datasets have also arrived. The idea is less about open data and providing a service to academia, but more about building a platform between researchers who know data analysis methods on companies that have interesting data.

These recent approaches are often organized around competitions, where the datasets and prize money are provided by companies. One example is kaggle (kaggle.com), where companies can set up their own competition. Such websites became quite popular after the famous Netflix Prize challenge (www.netflixprize.com). Netflix, a provider of streaming video, set up a competition where the person who could improve over Netflix existing recommendation algorithm would win one million dollars.

However, the Netflix Prize also highlights some of the dangers of competitions based on live business data. Netflix was eventually sued over privacy concerns. Using competition data, researchers seemed to be able to de-anonymize the datasets by correlating the data with other sites. Netflix finally chose not to run a second competition [11].

Ultimately, such changes can also be seen as a cheap way for companies to outsource data analysis work to graduate students in machine learning and related fields. For competitions with many participants, the final prize money might be significantly less than what would have to be paid for the joint work of all participants.

13.7 Future Challenges

We have discussed an approach to support reproducible research in the area of machine learning based on adopting concepts and processes from open-source software and extensions. We believe that the combination of open-source software, open data, and open access leads to an environment where researchers can efficiently exchange their results and reuse the work of others. Still, a large number of challenges still exist that we will discuss in the following.

13.7.1 Interoperability and Standards

Due to the reasons mentioned earlier, it is desirable to have independent units of data, software, and computational resources that interact with one another. Furthermore, within a particular application pipeline, different parts of the pipeline such as the feature construction and classifier training may be provided by code from different software projects. One major challenge when building a long workflow is to ensure that when replacing a feature construction method with a novel approach, the whole pipeline still functions as expected. This requirement goes beyond simple replicability, but it is necessary in order to build large complex systems capable of solving real-world problems. To achieve this, the community would have to agree on certain standards or protocols of communication between the different parts of a data processing pipeline.

As mentioned before, mloss.org and mldata.org are only first steps toward the goal of open science in machine learning. We briefly review several other projects that work toward the same goals.

In the area of statistics, a lot of integration has already been realized in the form of the R programming language (r-project.org). R is an open-source reimplementation of the commercial S programming language and is similar in scope to other data analysis centered programming language environments like the commercial MATLAB® or the Python-based scipy. It provides specialized data types for dealing with all kinds of data and comes with a large library of standard statistical and data analysis functions, as well as libraries for plotting and visualization. In addition, it has a central package repository called CRAN (cran.r-project.org) which makes it very easy for researchers to publish their code and for others to install and use it (Figure 13.4).

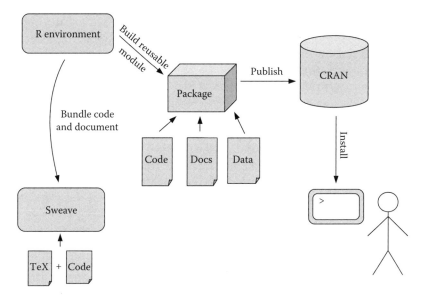

FIGURE 13.4
Overview of the R environment. R provides a very rich environment for data analysis, but one of the main strengths of the system is the central package repository called CRAN, which allows users to publish their code together with documentation and datasets easily, which can then be installed painlessly by other users. Another component are libraries like Sweave or knitr, which let users combine code and the LaTeX code used to typeset document to achieve a very high level of integration and reproducibility.

In statistics, R is the de facto standard, meaning that practically all papers also publish their methods in high-quality code, often including datasets as well. R also comes with very good documentation support, generating code that can be used with the LaTeX typesetting system, including example code snippets. It is also possible to package datasets together with the code that is a very good way to publish moderately sized datasets.

Finally, there are also the Sweave project, (www.statistik.lmu.de/~leisch/Sweave/) and knitr (yihui.name/knitr) which are systems where you can combine R code with the LaTeX code to typeset your paper such that the paper itself is turned into the code to produce your analysis results, leading to a very high level of reproducibility.

The success of R hinges on the homogeneity of the research community. For machine learning, the set of tools, programming languages, and approaches has always been too diverse to be integrated in the same tight fashion easily. For example, for real-time, or large-scale applications, the performance of R is insufficient. Therefore, while the R example shows the benefit of a tightly integrated infrastructure adopted by the majority of the community, achieving this complexity for other areas will typically be much more challenging.

For machine learning, one would have to integrate different programming languages like MATLAB, Python, C, or Java, support different data formats and data storage backends like databases, files, web services, and also different operating systems.

One way to approach this problem is to develop formal abstractions and descriptions to encode feature preprocessing and other operations and to provide an interface that others can plug into. Two examples of currently active projects are the ExpML project by Vanschoren et al. [33], who have developed an XML schema for doing exactly this. The goal is to provide support for this XML schema in the major existing machine learning platforms such that experiment descriptions can be automatically executed. The project is also working on setting up an experiment repository in the same sense as mldata.org and mloss.org. Other examples are tunedit.org and mlcomp. org. These websites provide computing facilities for people to run their methods on datasets and to collect benchmark results for a large number of algorithms.

Another project is the "Protocols and Structures for Inference" project by Mark Reid (psi.cecs.anu.edu.au), which also defines an interface language for common machine learning interactions. Here, the focus is less on reproducibility and open data, but more on laying the groundwork for improved interoperability between the different pieces of code.

13.7.2 Automation vs. Flexibility

From our experience building mldata.org, we observed the following unfortunate trade-off between automation and flexibility: We built an general representation of tabular style data in HDF5 that captures many different possible feature types, such as categorical, real valued, or strings. Using this HDF5 representation, we could easily automate conversion between several popular machine learning formats such as csv files, libsvm formats, and MATLAB binaries. However, this led to a large proportion of the datasets on mldata.org being of tabular form, and many users assumed that this was the only acceptable structure.

On the other hand, mldata.org also accepts any file format as a dataset. This flexibility means that we are unable to automatically convert between formats that are convenient for different programming languages, and the dataset is then less appealing to users.

In general, there is always the danger to create something so complex and complicated to make the system practically unusable. A formal description of a machine learning experimental setup quickly evolves to become a full domain-specific language (DSL), just another programming language for the user to learn.

One way to approach this dilemma is to focus on common cases and simple examples first, to keep the system simple and user-friendly. However, there will always be cases that cannot be represented in such a system.

13.7.3 Nonstatic Data

In traditional machine learning settings, data are considered in a "batch," that is, the whole dataset is available and is fixed. However, in many recent application areas such as social network analysis, there is a stream of data, and the corresponding research area of online learning that continuously updates the predictor has emerged. Defining reproducibility in such a setting is challenging.

Furthermore, in the setting of reinforcement learning, the algorithm has a choice of which data to receive and may even intervene in the environment. Apart from highly contrived simulated examples, it is an open problem on how to define reproducibility in such a setting.

13.8 Outlook

An open letter to the US congress signed by 25 Nobel laureates in 2004 states: "Open access truly expands shared knowledge across scientific fields, it is the best path for accelerating multi-disciplinary breakthroughs in research." This sentiment has extended to open data in recent years, quoting the Open Knowledge Foundation [19]: "The more data is made openly available in a useful manner, the greater the level of transparency and reproducibility and hence the more efficient the scientific process becomes, to the benefit of society." In a data-driven field such as machine learning, the easy availability of methods and data are cornerstones of reproducibility and scientific progress.

We believe that open source goes way beyond simply making your source code available to others under a license that invites collaboration, but is in fact a whole process for open collaboration, not unlike science. Science has always favored open collaboration through publication of scientific results. Isaac Newton is attributed with the famous quote "If I have seen further it is by standing on the shoulders of giants.", which reflects the importance of sharing scientific results to accelerate scientific growth. The open science model poses an interesting inspiration to transform the scientific progress in the information age.

Acknowledgments

The authors thank Luis Ibanez and Lydia Knüfing for useful comments and criticisms, which resulted in significant improvements in the chapter.

References

1. D. Barber. *Bayesian Reasoning and Machine Learning*. Cambridge University Press, New York, 2012.
2. C. Bishop. *Pattern Recognition and Machine Learning*. Springer, New York, 2006.
3. J.B. Buckheit and D.L. Donoho. Wavelab and reproducible research. Technical Report, Stanford, CA, 1995.
4. P. Bühlmann and S. van de Geer. *Statistics for High-Dimensional Data*. Springer, Heidelberg, Germany, 2011.
5. O. Chapelle, B. Schölkopf, and A. Zien, eds. *Semi-Supervised Learning*. MIT Press, Cambridge, MA, 2006.
6. B. Efron and R.J. Tibshirani. *An Introduction to the Bootstrap*. Taylor & Francis, Boca Raton, FL, 1994.
7. J. Feller, B. Fitzgerald, S. Hissam, and K. Lakhani, eds. *Perspectives on Free and Open Source Software*. MIT Press, Cambridge, MA, 2007.
8. R. Gentleman. Reproducible research: A bioinformatics case study. *Stat Appl Genet Mol Biol*, 4(1):1034, 2005.
9. R. Gentleman and D.T. Lang. Statistical analyses and reproducible research. Technical Report 2, Bioconductor Project Working Papers, 2004. http://biostats.bepress.com/biconductor/paper2 (Accessed November 29, 2013).
10. T. Hastie, R. Tibshirani, and J. Friedman. *The Elements of Statistical Learning*. Springer Series in Statistics. Springer, New York, 2001.
11. N. Hunt. Netflix prize update. http://blog.netflix.com/2010/03/this-is-neil-huntchief-product-officer.html, March 2010. (Accessed November 29, 2013).
12. M. Hutter. *Universal Artificial Intelligence: Sequential Decisions Based on Algorithmic Probability*. Springer, Berlin, Germany, 2010.
13. D.E. Knuth. Literate programming. *Comput J*, 27(2):97–111, 1984.
14. J. Kovacevic. How to encourage and publish reproducible research. *IEEE International Conference on Acoustics, Speech and Signal Processing*, 4:iv-1273–iv-1276, 2007.
15. F. Leisch. Sweave: Dynamic generation of statistical reports using literate data analysis. Härdle, W. and Rönz, B. (Eds.) *COMSTAT, Proceedings in Computational Statistics*, Physika Verlag, Heidelberg, Germany, 2002.
16. S. Marsland. *Machine Learning: An Algorithmic Perspective*. CRC Press, Boca Raton, FL, 2009.
17. J.P. Mesirov. Accessible reproducible research. *Science*, 327:415–416, 2010.
18. T. Mitchell. *Machine Learning*. McGraw Hill, New York, 1997.
19. J.C. Molloy. The open knowledge foundation: Open data means better science. *PLoS Comput Biol*, 9(12):e1001195, 2011.

20. A. Morin, J. Urban, and P. Sliz. A quick guide to software licensing for the scientist-programmer. *PLoS Comput Biol*, 8(7):e1002598, 07 2012.
21. K.P. Murphy. *Machine Learning: a Probabilistic Perspective*. MIT Press, Cambridge, MA, 2012.
22. R. Nisbet, J. Elder IV, and G. Miner. *Handbook of Statistical Analysis and Data Mining Applications*. Academic Press, Amsterdam, the Netherlands, 2009.
23. Open Source Initiative. The Open Source Definition. http://www.opensource.org/docs/osd (Accessed November 29, 2013).
24. R.D. Peng. Reproducible research in computational science. *Science*, 334:1226–1227, 2011.
25. C.E. Rasmussen, R.M. Neal, G. Hinton, D. van Camp, M. Revow, Z. Ghahramani, R. Kustra, and R. Tibshirani. Delve-Data for Evaluating Learning in Valid Experiments. http://www.cs.toronto.ca/~delve (Accessed November 29, 2013).
26. E.S. Raymond. *The Cathedral and the Bazaar*. O'Reilly Media, 1999.
27. S. Russell and P. Norvig. *Artificial Intelligence: A Modern Approach*, 3rd edn. Prentice Hall, Upper Saddle River, NJ, 2009.
28. B. Schölkopf and A.J. Smola. *Learning with Kernels*. MIT Press, Cambridge, MA, 2002.
29. E. Schulte, D. Davison, T. Dye, and C.N Dominik. A multi-language computing environment for literate programming and reproducible research. *J Stat Software*, 46(3):1–24, 2012.
30. M. Schwab, M. Karrenbach, and J. Claerbout. Making scientific computations reproducible. *Comput Sci Eng*, 2(6):61–67, 2000.
31. S. Sonnenburg, M.L. Braun, C.S. Ong, S. Bengio, L. Bottou, G. Holmes, Y. LeCun et al. The need for open source software in machine learning. *J Mach Learn Res*, 8:2443–2466, 2007.
32. V. Stodden. The legal framework for reproducible research in the sciences: Licensing and copyright. *IEEE Comput Sci Eng*, 11(1):35–40, 2009.
33. J. Vanschoren, H. Blockeel, B. Pfahringer, and G. Holmes. Experiment databases—A new way to share, organize and learn from experiments. *Mach Learn*, 87(2):127–158, 2012.
34. L. Wasserman. *All of Statistics*. Springer, New York, 2004.
35. D.A. Wheeler. Open source software is commercial. Software Tech News, Data & Analysis Center for Software, Department of Defence, USA, 14(1):16–19, 2011.
36. I.H. Witten, E. Frank, and M.A. Hall. *Data Mining: Practical Machine Learning Tools and Techniques*, 3rd edn. Morgan Kaufmann, Amsterdam, the Netherlands, 2011.

14

RunMyCode.org: A Research-Reproducibility Tool for Computational Sciences

Christophe Hurlin, Christophe Pérignon, and Victoria Stodden

CONTENTS

14.1 Introduction

Research reproducibility can be vastly improved by the open availability of the code and data that generated the results. In this chapter, we present a new web-based tool that aims to improve reproducibility in computational sciences. The RunMyCode.org website gives published articles a *companion webpage* from which visitors can (1) download the associated code and data and (2) execute the code in the cloud directly through the RunMyCode.org website. This permits results to be verified through the companion webpage or on a user's local system. RunMyCode.org also permits a user to upload their own data to the companion webpage to check the code by running it on novel datasets.

We present the structure of the RunMyCode.org system in Figure 14.1. Researchers provide the code and data associated with their publication. Users can either use the data provided by the researchers or provide their own. Then the code and data are sent to the cloud. When the computation is done, the results are sent back to the user.

The RunMyCode concept can be viewed as an attempt to provide, on a large scale, an executable paper solution. The difference between this

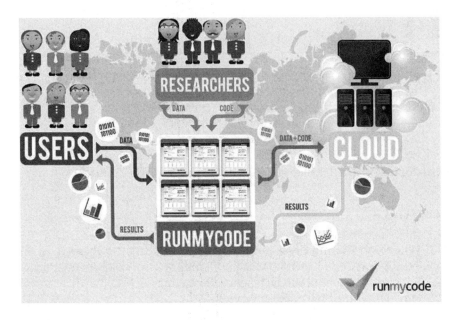

FIGURE 14.1
The RunMyCode system. Note: Researchers provide the code and data associated with their publication. Users can also provide their own data, which are sent to the cloud along with the computer code. When ready, the results are sent back to the user.

and the executable paper approach proposed by the scientific publishers (see, for instance, Elsevier's Executable Paper Grand Challenge, 2011, http://www.executablepapers.com) is that the companion webpage is not encapsulated within the text of a scientific publication. In that sense, a companion webpage can be considered as providing *additional* material for a scientific publication, in particular the digital objects that permit verification and replication of the published computational results.

Of course, being able to reproduce the main findings of scientific papers is important for the scientific community itself, but it also matters for the credibility of science in society. Furthermore, reproducibility is of primary importance for governments and corporations since it is a necessary condition to convert scientific ideas into economic growth. We summarize in Figure 14.2 how the RunMyCode website can improve transfer of technology from the academia to society (students, corporations, administrations, general public, etc.). A key feature of the website is to reduce the technical cost for users to access and use a new scientific technique.

RunMyCode has three main objectives. The first is to allow researchers to quickly disseminate the results of their research to an international audience through an online service. This should lead to a notably increase in the citations of certain academic articles. Second, RunMyCode aims to

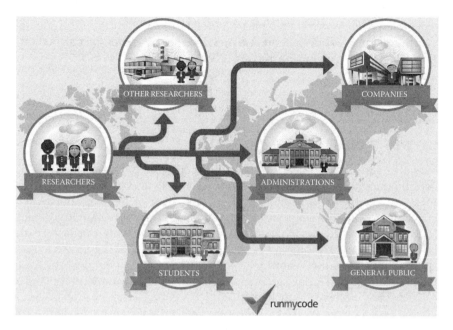

FIGURE 14.2
Improving transfer of technology. Note: The RunMyCode website aims to improve transfers of technology within academia (researchers to researchers and researchers to students), from the academia to companies, as well as from the academia to society (administrations, general public).

provide a very large community of users—potentially beyond the academic sphere—with the ability to use the latest scientific methods in a user-friendly environment, for their own data and parameter values. To date, such analyses were impossible for users without the necessary computing skills to implement the methods in specific software. Third, it allows members of the academic community (researchers, editors, referees, etc.) to replicate scientific results and to demonstrate their robustness.

RunMyCode is an international academic project founded by economics and statistics professors from Columbia University, HEC Paris, and University of Orléans (France) and engineers from CNRS (the French National Science Foundation). RunMyCode is incorporated as a non-for-profit scientific association and is funded by universities, national research agencies, and foundations.

The rest of our chapter is structured as follows. In Section 14.2, we explain why researchers should share their code and data and why they often do not. We explain in Section 14.3 why, on top of *sharing* code and data, making code running in the cloud is a further step toward full reproducibility. We then take economics as an example of computational science (Section 14.4) and we discuss the case of code and data sharing in this fields, as well as executable code. Section 14.5 focuses more specifically on RunMyCode and on its functioning, while Section 14.6 mentions several potential partnerships and further developments for the RunMyCode initiative.

14.2 Why (Not) Sharing Code and Data

There are many good reasons to share the code and data associated with a scientific paper. Lerner and Tirole (2002) show that in the context of open source, researchers can benefit from enhancement of their reputation and that of their potential value on the labor market. The availability of data and codes is related not only to the reproducibility issue but also to the dissemination and exploitation of academic research. Having access to such resources improves the visibility of articles and their impact on both scientific community and nonacademic sphere. A recent example is the V-Lab (*Volatility Lab*) website launched in 2012 by Nobel laureate Robert Engle at New York University in order to ease the diffusion of the systemic risk measures proposed by Engle and his coauthors.

However, in practice, most researchers are still reluctant to share their code and data. Borgman (2007) identifies four major factors preventing systemic disclosure of code and data: (1) lack of incentives (citations or promotion), (2) the effort required to clean data and codes, (3) the creation of a competitive advantage over other fellows, and (4) intellectual property issues. Similar impediments for reproducibility have been identified in previous work (Stodden 2010) in a survey of 723 American academic researchers. In her study, the main factors restraining researchers from making computer codes available are the time for documenting and preparing the codes (77% of subjects), the idea of having to answer questions from possible users (52%), and having no direct benefits (44%). A possible loss in future publications was also indicated as a subsequent factor by 30% of the researchers. Finally, in some research area, a significant fraction of research is conducted using proprietary data. For instance, Glandon (2010) report that 28% of the articles published in the top economics journal, the *American Economic Review*, used confidential data.

RunMyCode.org solves several of the problems given earlier, confronting computational scientists in 2007–2008 who wish to engage in reproducible research. It removes the difficulty of hosting the code and data, it removes the difficulty of installing and running (even correct) code on a local computer system, and by providing the ability for users to execute the code in the cloud, it minimizes the amount of support coders and authors are asked to supply. RunMyCode.org also provides suggested citations, to help encourage a reward system that encourages code and data release, by giving credit for these scientific contributions. RunMyCode.org provides a public date of creation of the companion webpage, helping to ensure primacy to those who release code and data and encourage attribution. Perhaps most importantly, RunMyCode.org provides a central field-independent platform to facilitate both code and data sharing and the verification of published computational results.

14.3 Why Make Code Executable in the Cloud?

We argue that sharing code and data would be a significant step toward research reproducibility. However, it may not be a sufficient one. A further step would be to make code running in the cloud. To make our point, we present a landmark experiment conducted by researchers in economics. McCullough et al. (2006) aimed to reproduce the results of the 266 papers published in the *Journal of Money, Credit and Banking* between 1996 and 2003. The replication team only had to use online material associated with the 266 papers available on the journal website, which had a data availability requirement. Out of the 266 papers, 193 of them contain an empirical section and, as such, should have data and/or code provided by their authors, in compliance with journal policy. In reality, 35% of the papers had no online material whatsoever, 5% had data but no code, and 4% had code written in languages not supported by the replication team. Other research confirms this is not a situation unique to economics (see Alsheikh-Ali et al. 2011, Tenopir et al. 2011, and Savage and Vickers 2009).

The main finding of this chapter is that a small fraction of the papers with available data and code were reproduced to their full extent. Hence, sharing code and data may not always be a sufficient condition for engaging in reproducible research. Indeed, only 14 articles (7% of the sample) have been reproduced. Several reasons can explain this extremely low reproducibility rate. First, the authors of the original papers were not always careful enough when preparing the final version of the code and data uploaded on the journal website. Hence, this material is hard to use and results hard to reproduce. Second, there is typically very little and often no explanation on how to use the online material (e.g., no readme file). This is due, in part, to the fact that the editorial boards provided no strict guidelines about the code and data submission process.

The case for executable script has recently been made in biostatistics (Peng 2011). Indeed, in the journal *Biostatistics*, each article receives a mark mentioned on the first page. "D" and "C" stand for available data and code, respectively, whereas "R" signifies a reproducible article. In the latter case, a "reproducibility review" (execution of the code on the original data) has been performed by the editor on the request of the author. The journal hence identifies four levels of reproducibility, from nonreproducibility to "gold standard": (1) publication only; (2) publication and code; (3) publication, code, and data; and (4) publication and executable code and data.

14.4 Example of Computational Science: Economics

RunMyCode.org was first launched in economics and there are several reasons for that. Over the last few decades, economics has become more empirical and data-driven. Furthermore, economics is nowadays a highly computational discipline, far ahead of many other social sciences. Numerical computation is now ubiquitous in modern economics: statistical analysis, estimation, optimization, simulation, numerical equation solving, and the entire spectrum of econometrics. Barrou (2008) reports that the fraction of theoretical papers published in the top economics journal, *American Economic Review*, dropped from 70% in the 1970s to 20% in the recent years. Further evidence is provided in the survey of Kim et al. (2006) of all the articles with more than 500 citations from top economics and management journals. They show that at the beginning of the 1970s, 77% of these papers were theoretical and 11% empirical. By the end of the 1990s, the proportions were reversed: 11% theoretical vs. 60% empirical.

Another reason that contributed to the development of an executable-code platform is the fact that scripts and data tend to be smaller than in many other computational sciences. There are some recent exceptions though in economics with datasets of several terabytes of high-frequency financial transaction data or shopper data at retailers.

Since the 2000s, mainly top-ranked economics journals, such as *American Economic Review* and *Econometrica*, have created data and code/script archiving systems. However, as noted by McCullough and Vinod (2003), sharing code and data have remained on a voluntary basis for a while. In 2004, the chief editor of the *American Economic Review*, Ben Bernanke, decided to make mandatory data and code submission after publication. Glandon (2010) studies the performance of this policy in 2007–2008 and shows that only 79% of the published papers could be replicated without contacting the authors. Another scientific policy recommendation would be to require the code and data associated with a scientific work prior to its publication (without making them publicly available yet).

While discussions among economics journals focused on disclosure of scripts and/or data, we are aware of only one paper advocating executable scripts. In a pioneering article, Phillips (2003), one of the best econometricians in the world, describes an Internet service for automatic forecasting similar in some respects with the RunMyCode companion website concept. Phillips anticipates that the future of economic forecasting is in automatic Internet-based econometric modeling, which he calls *Interactive Econometric*

Web Service (IEWS). Phillips imagined a web interface on which different forecasting methods are presented. The user is allowed to choose the parameters and options. The results are executed on a local server and displayed in the webpage as tables and graphs. He then summarizes the advantages of his IEWS:

> Perhaps the main advantage of econometric web services of this kind is that they open up good practice econometric technique to a community of users, including unsophisticated users who have little or no knowledge of econometrics and no access to econometric software packages. Much as users can presently connect to financial web sites and see graphics of financial asset prices over user-selected time periods at the click of a mouse button, this software and econometric methodology make it possible for users to perform reasonably advanced econometric calculations in the same way. The web service can be made available on a 24/7 basis so that people can perform online calculations in presentations and lectures.
>
> Phillips (2003), *The Law and Limits of Econometrics*, p. 25

Phillips' paper has been a major source of inspiration for the RunMyCode project. The companion website proposed by RunMyCode can be seen as a generalization of Phillips' IEWS.

14.5 How Does RunMyCode.org Work?

RunMyCode is based on the concept of a companion webpage associated with a scientific publication. It allows people to run online computer scripts associated with an article, the results being automatically displayed to the user as a SaaS (software as a service), or to download the script and demo data directly. The companion webpage is thought of as a frame of the scientific publication making it possible to both download the research resources associated with publications and to simply use them through the web to check the robustness, performance, and reproducibility of the results.

An example of a companion website is presented in Figure 14.3. A scientific paper's companion webpage on RunMyCode.org is structured as follows. The upper panel displays information about the paper, including a direct link to the pdf file and the abstract, and the authors. The intermediate panel contains information about the coders (i.e., the researchers who wrote the code and who may not be the original authors of the paper) along with a description of the goal of the code. The lower panel allows the user to upload the data, select models, and set parameters values. Finally, the green RunMyCode button launches the computation.

FIGURE 14.3
Example of a scientific paper's companion webpage on RunMyCode.org.

As shown in Figure 14.1, RunMyCode plays the role of an intermediary between the researchers offering the code (which may, in some cases, be different from the authors of the publication) and the users (researchers, students, public administration, private firms, etc.). RunMyCode allows researchers to create a custom companion webpage online without any particular computing skills. This is a six-step process, each of them requiring the author to give some information about the publication and the coauthors, as well as a clear description of the variables and input parameters of the computer code. The author can declare five types of inputs: scalar, vector, matrix, text, choice list, and file (in this case, he or she defines the type of file, such as an image file). For any other type of inputs, the author is asked to give particular recommendation to our technical team. In contrast, no information about the output is required: the companion webpage reproduces the output of the computer code (tables, figures, numerical values, text, image, etc.) as it would appear on the researcher's personal computer. The final task of the researcher is to preview and validate their companion webpage.

The RunMyCode back end can take scripts or code, where the code needs to be compiled before execution and scripts are interpreted at runtime only and need not to be compiled. Currently, it is possible to create a companion webpage from code written in C++, Fortran, MATLAB®, R, and RATS. More software will be added in the near future, especially Python.

Note that the creation of a companion webpage does not typically require any modification to the original script/code and as such requires no additional effort from the researcher. The source scripts are simply encapsulated and sometimes transformed into an executable on the RunMyCode system. For instance, MATLAB scripts are compiled and then run with the MATLAB Compiler Runtime (MCR). The MCR is a stand-alone set of shared libraries that enables the execution of compiled MATLAB applications that do not have MATLAB installed. For other software (for instance, specific econometric software such as RATS), the scripts cannot be transformed into an executable file. In this case, the script is simply used in batch mode. The source codes are compiled according to the recommendation provided by the author, and when the code uses some specific libraries, we use exactly the same libraries. RunMyCode runs in the Linux environment. If a code runs on a specific Windows system, we emulate a virtual machine (Windows) with the same environment as that used by the author.

During this process (called preproduction), we may introduce some additional instructions in the original script if necessary in order to (1) link it with the inputs provided by the companion websites and (2) format the results in a pdf file. Indeed, once the job is executed, a posttreatment is done from the raw results issued from the software. This posttreatment is done with LaTeX: all the numerical results (tables) and all the comments (text) produced by the codes are automatically saved in tex format. An automatic program compiles these results to produce a pdf file. The visual results (figure) are saved in an eps format and included in the LaTeX file during the posttreatment process. Currently, they are published in the same pdf file as the other numerical results. In the future, we plan to improve this mechanism in order to produce the results and the figure in html.

During this preproduction process, we check the code to ensure that the required inputs match the descriptions and constraints provided by the author. Note that this is not a scientific validation. We only check for typical bugs (infinite loops) for the duration of the computing process and for security (malicious codes). Note that RunMyCode is responsible for the security of the codes that are submitted to our cloud provider. But, the cloud provider has also its own security rules that are not specific to RunMyCode.

Once the website is created, it enters the validation stage. First, the authors or coders validate it. Then, the editorial team checks whether the topic complies with the editorial policy of the website, similar to arXiv, for example, or any peer-reviewed academic journal. Finally, a technical validation of the code is undertaken, which focuses on its robustness, security, CPU requirements, and computing time.

Once the validation step is completed, the code is uploaded on the cloud and the companion webpage goes online. Companion websites can be found directly on the web, through any search engine, or starting from the RunMyCode website. Each contributor within RunMyCode is given a unique profile called a "coder page." This permits the researcher to find and connect with people working on similar or other interesting problems. Most importantly, it offers various statistics on the visibility of their websites: number of visits, number of executions of the code, number of downloads, etc.

Developing the concept of executable papers is an important issue nowadays for the major scientific editors worldwide. PDF publications can no longer be considered as the ultimate stage in scientific research. For example, two major conferences called "Beyond the PDF" were organized in 2011 and 2013.* As another example, Reed Elsevier issued a call for tenders in 2010, for the executable paper concept. Its objective was to find ways to easily replicate the results of scientific publications. Nevertheless, to our knowledge, no functional form of the executable paper concept has been proposed so far and no published article describes or proposes such services in economics and management.

As an illustration, we show in Figure 14.4 how a specific result can be reproduced with a RunMyCode companion website. Consider a given study in which one of the key results is a plot of the value of a Y variable that depends on an X variable. With the companion website, one can reproduce the result published in the original paper using the same parameter values as in the paper (n = 100 in this example). However, as shown in Figure 14.4, one can also launch the computation using different parameter values (n = 50)

---

* See   https://sites.google.com/site/beyondthepdf/   and   http://www.force11.org/
  beyondthepdf2.

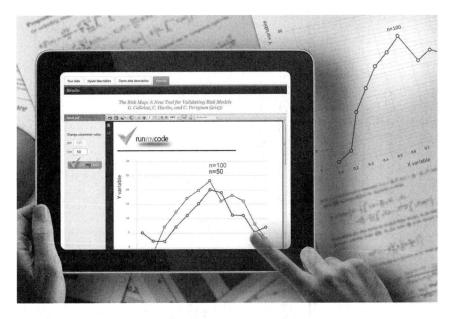

FIGURE 14.4
Reproduced and generalized computational results.

and see whether the key result is robust to a change in the value of one or several parameters.

Figure 14.5 gives a representation of the RunMyCode system.

RunMyCode is based on a cloud computing architecture type and a message routing mechanism built on message-oriented middleware (MOM). The message includes the data and all the parameters needed to run the script on the cloud. For all the applications, cloud facilities are provided by the French National Research Agency (CNRS)'s TGE Adonis. The management of the jobs is done through the distributed task manager (DTM) application provided by the TGE Adonis (CNRS). DTM is a lightweight tool for submitting and monitoring jobs through a local batch scheduler, gLite grid, and local Linux/Unix host. Jobs in DTM may consist of one or several tasks. The RunMyCode jobs are registered and then they are executed by DTM jobs agents in SGE or grid. Once the posttreatment is ended, the website receives the information and displays the results to the user. If the user is still on the companion website, he or she can display the results by clicking on the button "view" of his or her computing queue. If the user browses other sites, or if he or she is logged out, he or she can retrieve his or her results on the tab "past results."

There is no mechanism to check if the code halts in a reasonable time, since the length of the process may vary with the inputs or parameter choices provided by the user. In order to ensure that no process will hog the computational resources, we only fix a limit in terms of CPU time (10 h).

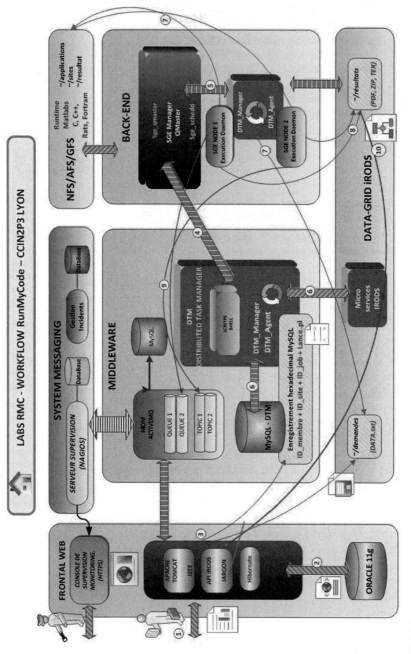

FIGURE 14.5
RunMyCode system workflow.

Currently, each job is submitted to a specific node of the computing cloud and so we do not use parallelization at this point. Because of this, the current architecture of RunMyCode generally does not provide better time performances that the user would have on his or her personal computer or personal system. On the contrary, the performance is generally worse due to the task scheduler, the check on the inputs, etc. But, improvement of compute time is on our working agenda (Box 14.1).

BOX 14.1   WHAT DO I NEED TO CREATE A COMPANION WEBSITE?

The making of a companion website is the following. Users will be able to generate automatically their own companion website from their computer codes. They follow the following process:

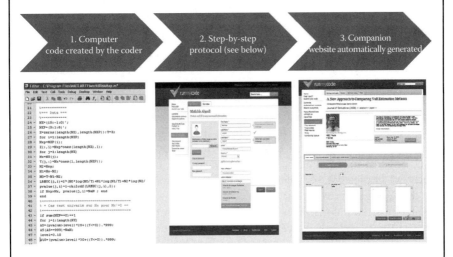

In the step-by-step protocol, the coder provides information about (1) the scientific paper, (2) the coders, (3) the code and the software, (4) the inputs (e.g., variables), and (5) the outputs. The first step requires very standard information about the publication: the name of the authors, the affiliations, the abstract, the DOI or the link to the publication (published article or working paper), some key words, etc. The second step consists in listing the authors of the codes or the scripts. In order to avoid confusion with the authors of the scientific publication, we introduce use the term "coders." Indeed, the coders may not be the authors of the publication, which is actually generally the case. The third step consists of declaring the main information about the code or the scripts. The coder has to upload all the required files (main codes and

(continued)

BOX 14.1 (continued)

subfiles or library), eventually in zip format, for the execution. We also ask for some information about the software used (the list is currently limited to MATLAB, C, R, Fortran, C++, Python, or RATS, although if the software can run on a Linux system, RunMyCode can probably support it), the version, the architecture (32 or 64 bit), and the compiler (for the codes only). The coder has also to provide a description of the goal of the code that will be displayed on the companion website. This description may be different from the abstract of the paper and may be designed to give all the required information to the future user of the companion website. The coder has also the possibility of uploading a pdf file if this description is longer than 800 characters. Finally, the coder could also provide a copy of the results (pdf file) obtained with the demo data.

    The fourth step is the most crucial. The coder is asked to describe all the inputs of the codes. For each input, the coder has to declare the type (scalar, text, vector, matrix, choice list, or file), the label that it will be displayed on the companion website, and the name of this variable in the code/script. The application checks in the main code if this name is present. Then, for each input, the coder has to provide a value (for the scalar or text types) or a set of demo data. The coder could also provide a text description of these inputs and these demo data. These descriptions will be displayed on the companion website. Then, the coder gets a first visualization of the input form of his or her future companion website: each type of input is associated to a particular object (box for the scalar and text, choice list, etc.). He or she has the possibility to design this form by dragging and dropping all these objects.

    The last step consists in declaring the outputs. This step is very limited, since by default, RunMyCode will reproduce the same presentation of the results (tables, figures, etc.) as that the user would obtain on his or her personal computer. All these inputs are included in a pdf file. So this last step is only devoted to the cases where the code produces some numerical data useful to the user. In this case, the coder declares the name and the label of all the corresponding variables in the code. The results contained in these variables will be stored in a csv file and can be downloaded by the user.

14.6 Partnerships and Expansion

To develop its operations, RunMyCode is currently partnering with scientific publishers, scientific association, editorial boards of scientific journals, conference and workshop organizers, pdf archives, and digital archiving services.

References

Alsheikh-Ali, A.A., W. Qureshi, M.H. Al-Mallah, and J.P.A. Ioannidis (2011) Public availability of published research data in high-impact journals. *PLoS ONE* 6(9): e24357. doi:10.1371/journal.pone.0024357.

Barrou, V. (2008) L'économie expérimentale: Un nouvel outil pour les SES?, Idées économiques et sociale, 3:48–56. doi: 10.3917/idee.153.0048.

Borgman, C.L. (2007) *Scholarship in the Digital Age: Information, Infrastructure, and the Internet*. MIT Press.

Elsevier's Executable Paper Grand Challenge (2011). http://www.executablepapers.com

Glandon, P. (2010) Report on the *American Economic Review* Data Availability Compliance Project, http://www.aeaweb.org/aer/2011_Data_Compliance_Report.pdf

Kim, E.H., A. Morse, and L. Zingales (2006) What has mattered to economics since 1970?. *The Journal of Economic Perspectives* 20(4): 189–202.

Lerner, J. and J. Tirole (2002) Some simple economics of open source. *Journal of Industrial Economics* 50(2): 197–234.

McCullough, B.D., K.A. McGeary, and T. Harrison (2006) Lessons from the JMCB archive. *Journal of Money, Credit and Banking* 38(4): 1093–1107.

McCullough, B.D. and H.D. Vinod (2003) Verifying the solution from a nonlinear solver: A case study. *American Economic Review*, 93(3): 873–892.

Peng, R.D. (2011) Reproducible research in computational science. *Science* 334: 1226–1229.

Phillips, P.C.B. (2003) Law and limits of econometrics. *The Economic Journal* 113: 26–52.

Savage, C.J. and A.J. Vickers (2009) Empirical study of data sharing by authors publishing in *PLoS Journals*. *PLoS ONE* 4(9): e7078. doi:10.1371/journal.pone.0007078.

Stodden, V (2010) The scientific method in practice: Reproducibility in the computational sciences. MIT Sloan School Working Paper 4773-10. Available at http://papers.ssrn.com/sol3/papers.cfm?abstract_id=1550193

Tenopir, C., S. Allard, K. Douglass, A.U. Aydinoglu, L. Wu et al. (2011) Data sharing by scientists: Practices and perceptions. PLoS ONE 6(6): e21101. doi:10.1371/journal.pone.0021101.

15

Open Science and the Role of Publishers in Reproducible Research

Iain Hrynaszkiewicz, Peter Li, and Scott Edmunds

CONTENTS

15.1 Evolution of Policies on Open Access and Open Data in the Life Sciences

When we read about the claims made in scientific papers, we tend to believe that they have been written by their authors in good faith. The process of science therefore demands the highest ethics and quality in order for the content of a scientific paper to be taken at face value. However, the increasing number of retractions in the scientific literature suggests that peer review is not of sufficient rigor to assess whether the results reported in papers can in fact be reproduced.[1] It is often only the results and conclusions of a study that are examined, while the methodology, raw data, and the source code used to generate the results of a paper are usually not fully evaluated.

15.1.1 Open-Access Publishing, BioMed Central, and the Literature as a Resource for Science

Openness enables reproducibility, and reproducible computational research requires openness in all products of research. Open data and code must be supported by full and accurate descriptions of the experiments as initially proposed (protocols) and as eventually carried out (methods and results).[2] Openness in scientific papers (journal articles) is encompassed by open-access publishing.

Open access to scholarly articles is generally achieved through two mechanisms. First, scholars or their institutions can "self-archive" and share peer-reviewed prepublication versions of papers, which have been accepted for publication in journals. This happened before, and after, digital scholarship was possible and is known as "green" open access. Second, scholars can publish their papers in open-access journals, known as gold open access,[3] which is the focus of this section. The first free-to-access online journals emerged soon after the introduction of the World Wide Web,[4] but characteristics of open-access publishing in the twenty-first century have helped the literature itself to become a scientific resource.

Open access to journal articles enables them to be read online without a subscription, but open access is about more than accessibility. Open access to journal articles is also about reusability, which means considering the format in which the literature is available and the copyright license under which it is published. This emerged from three overlapping definitions of open access resulting from three influential meetings (in Budapest, February 2002; Bethesda, June 2003; and Berlin, October 2003), which subsequently released public statements.[3] The Budapest definition states: "By 'open access' to this literature, we mean its free availability on the public internet, permitting any users to read, download, copy, distribute, print, search, or link to the full texts of these articles, crawl them for indexing, pass them as data to software, or use them for any other lawful purpose, without financial, legal, or technical barriers other than those inseparable from gaining access to the internet itself."[5] This is the most pertinent definition when deriving reproducibility from the published literature. The Budapest, Bethesda, and Berlin (BBB) definitions of open access unified policies on copyright in scholarly works and unified practices in how electronic literature should be formatted and structured. These policies and practices had already been put into practice, in 2000, by the first commercial open-access publisher BioMed Central and the full-text open-access repository funded by the National Institutes of Health (NIH), PubMed Central.

The idea for creating an online, open-access life-science publisher emerged in 1998 from a meeting between the publishing entrepreneur Vitek Tracz, chairman of the Science Navigation Group in London, United Kingdom, and David Lipman, director of the National Center for Biotechnology Information (NCBI) in the United States. Lipman's responsibilities include infrastructure for the implementation of data-sharing policies in genomics—databases such as GenBank—and the bibliographic database PubMed. After Lipman discussed the idea with the then director of the NIH Harold Varmus, a proposal emerged for E-BIOMED, an NIH-sponsored free, full-text research publishing platform.[6] However, a large research funder's potential conflict of interest in becoming, or being seen to be, a publisher meant that the NIH could only ever provide a repository for full-text open-access articles originally published elsewhere. As a result, BioMed Central was conceived as a publisher of online biology and medical journals to support deposition of content in the repository. The repository was launched, as PubMed Central, in February 2000.[7] BioMed Central began accepting its first submissions in May 2000, with an inclusive editorial and peer-review policy focusing on scientific accuracy rather than interest, including publication of negative results and single experiments.[8]

Both PubMed Central and BioMed Central publish the full text of articles in an open standardized XML format with a document type definition (DTD) to enable efficient filtering and querying of content. This approach, under open access, enables the rapid development of computational analysis tools—so the literature itself becomes a scientific resource.

15.1.2 Licensing the Literature for Reuse

Efficient reuse of published research requires that the appropriate legal tools—copyright licenses—be put in place by publishers and rights holders. Legal restrictions, engrained in traditional copyright transfer agreements, on the sharing and reuse of the products of scientific research are another barrier to reproducibility. In its first author license agreement, BioMed Central authors retained copyright in their work, with the publisher acting as a provider of layout, archiving, and peer-review coordination services. Authors were free to redistribute their work as they wished, with the only requirement being attribution of the original publisher. This was—and generally still is—in contrast to the traditional model of science publishing, where researchers typically work for several years on a piece of research and then hand exclusive rights to display and distribute that work to a publisher who controls access to the work. In 2004, BioMed Central's license agreement was made consistent with the Creative Commons Attribution License[9] (CC-BY), which has emerged as the gold standard for licensing journal articles under an open-access model in STM publishing.

There are several derivatives of the Creative Commons Attribution License, with CC-BY being the most liberal. The only requirement for sharing, redistribution, reproduction, remixing, reuse, and translation of content published under CC-BY is attribution of the original author who retains copyright. The use of CC-BY for papers fits with the reproducible research standard proposed by Stodden[10] and is compatible with scientific norms as citation practices ensure that reuses of scientific media, such as descriptive text within papers, and images, are credited to the original author(s). Less liberal derivatives of CC-BY, such as CC-BY-NC, which restricts commercial reuse, are discouraged for open-access publishing but are used by a number of publishers. As Mike Carroll, who sits on the board of Creative Commons, explains: "Granting readers full reuse rights unleashes the full range of human creativity to translate, combine, analyze, adapt, and preserve the scientific record." Commercial use restrictions also affect authors who, for example, could not upload images to Wikipedia if they published their research in a journal, which restricts commercial use. Also, commercial organizations can assist with the preservation of content (e.g., if a publisher went out of business and a new commercial publisher wished to republish content to make it available to readers). It is presumed that some publishers restrict commercial use to protect revenue streams from services such as the sale of reprints and the future development of commercially valuable text-mining applications.[11] However, publishers permitting commercial reuse of content and gaining commercially from CC-BY content are not mutually exclusive.

The Open Access Scholarly Publishers Association (OASPA), which includes many traditionally subscription-based publishers who have set up open-access journals, sets standards for content licensing. OASPA strongly encourages the use of CC-BY "to fully realize the potential of open access to research literature."[12] CC-BY-NC is the least liberal license permitted for membership of OASPA. It was reported in 2012 that 17% of the scholarly literature published in 2011, and indexed by the largest citation and abstract database Scopus, was published as open access.[13] Scholarly publishing is growing, but open access is growing faster than publishing under the subscription model.[14] These are promising developments for promoting barrier-free reproducible science, which uses the published literature as a resource.

15.2 Publisher-Community Policies Supporting Reproducible Research

15.2.1 Supplementary Materials

Online publishing in journals enables the publication of more than just digitized paper-based documents. Many journals include supplementary materials, which are referred to as additional files by BioMed Central. Despite the fact that supplementary materials can be limited to relatively small file sizes since publishers typically allow files of only 10–20 MB each to be included with online articles, reproducibility is enabled when these files contain data and code supporting the reported results in a paper. In principle, any file formats can be uploaded for publication, but formats that facilitate reuse—open formats that are not platform specific and are viewable using freely available tools—are generally preferred to proprietary file formats.[15]

Online supplementary materials have been a subject of debate in scholarly publishing, particularly in 2010 when the *Journal of Neuroscience* announced it would no longer accept supplementary material.[16] The main reason stated by the *Journal*—not wanting to overburden peer reviewers—was honorable, but misguided when considered in the context of online-only journals. Although the vital services provided by a limited number of peer reviewers should be used as efficiently as possible, the expectation that every reviewer should reanalyze a dataset provided as an additional file is unrealistic. Journal editors often invite reviewers with specific expertise pertaining to certain parts of a paper, such as a particular statistical technique. Making data and code available as supplementary material promotes transparency and reproducibility, enabling reviewers to analyze data if they or the editor feels it is essential to editorial decision making.[17] In response to this debate, BioMed Central amended the peer reviewer guidelines on all its journals to clarify this expectation of peer reviewers receiving manuscripts with additional files. Reviewers are not expected to reanalyze all supporting data unless the editor or reviewer feels a more detailed analysis is necessary.[18]

Another reason for not publishing supplementary materials was put forward by *Lab Times*[19]—preventing detailed methods and important tables being removed from the main article—However, online-only (open-access) journals rarely have restrictions on the length of research articles, the number of references, figures, and tables.

Supplementary materials do not replace the need for data archiving in specialized repositories since supplementary materials can have significant limitations on size and rarely enable datasets and other digital research objects to be independently harvestable, discoverable, and citable. However, data repositories do not yet exist for all experimental data types and scientific domains, so journals can assure online permanence of content by playing an important "stopgap" role[20] by making available reproducible materials.

15.2.2 Journal and Publisher Policies on Data Sharing

Science policies, such as for the availability of data in public databases, can be enforced iteratively by funders, peer reviewers, editors, and the wider scientific community. However, journals can act as a last line of enforcement. Given scientists must "publish or perish," the prospect of being rejected for publication can in principle be a powerful way to change authors' behavior— such as with regard to data sharing.

The data availability policies for different types of data and materials in the 50 journals with the highest impact factor in 2007 have been catalogued (Figure 15.1) and their effectiveness evaluated.

Of the 50 journals, 44 had statements about data sharing in their information for authors. Ioannidis and colleagues looked at the first 10 research papers published in each journal in 2009 and checked if the supporting data were subject to a data-sharing policy. Of 351/500 papers, which were subject to a data-sharing policy, 208 of these did not fully adhere to the policy. Nondeposition of microarray data in a public database was the most common violation. They found that 47 papers deposited full primary raw data online, but these included none of those papers not subject to data availability policies.[21] Another study, assessing the sharing of psychological data subject to a policy of authors agreeing to share data with other scientists on request had a 25% rate of compliance.[22] Improvements in how community and journal policies on data availability are enforced are clearly needed, although the results of the Ioannidis study support previous findings that journal policies on data sharing lead to at least some increase in data sharing.[23]

There are several different approaches to journal data-sharing policies:

1. **Data sharing implied by submission (e.g., BioMed Central journals):** The minimum requirement for BioMed Central's journals is that submission of a manuscript implies "readily reproducible materials described in the manuscript, including all relevant raw data, will be freely available to any scientist wishing to use them for noncommercial purposes."[24] Where databases exist and communities require it, such as for genetic sequence data, public data sharing as a condition of publication applies.

2. **Data sharing as a condition of publication (e.g., *Nature*, *PLOS*):** *Nature* requires that authors "make materials, data and associated protocols promptly available to readers without undue qualifications in material transfer agreements" and that supporting data be available to editors and peer reviewers. *Nature* also specifies how it deals with infringements to the policy, which includes publishing corrections or refusing publication.[25] *PLOS*'s policy states: "Publication is conditional upon the agreement of the authors to make freely available any materials and information described in their

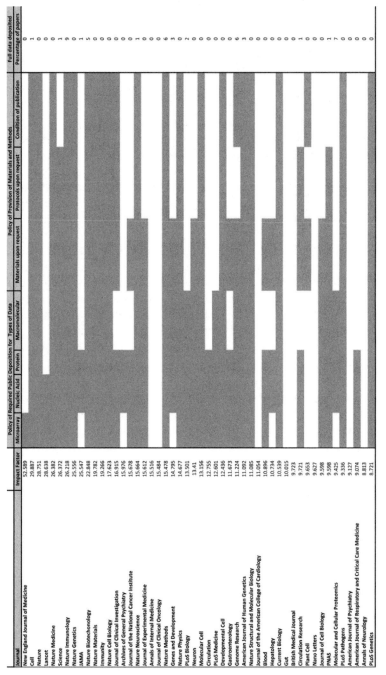

FIGURE 15.1

Breakdown of journal policies for public deposition of certain data types, sharing of materials and/or protocols, and whether this is a condition for publication and percentage of papers with fully deposited data. (Available under CC-BY, reproduced from Alsheikh-Ali, A.A. et al., *PLOS ONE*, 6(9), e24357, 2011.)

publication that may be reasonably requested by others for the purpose of academic, non-commercial research."[26] Compliance with the policy is taken into account in editorial decisions, and corrections will be published or publications withdrawn if noncompliance is discovered retrospectively. Also, *PLOS* encourages readers to contact them if they encounter difficulty in obtaining materials supporting published papers and have retracted at least one paper in cases where data reanalysis revealed less than complete support for the published conclusions.[27]

3. **Reproducible research, or data-sharing, statements in published papers (e.g., *Annals of Internal Medicine, BMJ*):** Since 2007, *Annals of Internal Medicine* has required all authors of original research to state in their published article *their willingness to share* their study protocol, statistical code used to generate the results, and the dataset from which the results were derived.[2] In 2010, the *BMJ* emulated this policy[28] (which itself was inspired by the *American Journal of Epidemiology*[29]) and began requiring data-sharing statements in published research papers. In these cases, sharing of materials is not required, but transparency about whether the materials are available is required—so readers and reviewers can take this into account when judging the merits of the article. *BMJ* announced in October 2012, however, a stronger policy for clinical trials of drugs or devices where it would only publish these studies when there is evidence of data sharing.[30]

4. **"Availability of supporting data" statement and link to dataset(s) (BioMed Central journals):** As of 2012, nearly 50 BioMed Central journals encourage or require authors to include a persistent link in their papers to the data supporting the results of their study. This standard article section, "Availability of supporting data," aims to address several challenges in linking data to publications—functionality, credit, and consistency. While statements in papers about what reproducible materials are available are useful for readers and reviewers, BioMed Central's approach focuses instead on evidence of data sharing, where it exists, by providing functional links between papers and datasets. The aim is to enhance the scientific record by enabling easier discovery of reproducible materials. The policy also encourages persistent identifiers for datasets to be formally cited in the article's reference list, helping to increase the potential for gaining academic credit for data sharing through citation. Linking articles to supporting data has happened since papers and data have been available online, but the approach—including where links are placed in papers—can differ between journals and, sometimes, between different papers in the same journal. The "Availability of supporting data" section accommodates

different types of repository and persistent identification formats provided a unique link to the dataset is provided in the http:// format. Digital object identifiers (DOIs), handles and GenBank identifiers, for example, can all be expressed in this format. However, the majority of the 50 journals encourage rather than require authors to link their papers to supporting data. For a journal to require permanent links to data for every article, there must be an appropriate data repository for every type of data, which conceivably could be described in the journal—a challenge for broad-scope journals in life sciences.

5. **Open data as a condition of publication:** In 2011, a number of evolutionary biology journals, including *The American Naturalist*, *Evolution*, and *Molecular Ecology*, collectively adopted a Joint Data Archiving Policy (JDAP). This policy requires, "as a condition for publication, that data supporting the results in the paper should be archived in an appropriate public archive." The Dryad repository[31] rapidly emerged as the repository of choice for implementing the JDAP. Dryad is also partnered with journals in other areas of life science research when there is a need for "idiosyncratic," "orphan," or unstructured data, which are not accommodated by structured databases such as GenBank.[32] Under the JDAP, all data supporting accepted peer-reviewed papers are deposited in Dryad including spreadsheets, images or maps, alignments, and character matrices. Files up to 1 GB in size can be uploaded through Dryad's web-based submission system. All data packages are released to the public domain under the Creative Commons "no rights reserved" CC0 waiver, so they can be reused with the minimum of restrictions (as "open data"; see following text). Exceptions to conforming to the JDAP can be granted for sensitive information—including locations of endangered species and human subjects data.

6. **Open data as a condition of submission (e.g., *F1000Research*):** The journal *F1000Research*, which covers all of life sciences and began publishing in 2012, has a policy that "all primary research articles should include the submission of the data underlying the results." Similar to other approaches earlier, data must be deposited in a public repository where one exists. Unlike other journals, however, where there is no public repository for data, the journal requires that authors transfer their supporting data to the journal. This is a mandatory policy with the only exception being where patient privacy may be put at risk by data sharing. Data can then be deposited, on the authors' behalf, in the FigShare[33] general data repository and integrated—with data file viewers in line—with their published article. Like Dryad, FigShare (and *F1000Research*) uses the CC0 waiver for public datasets.

Of the various approaches to journals implementing data sharing, evidence published in 2013[34] found that the most effective way for a journal to ensure datasets supporting publications are available is to have a mandatory data availability policy and require a data-sharing statement. This is more effective than having no policy, a policy of recommending data sharing, or having a mandatory policy but not requiring a data-sharing statement.

15.3 Field-Specific Policies, Standards, and Challenges

15.3.1 Data Sharing in Genomics

Genomics is often highlighted as an area of biology that leads the way in data access and standardization. The growth of this field has been driven by the huge technological advances in DNA sequencing and the massive investment in the human genome project (HGP) and other "megasequencing" projects. This huge pool of genomics data is made freely available to the community by a data-sharing infrastructure called the International Nucleotide Sequence Database Collaboration (INSDC).[35] It evolved from the bulletin boards and e-mail lists that were forums for electronic communication of sequencing information in the early 1980s. NCBI at the US National Library of Medicine,[36] the European Nucleotide Archive (ENA) EMBL-Bank at the European Bioinformatics Institute (EBI) in the United Kingdom, and the DNA Database of Japan (DDBJ) in 1987 formed a three-party partnership that persists to this day.[37] Combining forces has allowed the databases to federate, standardize, and mirror nucleotide sequencing data. This gave them more power to lobby publishers (see open letter to journal editors from the INSDC[38]) and data producers for mandatory data submission policies. This has been particularly successful in getting most publishers to require in their instructions for authors that deposition of the types of data that the INSDC databases handle—in particular raw sequencing data and genome assemblies—is a requirement of publication.

While there were arguments over the free access to genomics data during the battles between the public and commercial HGPs, the data were eventually released into the public domain.[39] Since the HGP was publicly funded and to prevent any particular center from establishing a privileged position in the exploitation and control of sequence information, data were required to be made public as soon as possible. With the public repository infrastructure already available from the INSDC, it was logical to do this through their databases.[40]

Rules to formalize and speed up these genomics data-sharing conventions were formalized at a meeting in Bermuda in 1996. But as sequencing continued to get faster and cheaper for subsequent genomes after the HGP, the risk of scientists with no role in producing original sequences publishing a paper before the sequencing center releasing the data increased.[41] Conflicts were building between sequencing centers and the communities wanting to use these data over the timing of the release of data, and there were also concerns whether journals may prejudice the publication of complete, annotated genomes after a preliminary release of data.[42] To address these fears and incentivize the continued release of data, a compromise was made regarding the assignment of credit and priority to publish. To do this, all of the various stakeholders involved including representatives of sequencing centers, data users, journals, and funders met in Fort Lauderdale in 2003 to update these policies.

The Fort Lauderdale agreement,[43] rather than a set of binding rules, was set up as an "agreement" in a similar way that giving scientific attribution and credit through citation is more of a courtesy than anything legally binding. Making this a system of "tripartite responsibility," it asked data producers, users, and funders to follow these guidelines and asked the journal editors present to assist in its implementation. Those carrying out whole-genome shotgun sequencing projects were asked to deposit raw data within 1 week of production and deposit whole-genome assemblies in a public nucleotide sequence database as soon as possible after the assembled sequence met a set of quality evaluation criteria. As an incentive and protection for the data producers to do this, data users were permitted to use the unpublished data for all purposes, with the sole exception of publication of the results of a complete genome sequence assembly or other large-scale analysis in advance of the sequence producer's initial publication. Following these guidelines, most of the large funders and centers producing sequencing data worked these instructions into their data release policies, and most scientific journals now make efforts to ensure that genomics research can only be published if the supporting sequence data were available in one of the INSDC databases. This was followed in 2009 by the Toronto International Data Release Workshop that recommended extending the policies for genomics to the increasing number of fields producing large-scale data.[44]

Post Fort Lauderdale and Toronto, adherence and compliance to these guidelines has been mixed, but these agreements are still seen as a success compared to the rest of biology. Engaging all of the various stakeholders enabled a carrot and stick approach, with data centers incentivized to release their data early with the protection of having priority to publish the first study and funders and journals insisting upon compliance by including it in their policies. As more of an informal "gentleman's agreement" than a legal framework, it has been hard to give teeth to these policies, although compliance of published articles depositing sequencing data in GenBank has still been estimated to be up to 90% in 2006.[45]

15.3.1.1 Sequencing Data Standardization

As well as making data publicly available, it is also essential to provide enough information to guide data integration, comparative studies, and knowledge generation. The interoperability and usability of data is essential to enable its reuse, and the development of community-wide standards for capturing and exchanging metadata is important to unite groups and enable collective change. One group that has taken responsibility to help enable this in genomics is the genomic standards consortium (GSC). Having rich and standardized contextual information is particularly important for environmental metagenomics, and the GSC has been particularly involved in producing standards for the wider genomics community to be able to combine and integrate their datasets. The GSC was established in 2005 to produce minimum information standards and checklists for describing genomes and metagenomes.[46] It has since become an open-membership working body with the goal of standardizing the description of genomes and the exchange and integration of genomic data.[47] Following on from their "minimum information about a genome sequence" (MIGS) and "minimum information about a metagenome sequence" (MIMS) checklists, the GSC produced the minimum information about any (x) sequence (MIxS) framework that builds upon their previous checklists, but on top of a sharing of a central set of core descriptors, which allows communities to build and add specific modules on top of it.[48] The use of MIxS standards seems to have the most use and uptake in the environmental metagenomics community, but the GSC have worked closely with the INSDC and a number of other databases to recognize these standards and support submission of compliant datasets. The GSC has also set up its own journal, *Standards in Genomic Sciences*, which was the first journal to require MIGS for the publication of all genome paper.[49]

15.3.1.2 Source Code

Postgenomic science typically contains a data-processing component, which is responsible for the analysis of data after it has been generated by assays of biological samples obtained from laboratory experimentation.[50] The reality of science in modern-day biology is that such data analyses cannot be reproduced based on the information made available in the published paper. This was comprehensively shown by Ioannidis et al.[51] who evaluated a set of 18 microarray studies and found only 2 studies that could be reproduced in principle.[51] In addition to raw data not being made available, the published information provided insufficient knowledge, such as software version and parameterization details, about how data analyses were carried out and subsequently stopped results from being reproduced.

Bioinformatics data analyses involve a series of processing steps on data involving the use of specific functions or command line applications. The reproducibility of such data analyses can be facilitated if the source code of applications and how they are combined in a script are made available during peer review. To this end, there have been a number of calls for publishers to take on this responsibility to facilitate reproducibility of research findings reported in their journal papers.[52,53] Some publishers have started to address these issues including the enforcement of source code availability with manuscript submissions. For example, *Biostatistics* published by *Oxford Journals* publishes papers on statistical methods with applications to human health and disease. *Biostatistics* encourages authors of their accepted papers to make their work reproducible by others by enabling them to submit their code (and data) to the journal to be hosted as supporting online material. Furthermore, authors can request a reproducibility review of their analyses, which involves the "associate editor for reproducibility" running the submitted code on the data and verifying that the code produces the results published in the article.[54] Articles with accompanying data or code receive a "D" or "C" kite-mark, respectively, while those that have passed the reproducibility review receive an "R." *PLOS* advocates making source code accessible in an open-source manner in research articles involving the use of software.[26] Conditions of this policy include ensuring that methods are described with a level of detail such that the results can be reproduced by reviewers and readers. This involves a requirement for documentation and a dataset to be made available for using the software with example parameterization. *PLOS* also asks for source code to be deposited in an open software archive, such as SourceForge or GitHub, and be included as part of the submission with an open-source license. A similar policy is enforced by the *Journal of Open Research Software*, which is published by Ubiquity Press in collaboration with the Software Sustainability Institute. This journal publishes peer-reviewed reports of research software and where to find it in a public code repository under an Open Source Initiative (OSI)-compliant license, or the Creative Commons CC0 waiver.

BioMed Central supports the publication of source code along with scientific papers through its editorial policies and in particular through its journals *Source Code for Biology and Medicine*[55] and *GigaScience* (discussed in more detail later). These journals hope that this form of code dissemination can lead to shortened times required for solving computational problems for which there is limited source code availability or software resources and, in time, serve as a repository for source code with applications in the life sciences. While the standard policy of many BioMed Central journals encourages rather than requires OSI compliance, *Source Code for Biology and Medicine* launched an ongoing collection of software papers[56–58] in 2012, which aims for the highest standards of reproducibility. This "Open Research Computation" collection[59] requires source code to be made available under an OSI-compliant license, and the peer-review process, developed by the series' editor Cameron Neylon, assesses the quality of documentation and testing of the software and aims for a very high level of unit test coverage.

15.3.2 MIBBI, BioSharing, and Data Interoperability

After sequencing, the functional genomics and transcriptomic communities have been particularly successful in formulating and building community standards and data-sharing resources. With the growth of the use of microarray-based technologies and platforms in the late 1990s, it was important to develop standards and infrastructure to enable sharing and reuse of these data. The Microarray Gene Expression Data Society (now the Functional Genomics Data Society) filled much of this role, publishing their "minimum information about a microarray experiment" (MIAME) standard in 2001.[60] The MIAME checklist has been a model for other communities, and after the development of many other "minimal information" standards for other communities such as Proteomics (MIAPE: the minimum information about a proteomics experiment), the MIBBI portal was created.[61] In 2012, the MIBBI portal comprised 35 bioscience projects. MIAME has been the most widely adopted, with over 70% of journals with the highest impact factors including the requirement to deposit data in a MIAME compliant format in a public repository.[21] Standardized, simple tab-delimited, spreadsheet-based formats such as MAGE-TAB[62] have aided submission to databases, and in the decade after their creation, huge amounts of data reuse have been enabled, through meta-analyses.[63]

Outside of the omics fields, rates of data deposition are much lower with, for example, only 4% of evolutionary biology studies using the TreeBASE repository for phylogenetic trees,[64] although most of these fields have not (until with the arrival of general purpose repositories such as Dryad[31]) had stable homes for their data. One reason may be the diffuse and confusing infrastructure for data sharing, with thousands of biological databases and hundreds of terminologies and reporting guidelines available.[65] Aiming to be a "one-stop shop" and centralized portal for bioscience data policies, reporting standards and links, and building from the MIBBI projects, BioSharing[66] has extensive web-based catalogues and a communication forum to connect the research community, funding agencies, and publishers. Journals, including *BMC Research Notes* through a special issue on data standards in life sciences,[65] are participating in the development of standards and tools to support data sharing.

While MIBBI-style checklists[61] help standardize data collection in particular fields, moving between fields and techniques data interoperability becomes increasingly difficult and hinders multidisciplinary research. One attempt at dealing with and integrating the increasing number of categories of comparative omics data types is the biological observation matrix (BIOM) format: a file format for representing arbitrary observations by sample contingency tables with associated sample and observation metadata.[67] A much broader community effort is the ISA ("Investigation–Study–Assay") framework, a group of open-source tools and formats to aid standards-compliant collection, curation, local management, and reuse of datasets.[68] With a list of conversion tools and templates, ISA-Tab files can be submitted to a growing number of international public repositories such as PRIDE (proteomics), ArrayExpress (transcriptomics), Metabolights (metabolomics), and the ENA (sequencing data).[69]

15.4 Toward Reproducible Research Licensing of Content in Open-Access Journals

Legal restrictions on data sharing and reuse, including copyright and materials transfer agreements, are complex and internationally heterogeneous and are barriers to reproducibility. Being able to build on previous findings and reuse data to drive new discoveries freely, without legal or other impediments, ensure society gains the maximum benefit from scientific endeavors. These ideals are set out in the Panton Principles for Open Data in Science, which were published in February 2010.[70] Under these principles, data should be placed explicitly in the public domain with a public domain license or waiver of rights, such as Creative Commons CC0, which permits all reuses including commercial use and preservation.

CC0 is recommended for data rather than attribution licenses such as CC-BY. Licenses that legally require attribution can be prohibitive when integrating data from very large numbers of different sources, as unmanageable legal requirements to provide attribution in the form of links can be created. CC0 helps address this "attribution stacking" problem by enabling rights holders to waive legal requirements for attribution. Rather than being a license—a means for an author or rights holder to assert their rights over works—a waiver is a mechanism for a rights holder to give up their rights. Moreover, CC0 is universal and irrevocable (widely recognized as covering all types of data and all legal domains, in perpetuity), interoperable (it is human and machine readable ensuring unambiguous expression of rights), and simple. Waiving rights in data means making individual requests for reuse and transfer agreements are unnecessary, increasing efficiency as scientists can focus on science rather than legal matters.[71]

Whether copyright actually applies to data is questionable and depends on the jurisdiction. Copyright cannot generally be expressed in facts—and data are numerical representations of fact—only the ways in which they are presented. This holds for US law and, with the exception of databases, in much of the EU also. In contrast, in Australia, copyright could exist in data, where the law focuses on originality rather than creativity. These international legal differences lead to ambiguity about the legal status of content, which might ultimately only be resolved in case of legal challenge in court. Being explicit about the legal status of data through an appropriate license or waiver at the outset avoids these potential problems. Other legal tools for dedicating data to the public domain include the Open Data Commons Public Domain dedication and License (PDDL), which is compatible with CC0.[72] Analogous to the BBB definitions on open access to papers, open data are about more than accessibility. "Open data" must be "freely available on the public internet permitting any user to download, copy, analyse, re-process, pass them to software or use them for any other purpose without financial, legal, or technical barriers other than those inseparable from gaining access to the internet itself."[73] To achieve this in journal publishing, CC0 or an equivalent legal tool must be applied to the data within (e.g., in tables) or included as additional files with journal articles.

BioMed Central was among the first public supporters of the Panton Principles for Open Data in Science, but putting them into practice at an established publisher had to be done in careful consultation with the scientific community. Between 2010 and 2012, BioMed Central published several statements and contacted hundreds of its journal editors to seek their opinions on applying CC0 to data in peer-reviewed journals. Also, in 2011, BioMed Central formed a publishing open data working group comprised of authors, editors, funders, librarians, legal experts, and other publishers—stakeholders in publishing scientific research data. The consensus of the working group was agreement that a variable license agreement—with CC0 applying to data and CC-BY to papers—could be implemented from submissions received after a specific date.

In September 2012, BioMed Central published in *BMC Research Notes* a paper describing and making the case for implementing CC0 for data in its journals,[74] which described the practical, legal, technical, and cultural implications. This included a new license statement for all published articles, as a model that could be adopted by many publishers. This new license statement, which is both human and machine readable, includes the CC-BY license and, for data, the CC0 waiver. To help determine scientists' opinions on changing the authors' default license agreement, BioMed Central held a 2-month public consultation[75] in September–November 2012. Respondents to the consultation ($n = 42$) were six to one in favor of implementing CC0, although there were a number of questions and concerns to emerge from the consultation (see Table 15.1).

15.4.1 Defining Data

Included in BioMed Central's publishing open data working group meeting were representatives from *Nature* Publishing Group and Faculty of 1000, publishers that, in 2013 and 2012, respectively, implemented CC0 policies for some of their publications. The *EMBO Journal* releases any files labeled "Source Data," "Dataset," or "Resource" under CC0.[76] Faculty of 1000's *F1000Research* journal, which began publishing in 2012, publishes open-access articles under a CC-BY license and makes all data associated with articles available under CC0. This approach was implemented in September 2013 by BioMed Central, and was supplemented with author guidelines, practical examples, and frequently asked questions informed by the outcomes of their public consultation.[77]

The *EMBO Journal* policy makes specifically tagged parts of the published work available under CC0. This makes the legal status of some content clearer but means that data within published articles—such as numerical

TABLE 15.1

Questions Raised in Response to BioMed Central's Public Consultation on Creative Commons CC0 for Data Published Peer-Reviewed Open-Access Journals and the Publisher's Responses to Them

Question/Concern	Summary Response
Will commercial organizations benefit from use of public domain data?	The CC-BY license already permits commercial use. There are wider benefits to the economy from commercial organizations gaining from open data. Companies including GSK have released some of their data under CC0 to stimulate scientific innovation.
Will plagiarism increase?	Plagiarism has increased with digital access to content regardless of content licenses. Processes (e.g., peer review) and tools (e.g., CrossCheck), which detect plagiarism, are agnostic of content licenses.
Will patient privacy be put at risk?	Changing the license of content does not change the accessibility of human subjects data, which are already being published open access and therefore must already be anonymized before publication.
Will articles receive fewer citations?	Attribution (a legal requirement of copyright) and citation (a scholarly cultural norm that ensures scientists receive credit for their discoveries) can sometimes be achieved in the same way, but the practices serve different purposes (see table in [http://www.biomedcentral.com/1756-0500/5/494] for practical examples). Removing the legal right of attribution—engrained in a CC-BY license—for reproducing, adapting, or copying a scholarly work does not remove the cultural expectation that scientists should cite one another's work when building on previous findings.
What is the incentive for the original data owner to make their data open?	Although it has not been empirically studied, public domain dedication maximizes the potential for data discovery and reuse, suggesting that open licensing might increase individual credit and citations. Sharing data underlying scientific journal articles increases citation share, increases reproducibility of results, and is associated with authors producing more publications. Data supporting publications and placed in the public domain in fields lacking combinable datasets promote collaboration and furthers scientific progress.
Will authors need to publish more data than they did previously?	A change in license will not require authors to publish more of their data. It only affects data that authors (already) choose to submit to journals for open-access publication and does not require release of any other data or a change in license of any data not submitted to the journal. Authors, editors, and their communities remain in control of what content they publish.
What if authors are not allowed, by their funders or employers, to use CC0 for any of their published work?	Where there are legitimate reasons for authors being unable to apply CC0 to their published data, then it is possible to opt out and use a nonstandard license. This already happens in journal publishing, such as when figures, tables, or charts are reproduced, with permission, in journal articles from other sources. Some research funders have agreements with publishers to use a nonstandard copyright statement in open-access articles.

Source: Hrynaszkiewicz, I., Busch, S., and Cockerill, M. J. Licensing the future: Report on BioMed Central's public consultation on Open Data in peer-reviewed journals. BMC Research Notes (in press).

tables, bibliographic data, and machine-harvestable data that could be obtained by text mining—are not covered by the CC0 policy. Data are difficult to define, with liberal definitions specifying data are anything, which can exist digitally to, the more relevant for science, "qualitative or quantitative attributes of a variable or set of variables. Data are typically the results of measurements and can be the basis of graphs, images, or observations of a set of variables," according to Wikipedia. The approach to implementing CC0 favored by BioMed Central and *F1000Research* applies CC0, more generally, to all "data" without the use of specific tags. This gives those publishing and reusing data flexibility and recognizes the differing definitions of, and needs for, data, in different types of research.

While there are a number of file types, which obviously pertain to data, comprehensively defining them is not feasible. However, practical examples are beneficial for interpreting the policy, and Table 15.2 (reproduced with permission from BioMed Central's proposed author guidelines on its open data policy) provides some examples of data associated with journal articles.

15.5 Community–Publisher Collaborations and Tools for Reproducible Research

15.5.1 LabArchives and BioMed Central

Considering the ease with which anyone can share and publish content on the web using freely available tools independently of publishers, publishers need to innovate to continue to add value to scientific communication. Integrating with scientists' online data management workflows and tools is a way to speed the dissemination of research, and workflow publication has been proposed as a role of publishers in the future.[78] Scientists are increasingly using electronic, often cloud-based, applications to store, manage, and share with collaborators their data and documents such as in electronic lab notebooks (ELNs). BioMed Central partnered with the ELN provider LabArchives in 2012 to enable more reproducible research by making more datasets supporting peer-reviewed publications available, openly licensed and permanently linked to publications.[79] As part of this partnership, BioMed Central authors are entitled to a free version of LabArchives' ELN, which includes 100 MB of file storage (the standard free edition included 25 MB of storage). The BioMed Central authors' edition of the software[80] acts as a personal or private file store and as a personal data repository for publishing scientific data and partially integrates scientific data management with manuscript preparation and submission to journals. Integration is achieved through the inclusion of manuscript templates conforming to journal style and links through to journal submission systems, from within the ELN.

TABLE 15.2

Examples of Data Associated with Journal Articles

Format	Explanation
Material submitted as additional files (supplementary material)	
Domain-specific datasets, supplied as additional files	Many domain-specific standards exist for the sharing of scientific datasets. Biosharing.org provides a useful catalogue of such standards. Many of these standards are based around XML, such as Gating-ML for flow cytometry experimental descriptions or MAGE-ML for microarray gene expression data.
Comma-separated values	CSV is a simple open tabular format used commonly for columnar data.
XLS/XLSX	XLS and XLSX are file formats used by Microsoft Excel. XLSX is a more modern, XML-based file format, and unlike the proprietary XLS format, it is an ISO standard (ISO/IEC 29500).
RDF	RDF is standard for representing knowledge and conceptual relationships using subject–predicate–object expressions (triples), which are widely used in modeling biological systems.
Material contained within the full text of papers	
Tables	Individual data elements, predominantly numbers, organized in columns and rows are a representation of facts and should be considered data.
Bibliographic data	Factual information, which identifies a scientific publication including authors, titles, publication date, and identifiers, should be considered open. Applies to individual articles and their reference lists.
Graphs and graphical data points	Software can harvest data points underlying graphs and charts, and graphs and other figures are often visual representations of data.
Frequency of specific words, names, and phrases in article text and their association to others	This information is frequently identified through text mining, for example, the frequency of particular gene and protein names and their potential associations with one another.

LabArchives can act as data publishing platform because datasets and other files can be shared publicly and assured permanence with the assignment of a DOI. LabArchives, also, helps enable maximum reuse by assigning the CC0 waiver to files that are assigned a DOI through its software. The assignment of DOIs means datasets are persistently identified and permanently linkable to journal publications and independently citable.

15.5.2 GitHub and BioMed Central

In 2013, BioMed Central announced a collaboration with the social coding repository, GitHub.[81] Many scientists—particularly bioinformaticians, one of BioMed Central's largest author groups—were already using GitHub for a variety of scientific activities beyond sharing of code before this partnership. However, scientists' uses of GitHub were happening independently of input from the GitHub training team, who had little direct experience of scientists' uses of their services. Scientists regularly use GitHub to, for example, publicly share their papers, author documents collaboratively, and version control their work. This three-way collaboration between a journal publisher, code repository, and a group of scientists (authors) enables sharing, documentation, and definition of good practices of using GitHub for science. Innovative uses of GitHub can make science more reproducible and research more transparent, and documenting these use cases to encourage wider adoption will further enable reproducibility.[82]

15.5.3 Data Publication, Data Citation, and GigaScience

With all of the challenges and difficulties of making data publicly available in as usable form as possible, data producers rarely receive the credit they deserve for the time and effort spent creating these resources. There has been much talk from data producers about the need for new mechanisms of incentive and credit, and the Toronto data workshop covered some of these issues, stating that "Data producers benefit from creating a citable reference, as it can later be used to reflect impact of the data sets."[44] Journal editorials following this discussed the need for a means of accreditation and a standardized tag for data that could be searched and recognized by both funding agencies and employers, providing recognition for those who share and enabling tracking of the downstream use and utility of data.[83] Founded in December 2009, DataCite is an international partnership set up to build a global citation framework for research data, leveraging the DOI system best known for its use in unambiguously identifying online publications. DataCite aims to enable researchers to find, access, and reuse datasets with ease.[84] With the aim to increase acceptance of research data as legitimate, citable contributions to the scholarly record, this infrastructure has already enabled the formation of a number of data publication platforms. The environmental sciences have been publishing datasets with DOIs for over a decade in the PANGAEA data repository,[85] but since the launch of DataCite, new repositories such as Dryad,[31] Figshare,[33] and the *Giga-Science* database[86] have utilized DataCite's DOI services and infrastructure. Data have historically only been searchable via DataCite's search engine and Application programming interface (API), but with the launch of the Thomson Reuters data citation index in October 2012, it is now possible to track and follow the downstream use and citation of these datasets. This is critical for those that believe data generated in the course of research are just as valuable to the ongoing academic discourse as papers and monographs.[87]

15.5.4 GigaScience and Adventures in Data Citation

GigaScience is a journal published by the BGI and BioMed Central that was launched in 2011 to provide a home for large-scale studies in biology and biomedicine. The *GigaScience* database implemented a new publication format by utilizing BGI's computational resources. The database, GigaDB, is able to host supporting data, from publications in the journal, of over 100 GB in size—as well as release previously unpublished BGI datasets.[86] DataCite DOIs are used to integrate datasets into the references of articles, and this has allowed *GigaScience* to experiment with the release of BGI datasets in a citable form before publication of analyses of the datasets.[88]

GigaScience has a policy in which all materials including the source code used for data analyses are submitted along with a manuscript for review and reproduction. Furthermore, a data analysis platform has been developed to support readers in the reuse and reproduction of the data analyses published in their journal articles as executable pipelines (see Section 15.5.6). There is extra effort required from developers to make the source code reusable and this may prohibit their release. This issue is being addressed at *GigaScience* by exploring how the source code can also be issued with DOIs, thereby enabling reuse metrics to be calculated for code and credit received by their developers.

The first published dataset to be published by *GigaScience* was the genome of the deadly 2001 *E. coli* 0104:H4 that caused an outbreak that killed 50 people in Europe.[89] Researchers at the BGI collaborated with the University Medical Center Hamburg-Eppendorf to rapidly sequence the genome of the pathogen. Due to the unusual severity of the outbreak, it was clear that the usual scientific procedure of producing data, analyzing it slowly, and then releasing it to the public after a potentially long peer-review procedure would have been unhelpful. By releasing and announcing via Twitter the first 0104:H4 genomic data before it had even finished uploading to NCBI and promoting its use and release, a huge community of microbial genomicists around the world took up the challenge to study the organism collaboratively. Once a GitHub repository had been created to provide a home to these analyses and data, groups around the world started producing and posting their own annotations and assemblies within 24 h. Releasing the data under a CC0 waiver (used by GigaDB) allowed truly open-source analysis, and other groups and GitHub members followed suit in releasing their work in this way. This "crowdsourcing" approach substantially aided in limiting the health crisis, with strain-specific diagnostic primers disseminated within 5 days of the release of the sequence data, and the draft unassembled genome sequence data subsequently enabled the development of a bactericidal agent to kill the pathogen.[90] It also brought to light a potentially useful way of scientifically addressing similar outbreaks in the future. Additionally, results of these open-source analyses were published in the *New England Journal of Medicine* a few months later, showing that journals do not have a problem with the release of data in this citable manner, and data citation can complement traditional forms of academic credit.[91] When the Royal Society published their "Science as an Open Enterprise" report, this project, and the *E. coli* genome sequenced, was highlighted on the front cover of the report and cited as an example of "The power of intelligently open data."[92] A subsequent study on the genomics of ash dieback disease, a fungal pathogen rapidly destroying ash trees and woodland biodiversity in Europe, also followed this open-source/GitHub-based approach.[93]

15.5.5 Use of the Cloud as a Reproducibility and Reviewing Environment

The rise of cloud computing brings new opportunities to increase transparency for data-intensive science. The use of virtualization and virtual machines where the whole state of a computer can be saved and transported to another host computer allows operating systems, pipelines, and tools to be easily deployed and replicated wholesale by potentially any researcher without concern for the underlying hardware, which brings unprecedented opportunities to aid reproducible research. Another advantage is the ability to take and exchange "snapshots," where the computer system used by researchers to produce experimental results is copied in their entirety into a single digital image that can be exchanged with other researchers. Doing this, researchers are able to obtain precise replicas of a computational system used to produce published results and have the ability to restore this system to the precise state of when the experimental results were generated.[94]

The use of virtual machines and the cloud as a reproducibility environment has historically been limited by the cost of computing and storage that needs to be paid to the provider.[95] But as these costs continue to fall, a growing number of projects are starting to take advantage of this functionality. The best example to date has been from the ENCODE consortium, where in September 2012, they simultaneously published their first 30 papers cataloguing functional DNA elements in over 100 cell types, all based on terabytes of shared data and huge numbers of tools and pipelines developed from 1600 experiments.[96] Needing to coordinate and share these resources among nearly 450 authors in more than 30 institutions across the world was a major challenge, and the ability to share this information in an easily reproducible manner was a boon to the authors, reviewers, and eventually downstream users of these resources. Complex computational methods are very hard to track in all their detail, and the ENCODE virtual machine provides all of the methodological details explicitly in a package that will last at least as long as the open VirtualBox virtualization format they used.

Accompanying the supplementary material from the main ENCODE integrative analysis, publication in *Nature* was a set of code bundles that provide the scripts and processing steps corresponding to the methodology used in the analyses associated with the paper.[97] The analysis group established an ENCODE analysis virtual machine instance of the software containing the functioning analysis data and code, where each analysis program has been tested and run. Where possible, the virtual machines were used to reproduce stages of the analysis performed to generate the figures, tables, or other information; however, in some cases, this was not possible in steps involving highly parallelized processing within a specialized multiprocessor environment. In these cases, a partial example was implemented leaving it to the reader to decide whether and how to scale to a full analysis. It cost around $5000 dollars, paid to Amazon Web Services (AWS), to run during the writing and review process (although it was nowhere near fully utilized and could have been cheaper with more efficient code integration and testing). After publication, it was possible to examine the figures in the Amazon cloud, and the virtual machines remained freely available for interested parties to work with and run the data and tools used by the project.[98] The aim of providing these virtual machines was not to produce a portable and reusable piece of software for every aspect of this analysis, since bugs may be present in the code, but to provide a completely transparent way of sharing the methods. These unprecedented levels of transparency give users more confidence to utilize the tools when they have the code that actually executes and produces the published result in a controlled environment.

Demonstrating the utility and reproducibility of cloud computing for microbiome research, an application combining a number of metagenomics resources such as QIIME, the IPython collaborative notebook, and StarCluster for setting up preconfigured clusters on EC2 was made reproducible using an Amazon machine image (AMI) containing all the necessary biological libraries and IPython/StarCluster support. Including the Amazon machine identifier used for the analyses published in ISME Journal allowed anyone with an Amazon account to repeat their analysis or modify it to address related questions.[99]

Another example of using the cloud as a reproducibility platform was the Sequence Squeeze challenge organized by the Pistoia Alliance (a precompetitive alliance of research groups, pharmaceutical companies, and scientific societies) and Eagle Genomics.[100] With the aim of outsourcing innovation in a public competition to build better FASTQ compression algorithms—a task of urgency due to the explosive growth of sequencing data—the competition gathered over 100 entries from data experts from around the world. Using publicly available test data from the 1000 genomes project,[101] with sponsorship covering cloud charges, the contest was AWS based and entries were submitted as AWS S3 buckets. All entries needed to be under an open-source Berkeley Software Distribution (BSD) license with code available in Sourceforge. The winning entry from James Bonfield achieved a compression ratio of 11.4%, and the resulting tools from a number of the competing entries have already been published.[102,103]

15.5.6 Workflow Systems

Analyses of genomics data that use a sequential series of computational tools can be implemented using a generic programming language such as Perl or R. They may also be constructed using a desktop workflow application, which provides a graphical user interface for composing pipelines based on a palette of computational tools. Examples of open-source workflow applications include Taverna,[104] Kepler,[105] and Knime[106] and commercial variants, for example, Pipeline Pilot.[107] Workflows written using such software can be saved in a machine-readable format with information on how it has been parameterized, which allows it to be reexecuted to reproduce the results of an analysis for a given dataset.

Galaxy is a workflow system, which has become popular for the analysis of next-generation sequence (NGS) data.[108] It comes with a large set of NGS data analysis tools for use in constructing workflows, which is achieved with a web-based interface via drop-down menus. *GigaScience* is using Galaxy for delivering reproducibility by using this computational platform to reimplement the data processing described in *GigaScience* papers as executable workflows.[109] Research involving the use of Galaxy pipelines has been extensively published in journal articles, some of which have made their pipeline data available as online supplemental information in the form of published pages on Galaxy servers. For example, Miller et al.[111] investigated the evolutionary relationships of the polar bear with brown and black bears and have provided the Galaxy workflows for analyzing genomic data used to generate some of the results reported in their accompanying paper.[123,124] In accordance with its open-data policy, all *GigaScience* workflows will be made freely available using myExperiment, an online repository for workflows, which has recently been integrated with the Galaxy workflow system.[125] myExperiment has tentatively been used to host workflows to be cited in the scientific literature, thereby providing more explicit documentation of data analyses in a form that can be executable by a particular workflow system. For example, journals published by BioMed Central have published papers, which have included citations of workflows stored in myExperiment.[110,111]

Since workflow systems allow data analyses to be recorded and subsequently used to reproduce results, they can be expected to play an important role in facilitating data reproducibility in scientific papers. Mesirov[53] described a reproducible research system (RRS), a paradigm for a computer application that enables computational data analyses to be reproduced and embedded within papers.[53] An RRS consists of two components. First, a reproducible research environment (RRE) provides an integrated infrastructure for producing and working with reproducible research and has the functionality to automatically track the provenance of data, analyses, and results and to package them for redistribution. Second, a reproducible research publication environment (RRPE) component would act as a document-preparation system that is responsible for embedding computation into the document, allowing it to be accessible from the paper by its readers. A number of applications have explored this type of functionality such as the Collage Authoring Environment, which enables researchers to seamlessly embed chunks of executable code and data into scientific publications in a form of collage items and to facilitate repeated execution of such codes on underlying computing and data storage resources.[112] Utopia Documents is an exciting project, which has developed an interactive PDF reader. This application provides access to information from various online data sources, such as PubMed and Mendeley, about specific terms or phrases selected by the viewer. In the future, it is feasible to see PDFs of scientific papers linking out to workflows and data such as those created by *Giga-Science*, to view how results of such papers have been generated as and when reading the document.

15.6 Role of the "Reproducible Research Publisher" of the Future

With more and more areas of research following genomics' lead and becoming increasingly data-intensive, there are enormous challenges, and equally enormous opportunities, to change publishing from being based on static papers to more interactive and dynamic *in silico* packages. A body of research consists of the scientific paper plus its data, methods, and tools, together with the people who have undertaken this research. These components are resources that collectively make up a research object.[113] The EU-funded Workflow4Ever project[114] is addressing how such resources can be described and aggregated electronically, shared, and discovered. A set of software tools will be developed by this project for creating, managing, and storing research objects. In addition, a collection of best practices will be delivered for the creation and manipulation of research objects. The "reproducible research publisher" of today, and the future, should aspire to collaborate and integrate with initiatives such as Workflow4Ever.

The increasing size and rate of growth of the body of published research,[115] combined with the ever-increasing amounts of data required to support publications, brings challenges to the hosting and filtering of content. The rise of open access and open data provides new opportunities to mine and analyze the literature and, coupled with semantic and data mining techniques,[116] make it more possible to discover and reuse information. However, providing data and text mining enhancements of value to both readers and *reusers* (researchers and their software) of the literature is challenging for commercial publishers' business models. Products such as BioMed Central's Cases Database[117]—a semantically enriched database of peer-reviewed medical case reports aimed at researchers, clinicians, regulators, and patients—show how semantic enrichment of the literature can work commercially, however.

Only in the second decade of the twenty-first century are groups starting to truly embrace online publication and break out of historically static formats, with interactive PDFs[118] and integration of data with research articles. Improvements made to searching and filtering published research combined with increased data publication demonstrate that it is also possible to change the perception and size of what is a publishable unit. On top of publishing data separately from analyses, it should also be possible to publish workflows and computerized methods as citable objects, allowing different combinations of executable, citable objects to be run against each other to produce novel results. Publishing smaller self-contained objects provides new forms of credit and incentives to release work quicker, allowing microattribution[119] or even nanopublications.[120]

We envision that this increasing computerization of papers will promote data reproducibility and new forms of knowledge discovery. It will become increasingly important for the system of peer review to keep up with this explosive growth of supporting data. Intelligent use of tools designed to visualize and assess data quality, integration with virtual machines and workflow systems, and easier-to-use systems promoting and aiding data interoperability and curation will be central to progress.[121]

To quote Cameron Neylon, who set out the challenges ahead in 2012,

> We need more than just reproducible computational research, we desperately need a step change in our expectations and in the incentives for communicating research in a reproducible form more generally. We need educators and the materials to support them in raising awareness and experience. And we need the development of policy and standards that help us move towards a world where reproducibility and replicability are minimum standards not aspirations.[122]

Scholarly communication is a slow-moving field with paper-based journals remaining and digitized documents usually representing little advance over letters exchanged between scholars in the seventeenth century. Pragmatically, we need stepwise changes in the scholarly communication system that involve developing existing platforms and integrating them with other commonly used tools for science. We need to better integrate the process of *doing* science with the process of communicating and evaluating science. There are promising collaborations between publishers and services upstream of paper submission (LabArchives, Figshare, GitHub) and downstream of paper publication (Mekentosj's Papers software was acquired by Springer, owner of BioMed Central, in 2012). We also need to make the data already published in journals more reusable. Simple enhancements will have an impact, such as tagging and classification of, and searching within, published data files. These will aid human and machine search, discovery, and integration of supplementary data. The ability to retrospectively associate data objects with published papers—dynamically and transparently adjust the scientific record—and the ability to link research objects in a manner that goes beyond hyperlinking through linked data (resource description framework [RDF]-based) approaches[74] should also be considered.

However, technology is just part of the solution. Journal, funder, and community policies, while of varying effectiveness, as we have described, are still important. We therefore need better, more automated, and scalable ways to check compliance with journal and community policies on the availability of materials for reproducible research and more automated ways to attach the right licenses and metadata to published objects, across all content disseminated by the reproducible research publishers of the future.

15.7 Summary

Reproducible computational research is and will be facilitated by the wide availability of scientific data, literature, and code, which is freely accessible and, furthermore, licensed such that it can be reused, integrated, and built upon to drive new scientific discoveries without legal impediments. Scholarly publishers have an important role in encouraging and mandating the availability of data and code according to community norms and best practices and developing innovative mechanisms and platforms for sharing and publishing products of research, beyond papers in journals. Open-access publishers, in particular the first commercial open-access publisher BioMed Central, have played a key role in the development of policies on open access and open data and increasing the use by scientists of legal tools—licenses and waivers—which maximizes reproducibility. Collaborations, between publishers and funders of scientific research, are vital for the successful implementation of reproducible research policies. The genomics and, latterly, other omics communities historically have been leaders in the creation and wide adoption of policies on public availability of data. This has been through policies, such as Fort Lauderdale and the Bermuda Principles; infrastructure, such as the INSDC databases; and incentives, such as conditions of journal publication. We review some of these policies and practices and how these events relate to the open-access publishing movement. We describe the implementation and adoption of licenses and waivers prepared by Creative Commons, in science publishing, with a focus on licensing of research data published in scholarly journals and data repositories. Also, we describe how some publishers are evolving the copyright system to ensure that published data are in the public domain under the Creative Commons CC0 waiver. Other cases where CC0 has been successfully implemented in science are discussed in particular by BGI, the world's largest genomics organization. BGI have developed an advanced platform for publishing executable research objects—including large data packages and code—which is integrated with open-access article publishing through *GigaScience* and its database, GigaDB. We look at journal and publisher policies, which aim to encourage reproducible research, and the comparative influence and success of these policies. We discuss specific problems faced in data sharing and reproducible research such as data standardization and some of the solutions. Finally, we review the state of the art in scientific workflows and large-scale computation platforms—including Galaxy and myExperiment—and how current and future collaborations between the scientific and publishing communities utilizing these innovative tools will further drive reproducibility in science.

References

1. Fang, F. C., Steen, R. G., and Casadevall, A. Misconduct accounts for the majority of retracted scientific publications. *Proceedings of the National Academy of Sciences of the United States of America* **109**, 17028–17033 (2012). http://www.pnas.org/content/early/2012/09/27/1212247109. (Accessed date 28th February 2013.)
2. Laine, C., Goodman, S. N., Griswold, M. E., and Sox, H. C. Reproducible research: Moving toward research the public can really trust. *Annals of Internal Medicine* **146**, 450–453 (2007).
3. Suber, P. What is open access? Open Access (2012). http://mitpress.mit.edu/sites/default/files/titles/content/9780262517638_sch_0001.pdf. (Accessed date 28th February 2013.)
4. Suber, P. Open-access timeline (formerly: FOS timeline). http://www.earlham.edu/~peters/fos/timeline.htm. (Accessed date 28th February 2013.)
5. Bethesda Statement on Open Access Publishing. http://www.earlham.edu/~peters/fos/bethesda.htm. (Accessed date 28th February 2013.)
6. ebi-PubMed Central: An NIH-operated site for electronic distribution of life sciences research reports. http://www.nih.gov/about/director/pubmedcentral/ebiomedarch.htm. (Accessed date 28th February 2013.)
7. Interview with Vitek Tracz: Essential for Science. http://www.infotoday.com/it/jan05/poynder.shtml. (Accessed date 28th February 2013.)
8. Butler, D. BioMed Central boosted by editorial board. *Nature* **405**, 384 (2000).
9. Haughey, M. Biomed Central using Creative Commons—Creative Commons (2004). http://creativecommons.org/weblog/entry/4077. (Accessed date 28th February 2013.)

10. Stodden, V. Enabling reproducible research: Licensing for scientific innovation. *International Journal of Communications Law and Policy* **13**, 2–15 (2009).
11. Carroll, M. W. Why full open access matters. *PLoS Biology* **9**, e1001210 (2011).
12. Redhead, C. Why CC-BY? The Open Access Scholarly Publishers Association. (2012). http://oaspa.org/why-cc-by/. (Accessed date 28th February 2013.)
13. Laakso, M. and Björk, B.-C. Anatomy of open access publishing: A study of longitudinal development and internal structure. *BMC Medicine* **10**, 124 (2012).
14. Pollock, D. *An Open Access Primer—Market Size and Trends*. Outsell Inc., Burlingame, CA (2009). http://www.outsellinc.com/b2b/products/ 873-an-open-access-primer-market-size-and-trends. (Accessed date 28th February 2013.)
15. *Genome Biology*. Instructions for authors. Software, preparing-additional-files, preparing-additional-files, preparing-additional-files. http://genomebiology.com/authors/instructions/software# preparing-additional-files. (Accessed date 28th February 2013.)
16. Maunsell, J. Announcement regarding supplemental material. *Journal of Neuroscience* **30**, 10599–10600 (2010).
17. Hrynaszkiewicz, I. and Cockerill, M. In defence of supplemental data files: Don't throw the baby out with the bathwater (2010). http:// blogs.openaccesscentral.com/blogs/bmcblog/entry/in_defence_of_ supplemental_data. (Accessed date 28th February 2013.)
18. *BioData Mining*. Instructions for reviewers. http://www. biodatamining.org/about/reviewers. (Accessed date 28th February 2013.)
19. Let's Get Rid of that "Supplemental Data" Madness… *Lab Times* 3 (2010). http://www.lab-times.org/labtimes/issues/lt2010/lt06/lt_ 2010_06_3_3.pdf. (Accessed date 28th February 2013.)
20. Piwowar, H. A. Supplementary materials is a stopgap for data archiving (2010). http://researchremix.wordpress.com/2010/08/13/ supplementary-materials-is-a-stopgap-for-data-archiving/. (Accessed date 28th February 2013.)
21. Alsheikh-Ali, A. A., Qureshi, W., Al-Mallah, M. H., and Ioannidis, J. P. A. Public availability of published research data in high-impact journals. *PLOS ONE* **6**, e24357 (2011).
22. Wicherts, J. M., Borsboom, D., Kats, J., and Molenaar, D. The poor availability of psychological research data for reanalysis. *American Psychologist* **61**, 726–728 (2006).
23. Piwowar, H. A. and Chapman, W. W. A review of journal policies for sharing research data (2008). http://precedings.nature.com/ documents/1700/version/1. (Accessed date 28th February 2013.)

24. BioMed Central. Availability of supporting data. http://www.biomedcentral.com/about/supportingdata. (Accessed date 28th February 2013.)

25. Availability of data and materials: Authors and referees @ npg. http://www.nature.com/authors/policies/availability.html. (Accessed date 28th February 2013.)

26. *PLOS ONE*: Accelerating the publication of peer-reviewed science. http://www.plosone.org/static/policies.action#sharing. (Accessed date 28th February 2013.)

27. Retraction Watch: Study links failure to share data with poor quality research and leads to a *PLOS ONE* retraction. http://retractionwatch.wordpress.com/2013/01/30/study-links-failure-to-share-data-with-poor-quality-research-and-leads-to-a-plos-one-retraction/. (Accessed date 28th February 2013.)

28. Groves, T. The wider concept of data sharing: View from the *BMJ*. *Biostatistics (Oxford, England)* **11**, 391–392 (2010).

29. Peng, R. D., Dominici, F., and Zeger, S. L. Reproducible epidemiologic research. *American Journal of Epidemiology* **163**, 783–789 (2006).

30. Godlee, F. Clinical trial data for all drugs in current use. *BMJ* **345**, e7304 (2012).

31. Dryad home. http://datadryad.org. (Accessed date 28th February 2013.)

32. Whitlock, M. C., Mcpeek, M. A., Rausher, M. D., Rieseberg, L., and Moore, A. J. Data archiving. *American Naturalist* **175**, 145–146 (2010).

33. FigShare. http://figshare.com. (Accessed date 28th February 2013.)

34. Vines, T. H. Andrew, R. L., Bock, D. G., Franklin, M. T., Gilbert, K. J., Kane, N. C., Moore, J.-S. et al. Mandated data archiving greatly improves access to research data. *FASEB Journal* **27** April (2013). doi:10.1096/fj.12-218164.

35. International Nucleotide Sequence Database Collaboration (INSDC). http://www.insdc.org/. (Accessed date 28th February 2013.)

36. DOE Genome Informatics—Announcement II. http://www.bio.net/bionet/mm/bionews/1994-January/000877.html. (Accessed date 28th February 2013.)

37. Cochrane, G., Karsch-Mizrachi, I., and Nakamura, Y. The International Nucleotide Sequence Database Collaboration. *Nucleic Acids Research* **39**, D15–D18 (2011).

38. Open Letter to Journal Editors from the INSDC. http://www.insdc.org/sites/insdc.org/files/documents/open_letter.txt. (Accessed date 28th February 2013.)

39. Kaiser, J. Genomics. Celera to end subscriptions and give data to public GenBank. *Science New York* **308**, 775 (2005).

40. Marshall, E. Bermuda rules: Community spirit, with teeth. *Science* **291**, 1192 (2001).

41. Roberts, L. Genome research. A tussle over the rules for DNA data sharing. *Science New York* **298**, 1312–1313 (2002).
42. Macilwain, C. Biologists challenge sequencers on parasite genome publication. *Nature* **405**, 601–612 (2000).
43. National Human Genome Research Institute. Data Release Policies (February 2003). http://www.genome.gov/10506537. (Accessed date 28th February 2013.)
44. Toronto International Data Release Workshop Authors Prepublication data sharing. *Nature* **461**, 168–170 (2009). (see http://www.nature.com/nature/journal/v461/n7261/full/461168a.html)
45. Noor, M. A. F., Zimmerman, K. J., and Teeter, K. C. Data sharing: How much doesn't get submitted to GenBank? *PLoS Biology* **4**, e228 (2006).
46. Field, D. and Hughes, J. Cataloguing our current genome collection. *Microbiology (Reading, England)* **151**, 1016–1019 (2005).
47. Field, D. et al. The Genomic Standards Consortium. *PLoS Biology* **9**, e1001088 (2011).
48. Yilmaz, P. et al. Minimum information about a marker gene sequence (MIMARKS) and minimum information about any (x) sequence (MIxS) specifications. *Nature Biotechnology* **29**, 415–420 (2011).
49. SIGS instructions to authors. http://www.standardsingenomics.org/index.php/sigen/pages/view/SIGS_i2a. (Accessed date 28th February 2013.)
50. Illuminating the black box. *Nature* **442**, 1 (2006).
51. Ioannidis, J. P. A. et al. Repeatability of published microarray gene expression analyses. *Nature Genetics* **41**, 149–155 (2009).
52. Begley, C. G. and Ellis, L. M. Drug development: Raise standards for preclinical cancer research. *Nature* **483**, 531–533 (2012).
53. Mesirov, J. P. Computer science. Accessible reproducible research. *Science (New York)* **327**, 415–416 (2010).
54. Peng, R. D. Reproducible research and biostatistics. *Biostatistics (Oxford, England)* **10**, 405–408 (2009).
55. *Source Code for Biology and Medicine.* About http://www.scfbm.org/about. (Accessed date 28th February 2013.)
56. Barton, M. D. and Barton, H. A. Scaffolder—Software for manual genome scaffolding. *Source Code for Biology and Medicine* **7**, 4 (2012).
57. Grosse-Kunstleve, R. W., Terwilliger, T. C., Sauter, N. K., and Adams, P. D. Automatic Fortran to C++ conversion with FABLE. *Source Code for Biology and Medicine* **7**, 5 (2012).
58. Ramirez-Gonzalez, R. H., Bonnal, R., Caccamo, M., and Maclean, D. Bio-samtools: Ruby bindings for SAMtools, a library for accessing BAM files containing high-throughput sequence alignments. *Source Code for Biology and Medicine* **7**, 6 (2012).
59. Open Research Computation collection thematic article series in *Source Code for Biology in Medicine.* http://www.scfbm.org/series/ORC. (Accessed date 28th February 2013.)

60. Brazma, A. et al. Minimum information about a microarray experiment (MIAME)—Toward standards for microarray data. *Nature Genetics* **29**, 365–371 (2001).
61. Taylor, C. F. et al. Promoting coherent minimum reporting guidelines for biological and biomedical investigations: The MIBBI project. *Nature Biotechnology* **26**, 889–896 (2008).
62. Rayner, T. F. et al. A simple spreadsheet-based, MIAME-supportive format for microarray data: MAGE-TAB. *BMC Bioinformatics* **7**, 489 (2006).
63. Engreitz, J. M. et al. ProfileChaser: Searching microarray repositories based on genome-wide patterns of differential expression. *Bioinformatics (Oxford, England)* **27**, 3317–3318 (2011).
64. Stoltzfus, A. et al. Sharing and re-use of phylogenetic trees (and associated data) to facilitate synthesis. *BMC Research Notes* **5**, 574 (2012).
65. *BMC Research Notes* thematic series on "Data standardization, sharing and publication." *BMC Research Notes* (2012). http://www.biomedcentral.com/bmcresnotes/series/datasharing. (Accessed date 28th February 2013.)
66. BioSharing. http://biosharing.org. (Accessed date 28th February 2013.)
67. McDonald, D. et al. The Biological Observation Matrix (BIOM) format or: How I learned to stop worrying and love the ome–ome. *GigaScience* **1**, 7 (2012).
68. Sansone, S.-A. et al. Toward interoperable bioscience data. *Nature Genetics* **44**, 121–126 (2012).
69. ISA tools. Welcome. http://isatab.sourceforge.net/
70. Murray-Rust, P., Neylon, C., Pollock, R., and Wilbanks, J. Panton Principles, Principles for open data in science (2010). http://pantonprinciples.org/. (Accessed date 28th February 2013.)
71. Schaeffer, P. Why does Dryad use CC0? *Dryad News and Views* http://blog.datadryad.org/2011/10/05/why-does-dryad-use-cc0/. (Accessed date 28th February 2013.)
72. Open Knowledge Foundation Conformant Licenses. http://opendefinition.org/licenses/. (Accessed date 28th February 2013.)
73. BioMed Central policies. Access to articles. http://www.biomedcentral.com/about/access/#opendata. (Accessed date 28th February 2013.)
74. Hrynaszkiewicz, I. and Cockerill, M. J. Open by default: A proposed copyright license and waiver agreement for open access research and data in peer-reviewed journals. *BMC Research Notes* **5**, 494 (2012).
75. Hrynaszkiewicz, I. Help put the open in Open Data and Open Bibliography (2012). http://blogs.biomedcentral.com/bmcblog/2012/09/10/put-the-open-in-opendata/. (Accessed 28th February 2013.)
76. *EMBO Journal*: Guide for authors. http://www.nature.com/emboj/about/authors.html. (Accessed 28th February 2013.)

77. Hrynaszkiewicz, I., Busch, S., and Cockerill, M. J. Licensing the future: Report on BioMed Central's public consultation on Open Data in peer-reviewed journals. *BMC Research Notes* **6**, 318 (2013), doi:10.1186/1756-0500-6-318.

78. Savage, C. J. and Vickers, A. J. Empirical study of data sharing by authors publishing in PLoS journals. *PLOS ONE* **4**, e7078 (2009).

79. Hrynaszkiewicz, I. LabArchives and BioMed Central: A new platform for publishing scientific data (2012). http://blogs.openaccesscentral. com/blogs/bmcblog/entry/labarchives_and_biomed_central_a. (Accessed 28th February 2013.)

80. LabArchives BioMed Central Edition. http://www.labarchives.com/ bmc. (Accessed date 28th February 2013.)

81. Hrynaszkiewicz, I. Social coding and scholarly communication - open for collaboration (2013) http://blogs.biomedcentral.com/bmcblog/ 2013/02/28/github-and-biomed-central/. (Accessed 28th February 2013.)

82. Ram, K. Git can facilitate greater reproducibility and increased transparency in science. *Source Code for Biology and Medicine* **8**, 7 (2013).

83. Credit where credit is overdue. *Nature Biotechnology* **27**, 579 (2009).

84. DataCite. http://www.datacite.org. (Accessed 28th February 2013.)

85. PANGAEA: Data Publisher for Earth & Environmental Science. http:// www.pangaea.de. (Accessed 28th February 2013.)

86. Sneddon, T. P., Li, P., and Edmunds, S. C. GigaDB: Announcing the *GigaScience database. GigaScience* **1**, 11 (2012).

87. Ball, A. and Duke, M. How to cite datasets and link to publications. In *How-to Guides*. Edinburgh, UK: Digital Curation Centre (2011). http:// www.dcc.ac.uk/resources/how-guides/cite-datasets. (Accessed 28th February 2013.)

88. Li, Y. et al. Single-cell sequencing analysis characterizes common and cell-lineage-specific mutations in a muscle-invasive bladder cancer. *GigaScience* **1**, 12 (2012).

89. Li, D., Xi, F., Zhao, M., Chen, W., Cao, S., Xu, R., Wang, G. et al. and the *Escherichia coli* O104:H4 TY-2482 isolate Genome Sequencing Consortium. Genomic data from *Escherichia coli* O104:H4 isolate TY-2482. BGI Shenzhen (2011). http://dx.doi.org/10.5524/100001. (Accessed date 28th February 2013.)

90. Scholl, D., Gebhart, D., Williams, S. R., Bates, A., and Mandrell, R. Genome sequence of *E. coli* O104:H4 leads to rapid development of a targeted antimicrobial agent against this emerging pathogen. *PLOS ONE* **7**, e33637 (2012).

91. Rohde, H. et al. Open-source genomic analysis of Shiga-toxin–producing *E. coli* O104:H4. *New England Journal of Medicine* **365**, 718–724 (2011). doi:10.1056/NEJMoa1107643.

92. Science as an open enterprise. The Royal Society Science Policy Centre Report (2012). http://royalsociety.org/uploadedFiles/Royal_Society_Content/policy/projects/sape/2012-06-20-SAOE.pdf. (Accessed 28th February 2013.)

93. MacLean, D. et al. Crowdsourcing genomic analyses of ash and ash dieback—Power to the people. *GigaScience* **2**, 2 (2013). doi:10.1186/2047-217X-2-2.

94. Dudley, J. T. and Butte, A. J. In silico research in the era of cloud computing. *Nature Biotechnology* **28**, 1181–1185 (2010).

95. Wilkening, J., Wilke, A., Desai, N., and Meyer, F. Using clouds for metagenomics: A case study. 2009 *IEEE International Conference on Cluster Computing and Workshops*, New Orleans, LA, pp. 1–6 (2009). doi:10.1109/CLUSTR.2009.5289187.

96. Nature ENCODE: Nature Publishing Group: A landmark in the understanding of the human genome. http://www.nature.com/encode/#/threads. (Accessed 28th February 2013.)

97. Bernstein, B. E. et al. An integrated encyclopedia of DNA elements in the human genome. *Nature* **489**, 57–74 (2012).

98. ENCODE Virtual Machine and Cloud Resource. http://scofield.bx.psu.edu/~dannon/encodevm/. (Accessed 28th February 2013.)

99. Ragan-Kelley, B. et al. Collaborative cloud-enabled tools allow rapid, reproducible biological insights. *The ISME Journal* **7**(3), 461–464 (2012). doi:10.1038/ismej.2012.123.

100. The Pistoia Alliance Sequence Squeeze competition. http://www.sequencesqueeze.org/. (Accessed 28th February 2013.)

101. Genomes Project Consortium, Abecasis GR, Altshuler D, Auton A, Brooks LD, et al. (2010). A map of human genome variation from population-scale sequencing. *Nature* **467**(7319), 1061–73.

102. Jones, D. C., Ruzzo, W. L., Peng, X., and Katze, M. G. Compression of next-generation sequencing reads aided by highly efficient de novo assembly. *Nucleic Acids Research* **40**(22), e171 (2012). http://nar.oxfordjournals.org/content/early/2012/08/14/nar.gks754.long. (Accessed date 28th February 2013.)

103. Hach, F., Numanagic, I., Alkan, C., and Sahinalp, S. C. SCALCE: Boosting sequence compression algorithms using locally consistent encoding. *Bioinformatics (Oxford, England)* **28**(23), 3051–3057 (2012). http://bioinformatics.oxfordjournals.org/content/early/2012/10/08/bioinformatics.bts593.short. (Accessed date 28th February 2013.)

104. Hull, D. et al. Taverna: A tool for building and running workflows of services. *Nucleic Acids Research* **34**, W729–W732 (2006).

105. The Kepler Project. https://kepler-project.org. (Accessed 28th February 2013.)

106. Knime. http://www.knime.org. (Accessed 28th February 2013.)

107. Pipeline Pilot. http://accelrys.com/products/pipeline-pilot. (Accessed 28th February 2013.)

108. Goecks, J., Nekrutenko, A., and Taylor, J. Galaxy: A comprehensive approach for supporting accessible, reproducible, and transparent computational research in the life sciences. *Genome Biology* **11**, R86 (2010).

109. *GigaScience* data analysis platform. http://galaxy.cbiit.cuhk.edu.hk. (Accessed 28th February 2013.)

110. Galaxy. Published Page. Polar-bears. https://main.g2.bx.psu.edu/u/webb/p/polar-bears. (Accessed 28th February 2013.)

111. Miller, W. et al. Polar and brown bear genomes reveal ancient admixture and demographic footprints of past climate change. *Proceedings of the National Academy of Sciences of the United States of America* **109**, E2382–E2390 (2012).

112. Nowakowski, P. et al. The collage authoring environment. *Procedia Computer Science* **4**, 608–617 (2011).

113. Belhajjame K. C. O. et al. Workflow-centric research objects: A first class citizen in the scholarly discourse. *Workshop on the Future of Scholarly Communication in the Semantic Web (SePublica2012)*, Crete, Greece (2012). http://users.ox.ac.uk/~oerc0033/preprints/sepublica2012.pdf. (Accessed date 28th February 2013.)

114. WF4Ever project. http://www.wf4ever-project.org. (Accessed date 28th February 2013.)

115. Gillam, M., Feied, C., Handler, J., and Moody, E. The healthcare singularity and the age of semantic medicine. *The Fourth Paradigm* 57–64 (2009). http://research.microsoft.com/en-us/collaboration/fourthparadigm/4th_paradigm_book_part2_gillam.pdf. (Accessed date 28th February 2013.)

116. Shotton, D., Portwin, K., Klyne, G., and Miles, A. Adventures in semantic publishing: Exemplar semantic enhancements of a research article. *PLoS Computational Biology* **5**, e1000361, (2009).

117. BioMed Central's Cases Database. http://www.casesdatabase.com. (Accessed 28th February 2013.)

118. Pettifer, S. et al. Reuniting data and narrative in scientific articles. *Insights* **25**, 288–293 (2012).

119. Patrinos, G. P. et al. Microattribution and nanopublication as means to incentivize the placement of human genome variation data into the public domain. *Human Mutation* **33**, 1503–1512 (2012).

120. Mons, B. et al. The value of data. *Nature Genetics* **43**, 281–283 (2011).

121. Bourne, P. E. What do I want from the publisher of the future? *PLoS Computational Biology* **6**, e1000787 (2010).

122. Neylon, C. et al. Changing computational research. The challenges ahead. *Source Code for Biology and Medicine* **7**, 2 (2012).

123. Galaxy. Published Page. polar-bears at https://main.g2.bx.psu.edu/u/webb/p/polar-bears.

124. Miller, W. et al. Polar and brown bear genomes reveal ancient admixture and demographic footprints of past climate change. *Proceedings of the National Academy of Sciences of the United States of America* **109**, E2382–90 (2012).

125. Goble, C. A. et al. myExperiment: A repository and social network for the sharing of bioinformatics workflows. *Nucleic acids research* **38**, W677–82 (2010).

Index